Modeling and Managing Shallow Lake Eutrophication

With Application to Lake Balaton

Editors:
L. Somlyódy and G. van Straten

With 156 Figures

Springer-Verlag
Berlin Heidelberg New York Tokyo

L. SOMLYÓDY
Research Centre for Water Resources Development, VITUKI
Budapest
Hungary

G. van STRATEN
Twente University of Technology
Enschede
The Netherlands

IIASA
International Institute for Applied
Systems Analysis
A-2361 Laxenburg
Austria

ISBN 3-540-16227-5 Springer-Verlag Berlin Heidelberg New York Tokyo
ISBN 0-387-16227-5 Springer-Verlag New York Heidelberg Berlin Tokyo

Library of Congress Cataloging in Publication Data
Main entry under title:
Modeling and managing shallow lake eutrophication.
Includes index.
1. Eutrophication--Hungary--Balaton, Lake.
2. Balaton, Lake (Hungary)
I. Somlyódy, L. (László) II. Straten, G. van (Gerrit van)
QH96.8.E9M59 1986 628.1'68 85-31737
ISBN 0-387-16227-5

Offsetprinting: Color-Druck, G. Baucke, Berlin; Bookbinding: Lüderitz & Bauer, Berlin
2161/3020-543210

Preface

In the late 1970s, the adverse effects of man-made eutrophication became manifest in many countries, which explains, perhaps, why there was such a broad interest when the former Resources and Environment Area of the International Institute for Applied Systems Analysis (IIASA) organized a workshop on the subject. There was such an enthusiasm among the participants that two further workshops were quickly organized, one on deep and the other on shallow lake eutrophication problems. The organization of these meetings was extremely stimulating, and the round table discussions among scientists from both West and East remain thought provoking for those who took part. The general feeling emerged that the complexity and multifaceted nature of the problem, even though perhaps not fully recognized at that time, clearly demanded a systems analysis approach. No wonder, then, that the request made by the Hungarian Member Organization of IIASA to adopt Lake Balaton as a "real life laboratory" for an IIASA case study fell on fertile ground, the more so since it appeared that shallow lake eutrophication had received less attention and was less well understood than that of deep lakes. And so the IIASA Lake Balaton Case Study began, with the appointment of Gerrit van Straten as the first leader of the project.

A number of years have passed since then, but the interest in eutrophication problems has not diminished. On the contrary, there is a pressing need for a sound, scientifically based analysis of various management options now under consideration in many countries, and there is a stronger demand to use the results of research in policymaking. This structural development, from research toward management application, is also reflected in the development of the Balaton case study. Initially, we had to concentrate on the organization of the necessary data and on the mechanisms of international communication and cooperation. The remainder of the first two-year period was used to initiate process-oriented research and to develop a set of simulation models. The results created the basis from which the study could gain its broader applicability, with evolvement toward its management setting realized in the second two-year period, this time with László Somlyódy as leader.

It is not an exaggeration to say that the period during which we worked on the case study became a crucial part of both our lives, and, during the many on-going professional and personal contacts made as the case study progressed, the challenging idea ripened to conclude our work with a book: this book. As such, this volume can be viewed as a final account of a successful IIASA case study, but from the very start of its compilation we were convinced that the study had spin-offs of much wider scope and interest, which, we feel, form the true justification for publication. A large part of the literature on eutrophication tackles the problem from one particular point of view, and places emphasis on one particular issue. In contrast, in this volume we emphasize the links between the various aspects, the need to build integrated management models on the basis of detailed research, but also the need to gear this detailed research into usefulness for application in a management context, where details are necessarily lost but the essentials must be preserved. So, in this volume, we do cover research at a disciplinary level, but we also discuss its integration in descriptive and predictive simulation models, and finally its management and policymaking implications. This, we feel, is a unique feature which distinguishes this volume from others in the field.

When the book project began, we faced the question of whether the two of us should be the only authors. For a number of reasons this idea was rejected and we decided to call on the help of some of the scientists who participated in the study. However, rather than using existing reports or publications we asked the contributing authors to write a chapter specifically for this volume, and we provided quite restrictive instructions in order to maintain the idea we had in mind. This procedure, as expected, proved time-consuming and sometimes painful. It required a considerable editorial effort and several feedbacks. But in the end we believe it was a rewarding exercise. It is to the judgment of the reader whether we succeeded in our endeavor to mould and reshape the material from multidisciplinarity into true interdisciplinarity.

The book consists of three parts. Part I, almost entirely written by ourselves, contains the methodology, and describes the paths from problem statement, through the approach, along modeling and its application for management, toward the impact on policymaking. In Part II the processes and subsystems are described. The contributions to this part can be considered as reports on each of the specific issues necessary for a complete understanding of the problem as a whole, given at a disciplinary level. Part III covers a number of selected topics. They were chosen because of originality or because of their explicit contribution to the success of the case study.

We are grateful for the large cooperative input we received from many scientists during the study, but there were also others who stimulated and supported the study. It happened more than once that we were given a little extra backup that helped us to overcome dead points. There is not space here to mention all who have contributed, but we are greatly endebted to Professor Oleg Vasiliev, former leader of the Resources and Environment Area, for his visionary view and his distant, but distinctive, support and enthusiasm. Without his tenacity at the initiation of the study, his international contacts, his kind but never offensive criticism, and his judgment in personal affairs, this book would not have emerged, simply because the study would not have commenced or would have died an early death. Had this happened, it would have deprived IIASA from what we believe to be a representative example of its ideals: a truly international cooperation between scientists from East and West, on the basis of mutual confidence and an unrestricted exchange of scientific experience. We still believe in these ideals, and we sincerely hope that this book may serve as evidence that such cooperation can be fruitful indeed and a benefit to the well-being of mankind, a cooperation worth cherishing and even expanding wherever and whenever possible.

László Somlyódy *Gerrit van Straten*

THE INTERNATIONAL INSTITUTE FOR APPLIED SYSTEMS ANALYSIS

is a nongovernmental research institution, bringing together scientists from around the world to work on problems of common concern. Situated in Laxenburg, Austria, IIASA was founded in October 1972 by the academies of science and equivalent organizations of twelve countries. Its founders gave IIASA a unique position outside national, disciplinary, and institutional boundaries so that it might take the broadest possible view in pursuing its objectives:

To promote international cooperation in solving problems arising from social, economic, technological, and environmental change

To create a network of institutions in the national member organization countries and elsewhere for joint scientific research

To develop and formalize systems analysis and the sciences contributing to it, and promote the use of analytical techniques needed to evaluate and address complex problems

To inform policy advisors and decision makers about the potential application of the Institute's work to such problems

The Institute now has national member organizations in the following countries:

Austria
The Austrian Academy of Sciences

Bulgaria
The National Committee for Applied Systems Analysis and Management

Canada
The Canadian Committee for IIASA

Czechoslovakia
The Committee for IIASA of the Czechoslovak Socialist Republic

Finland
The Finnish Committee for IIASA

France
The French Association for the Development of Systems Analysis

German Democratic Republic
The Academy of Sciences of the German Democratic Republic

Federal Republic of Germany
Association for the Advancement of IIASA

Hungary
The Hungarian Committee for Applied Systems Analysis

Italy
The National Research Council

Japan
The Japan Committee for IIASA

Netherlands
The Foundation IIASA–Netherlands

Poland
The Polish Academy of Sciences

Sweden
The Swedish Council for Planning and Coordination of Research

Union of Soviet Socialist Republics
The Academy of Sciences of the Union of Soviet Socialist Republics

United States of America
The American Academy of Arts and Sciences

Contents

PART ONE

Problem, Methodology, and Management

CHAPTER 1

Background to the Lake Balaton Eutrophication Problem

L. Somlyódy and G. van Straten

1.1. Introduction

Among the problems that have threatened freshwater lakes worldwide over the past 10–20 years, man-made eutrophication has been the most prominent. Eutrophication literally means "rendering rich in nutrients". Although some lakes are naturally rich in nutrients, the term eutrophication is usually taken to mean the unintended enrichment by human activity, particularly if the associated changes in the lake's ecosystem are unfavorable. Thus, in a narrow sense, increased loadings can be held immediately responsible for the generation of eutrophication phenomena. In a broader sense, however, eutrophication can be viewed as the response of a lake ecosystem to changes in the regional infrastructure; the deterioration in lake water quality is just the end result of processes that originate within the watershed. Consequently, if it is necessary to find remedies because the changes in the original functions of the lake are unacceptable, it is most logical to look for them primarily within the watershed region.

So, at first sight, it seems that it should be sufficient to concentrate on the watershed, its processes, and the ways in which they are influenced by human actions, in order to select the best management practices. Actually, the question may arise as to whether there is a need to consider the lake at all. Would the lake not automatically recover as soon as external loads are reduced? There are various reasons why such an attitude should be resisted. The first and most obvious is that the past can never be recreated. Developments within the watershed may have led to permanent and irreversible changes, so the kind of loads received by the lake will be different from before, even if the most stringent measures are taken. Second, there is an economic necessity to exploit the natural assimilation process. That is, nature is not untouchable; we can and may use and reshape nature, as long as we are aware of the limitations, so that it is probably not necessary to return the lake to its original unspoiled condition. On the other hand, in order to know the limitations, we must study the long-term behavior of the lake, and this again indicates that the lake is an essential element in the whole system. Thus, the conclusion is that a study of the process of eutrophication, with the aim of finding necessary and suitable management policies, requires a thorough analysis of both the watershed and the lake. The lake and its region are elements of a single system, and they cannot be treated in isolation.

Of the factors that alter the regional infrastructure, industrialization, urbanization, tourism, and agricultural development are the most important. As a result of these, large amounts of nutrients (mainly P and N compounds) are released to the environment and reach the lake through various artificial and natural processes (sewage discharge, runoff, erosion, atmospheric pollution, groundwater infiltration, etc.). Such increased loads induce water quality degradation, with such symptoms as algal blooms, discoloration, and floating debris. Taste and odor problems occur, and fish kills are common as a result of drastic changes in the ecosystem. All these signs may lead to restrictions of water use, including for example drinking water supply or recreational potential of the lake. This latter can endanger the utilization of the entire recreational area; that is, it limits the use of the region itself.

The water quality in the lake under certain load conditions is a function of complex, interrelated biological, chemical, and physical processes that are influenced by several external natural factors such as temperature, solar radiation, or wind. The study of these processes is necessary from the viewpoint of scientific understanding, which is a precondition of proper planning to improve water quality: an important practical issue.

At this point the reader will have obtained a sketchy picture of the various aspects one is likely to encounter when engaged in a study of eutrophication. In order to complete the picture the next step is to collect background material. Before going into further detail it is important to stress that this step should include consideration of both the watershed and its related processes, as well as the lake, its major internal processes, and their forcing functions. In addition, since we are interested in management, we should become familiar with the various management alternatives available. And, finally, all this depends upon data availability, which therefore should receive special attention since it is frequently necessary to take data uncertainty into account, as is discussed below.

In practice, no two lakes are the same, so that the background data collection process will differ from lake to lake. This book specifically deals with Lake Balaton in Hungary, so the details presented here are lake specific, but many elements and features important for Lake Balaton will also be applicable to other lake eutrophication studies. The objectives of this chapter are twofold: first, to describe the major features of Lake Balaton itself in order to provide a basis for the understanding of the remainder of this book, and second, to illustrate the various general aspects that may arise in any study of eutrophication, with Lake Balaton simply an example.

1.2. Characterization of the Watershed

The extension of the Lake Balaton watershed area (including the lake) is 5776 km^2 (Figure 1.1). The major subwatersheds, their areas, and elevations are listed in Table 1.1. The Zala River basin in the west mainly consists of hilly, alluvial land; the southern basin is dominated by low hills of sand and loess, with some marshland; the northern basin is characterized by higher hills of dolomitic limestone with evidence of volcanic activity.

About half of the land area is drained by the Zala River whose average discharge of 9 m^3/s represents 50% of the total water inflow to the lake. Tributaries of the Szigliget basin (basin II, see Figure 1.1) provide some 30% of the inflow. The surface areas of the four lake basins (A_{Li}) increase while those of the corresponding subwatersheds (A_{Wi}) decrease from west to east (see Table 1.2); the ratios A_{Wi}/A_{Li} for basins I–IV are roughly 72:11:3:1, suggesting much larger relative nutrient loads and less favorable water quality conditions at the western end of the lake. This is confirmed by actual measurement data, as discussed later.

Table 1.1. Subwatershed areas and elevations (after Baranyi 1975).

	Area (km^2)	Elevation (m)
Zala watershed	2622	340
Northern stream watersheds	820	47–711
Southern stream watersheds	1176	117–212
Direct shoreline watersheds	562	–
Lake Balaton	596	–
Total catchment area	5776	–

Table 1.2. Geometric data of the four Lake Balaton basins at average level (104.9 MASL) and the corresponding watershed surface areas.

Basin	Volume (10^6 m^3)	Depth (m)	Surface area of basins, A_{L_i} (km^2)	Surface area of corresponding subwatersheds, A_{W_i} (km^2)
I	82 (4.3%)	2.3	38 (6.4%)	2750 (53.1%)
II	413 (21.8%)	2.9	144 (24.4%)	1647 (31.8%)
III	600 (31.6%)	3.2	186 (31.1%)	534 (10.3%)
IV	802 (42.3%)	3.7	228 (38.1%)	249 (4.8%)
Total	1907 (100%)	3.2	596 (100%)	5180 (100%)

Soil erosion is a significant factor in changing conditions in the lake and its basins. Owing to geology and topography the soil loss has been relatively severe in the northern subwatershed and moderate in the southern subwatershed. According to various estimates the annual soil loss ranges between 4000–6000 tons/km^2 with lower and upper extremes of 2500 and

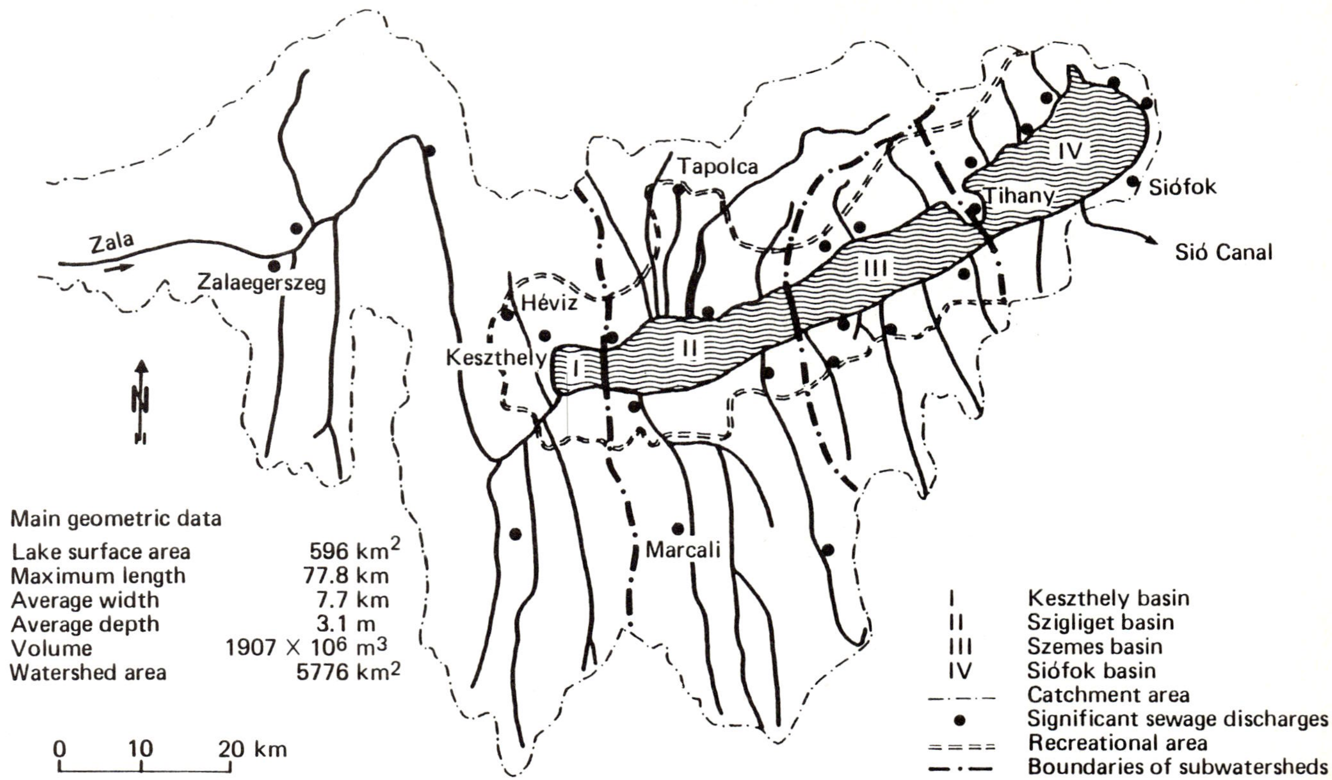

Figure 1.1. Lake Balaton and its watershed.

Table 1.3. Slope distribution of various land-use types (from Horváth and Kamarás 1976).

Land-use type	Percentage of total catchment area	Percentage of slope categories				
		<5	5–12	13–17	18–25	>25
Arable land	35	68	20	8	3	1
Meadows	8	100	–	–	–	–
Pasture	7	73	12	5	7	3
Orchards and vineyards	6	41	28	17	11	3

17000 tons/km^2. The erosion rate depends on many factors and varies considerably from site to site. Apart from natural factors such as slope, human factors such as the types of vegetation or crops grown, and land-use practices, have had a significant effect on erosion. Table 1.3 gives details of land-use types according to the steepness of slopes.

Generally, 1–3% of the applied fertilizer is washed from the soil and reaches the lake through runoff and erosion; the growing use of fertilizers thus increases the nutrient load and is one of the main causes of artificial eutrophication. Based on statistical records for the mid-1970s fertilizer use for the total catchment area was about 15000 tons/yr of P, which represents a six- to seven-fold growth since 1960 (see also Chapter 14). Taking the year 1950 as a reference, fertilizer use has thus increased 60–70 times.

Another spectacular change in agricultural production has been the wide-scale introduction of modern livestock breeding techniques. Up to 30 years ago there were no large-scale livestock farms in the region at all; today there are more than 40, with a total of some 100000 animals. From the liquid manure generated on such farms as much as 1000–1500 kg/d of P is discharged to the environment, some 5–20% of which may then reach the lake.

Another important factor characterizing the region of Lake Balaton is tourism. The permanent population is 405000, of which some 120000 live in close proximity to the lake. Tourism increased by a factor of 14 between 1950 and 1978; in 1978 the number of visitor days was about 8 million, double the local population (and also the sewage discharge) during the relatively short summer peak season, July and August. The drinking water supply increased fivefold between 1960 and 1978, so that today the demand is almost completely satisfied within the recreational area. The capacity of the sewerage and wastewater treatment system has also been improved, although not to the same extent as the development of drinking water supply.

All of these factors have contributed to the growing amounts of nutrients reaching the lake. Although direct observations are not available, it can be estimated that the load of the lake has increased by an order of magnitude during the past 20–25 years, and this has resulted in the artificial eutrophication of Lake Balaton.

The nutrient loads and their distribution among the lake basins are difficult to quantify. When performing source evaluations, all emissions (fertilizers, liquid manure, sewage discharges, etc.) and transmission processes in the watershed that contribute to the lake's external nutrient load have to be considered. Another approach, based on evaluation of data from the monitoring network of loads (tributaries, sewage discharges, atmospheric pollution, etc.) entering the lake directly, is more straightforward but does not provide information on the origin of the nutrients in question.

The uncertainties of this procedure are due mainly to infrequent observations; in general, only one sample per month is taken from each of the 20 major tributaries. For sewage discharges (see Figure 1.1 for the location of the most significant sources) even fewer observational data are available. Accordingly the data set does not completely reflect the influence of floods or peak wastewater releases even though, as shown in Chapters 6 and 14, floods account for 60–70% of the annual P and N loads. An exception is the largest tributary, the Zala River, on which daily observations were initiated by Joó in 1975. Based on these observations, past loads and their temporal changes in the most polluted segment of the lake can be quite accurately derived (the Zala River contributes 90% of the load of Keszthely basin). The long-term annual average total P load, L_{TP}, of the Zala River (1976–81) is approximately 80 tons/yr (the corresponding volumetric load of basin I is about 3 mg/m^{3}d), but in any one year it can reach 115 tons/yr depending on hydrological conditions. The total P load for the entire lake was estimated at 314 tons/yr (with an annual upper limit of approximately 500 tons/yr), corresponding to a lake-wide average of 0.5 mg P/m^{3}d (Chapter 6). The total nitrogen load, L_{TN}, is about ten times larger than that of P.

As seen from the estimates given above, a considerable proportion of nutrients (about 25%) reaches the lake through the Zala River and loads the Keszthely basin, the volume of which is only a tenth of basin IV. Consequently, the volume-related basin loads decrease from west to east inducing a typical longitudinal gradient in many water quality components (see below).

In general, the year-to-year changes in nutrient loads are determined by trend-like effects (related mainly to human activities within the watershed) and random changes are governed by natural processes. Although long-term changes in factors that influence the loads can be clearly demonstrated, conclusions on the loads themselves cannot be drawn because of the scarcity of observational data, the significant stochastic influence of the hydrological regime, and the relatively short records available.

1.3. Characterization of Lake Balaton

Geometry

The main geometrical parameters of the lake are shown in Figure 1.1. It is a shallow lake whose water depth is less than 1.5 m at 9% of the surface area, and less than 3 m at 30%. The deepest point, 11.6 m, is found at the Tihany peninsula. From hydrological and water quality aspects four separate basins can be distinguished (Baranyi 1974): the Keszthely, Szigliget, Szemes, and Siófok basins I–IV (see Table 1.2). This segmentation is used later for data collection and the development of mathematical models.

Temperature

Due to the shallowness of the lake large annual fluctuations can be observed in water temperature. On average, the lake is frozen for two months in winter, whereas in summer the water temperature may exceed 25 °C. Owing to strong wind action, thermal stratification almost never occurs (except perhaps during warm calm periods).

Hydrology

On an annual basis the total water inflow to Lake Balaton (957 mm, equivalent to 570×10^6 m^3/yr) balances evaporation losses (916 mm). The lake has just one outflow (671 mm) at the Sió control gate and is about equal to the precipitation (628 mm). The time required for full replacement of the water volume of the lake (or basin) is the exchange time (Baranyi 1979), which increases from west to east, with a long-term average value of one year for the Keszthely basin, and nine years for the Siófok basin. The filling time (volume/inflow) often used in water quality management practice is considerably shorter than the exchange time: 0.25, 0.72, 0.97, and 1.31 years for the four basins (about two years for the entire lake[1]).

Hydrodynamics

Water motion in the lake is characterized by the slow water throughflow (~ 0.05 cm/s) and the pronounced wind-induced circulation (the velocity ranges between 5–20 cm/s, with an extreme of 1 m/s at the Tihany Strait). The governing wind direction is N–NW. The mean monthly wind speed is about 2–5 m/s, but during storms it can exceed 25–30 m/s. The number of stormy days when the velocity is larger than 8 m/s is about 60 per year. The wind field is characterized by significant temporal and spatial nonuniformities (Béll and Takács 1974) due to the sheltering and deviating effects of the northern hills; the water is almost never found still. According to Muszkalay (1979) the number of longitudinal and transverse seiches reaches 1000 per year. For longitudinal wind conditions, the greatest difference in water level observed along the lake is around 1 m – an extremely high value. During such events the volume of basin IV can change by 10% for short periods of time.

Owing to the geometry of the lake and nonuniformities in the wind field a complicated circulation pattern is set up, which is characterized by large-scale eddies and gyres. These have been demonstrated by physical model experiments (Győrke 1975) and satellite photographs, and they seem to isolate the bays and basins of the lake rather than to strengthen mixing. No quantitative results on spatial and temporal water velocity changes are available for the whole lake, and again, the extent to which they influence water quality is unknown. From water quality observations it can only be stated that

[1]Owing to throughflow the sum of inflows of the four basins is larger than the inflow of the entire lake. This is why the filling time of the lake exceeds that of the individual basins (or their arithmetic mean).

intensive wind-induced water motion is not strong enough to level out the typical longitudinal water quality gradient (see below).

Apart from the components of water motion discussed briefly above, shallow waves should also be mentioned, which may reach up to 1 m in height (Muszkalay 1973). This phenomenon can play a role in sediment resuspension and the net settling of suspended material.

Water Quality

The chemical properties of the lake water reflect the mineral composition of the watershed. The $Ca(HCO_3)_2$ and $Mg(HCO_3)_2$ content is rather high, so that the m-alkalinity is also high (4–5 mval/l) and the lake is buffered effectively at pH 8.3 to 8.7. The biogenic lime precipitation, which is also influenced by the activity of the ecosystem, is considerable.

Similar to the changes in Ca^{2+} content, most water quality components show decreasing concentration profiles from Keszthely to Siófok. This gradient is illustrated for total P and chlorophyll-a (Chl-a) in Figure 1.2 on the basis of the regular observations made by VITUKI[2] (Tóth 1974). The same tendency can be found for total dissolved P, particulate P, suspended solids concentration, or Secchi disk depth. The explanation for the gradient lies in the volume-related loads, which decrease from west to east. For illustration the volume-related total P loads (see Chapter 6) of the basins are also given ($L^*_{TP} = L_{TP}/V$). The longitudinal (west to east) gradient, its temporal changes, and local nonuniformities are clearly reflected in Figure 1.3, which contains two Chl-a profiles (continuous measurements): one for June when the algal biomass is close to its summer minimum and the water quality of the lake is relatively uniform, and one for August when peak biomass values usually occur and considerable spatial changes can be observed.

In the early 1960s water quality throughout the lake was approximately uniform (see Chapter 8), but this has deteriorated as a result of changes in the region with a time lag of a few years. Primary production (in terms of C) was estimated at about 0.25 g C/m^{2}d (Böszörményi *et al.* 1962), but ten years later this had risen to 0.6 g C/m^{2}d at Tihany, and a gradient can now be observed. Primary production at Keszthely in 1973 exceeded 13 g C/m^{2}d (a hypertrophic value), 3 g/m^{2}d at Szigliget in 1974, 2 g/m^{2}d at Szemes in 1976, and 1.5 g/m^{2}d at Tihany in 1977 (Herodek and Tamás 1973, 1978; see also Chapter 8). Algal biomass has also undergone similar changes: 8 g/m^3 at Keszthely and 1 g/m^3 at Tihany in 1965. The corresponding values in 1977 were 60 g/m^3 and 7 g/m^3, respectively (Vörös 1979). Data for Chl-a show the same trends (Felföldy 1981). As is apparent from the data, artificial eutrophication has proceeded not only with time but also in space from west to east.

The changes over the last 13 years can be better traced using regular measurements by VITUKI at nine sampling locations (see Figure 1.2). Figure 1.4 illustrates the changes in the annual mean and extremes of Chl-a concentrations for the four basins (sampling locations 1, 2, 5, and 8). The maximum

[2]Research Centre for Water Resources Development, Budapest, Hungary.

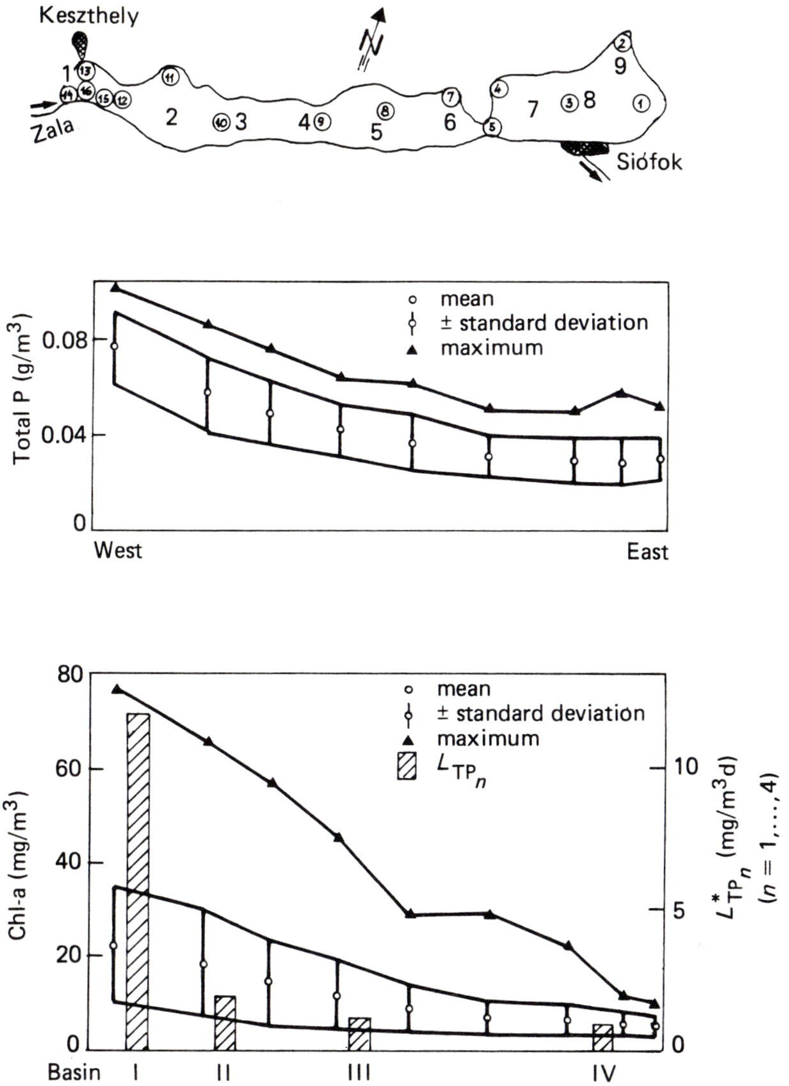

Figure 1.2. Longitudinal distribution of water quality and P loads, 1976–78. 1–9 and 1–16 (encircled; point 6 not included because not used after the initial investigations) are sampling locations of VITUKI and the Central Transdanubian District Water Authority, respectively.

values observed for basins I–III during May–September are also given. From a linear trend analysis the annual deterioration in water quality in terms of the mean is 10% on average (related to 1977). From Figure 1.4 it appears that the mean Chl-a concentration in the lake has increased by a factor of between 5 and 10 over the past 10 years, which is a striking value.

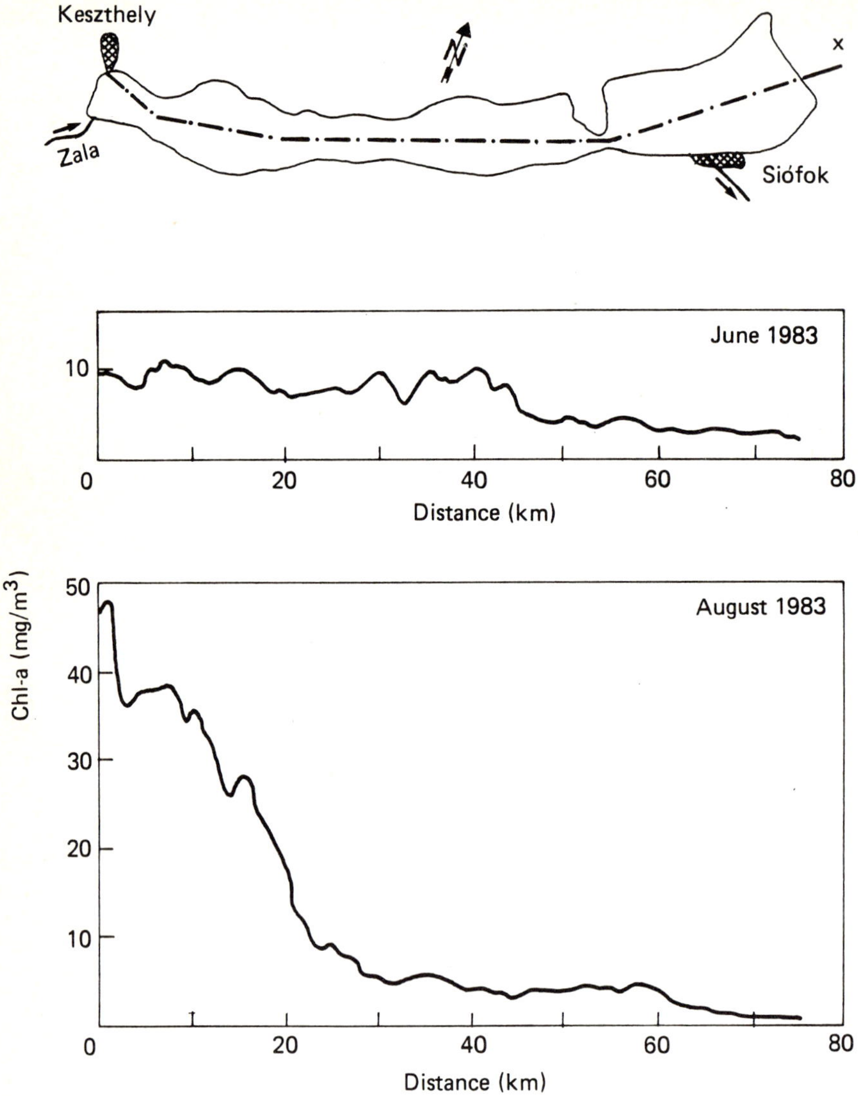

Figure 1.3. Two longitudinal profiles for Chl-a (Hoffmann *et al.* 1984).

Figure 1.4 also illustrates the trophic classes (in terms of the peak Chl-a concentrations) proposed in the OECD (1982) eutrophication study. Accordingly, the Keszthely basin became eutrophic in 1973–74, hypereutrophic in 1977, and definitely hypertrophic during 1982. Basin II has been eutrophic since 1976, and is now also hypertrophic. The water of basin III started to become eutrophic around the end of the 1970s, while that of basin IV became eutrophic in 1982.

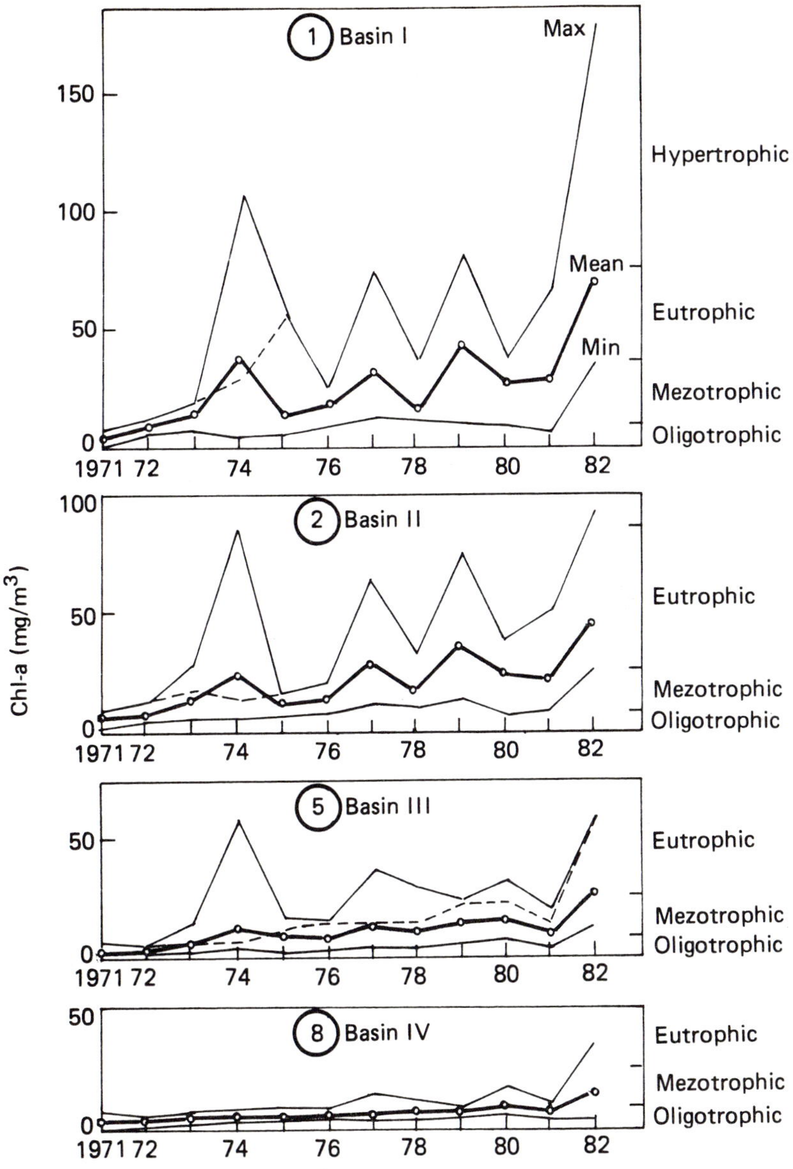

Figure 1.4. Changes in Chl-a profiles from 1971 to 1982 (sampling locations 1, 2, 5, and 8; see Figure 1.2). --- maxima for May–September.

Figure 1.4 clearly indicates how strongly the water quality of the lake is affected by meteorological factors, but it also reflects sampling problems. On average, only ten observations are made per year. Under such conditions the computation of the annual mean is rather uncertain, and even more so when dynamic properties are considered. For example, from Figure 1.4 the observation of an algal bloom in December 1974 resulted in a larger mean Chl-a value for the entire year than that determined for the period May–September.

Figure 1.5 compares the basins of Lake Balaton with those of other lakes. Following the structure of the empirical model of R.A. Vollenweider (OECD 1982) the maximum Chl-a concentration is plotted against the annual average total P concentration, P_λ (note that P_λ is computed from the actual total P load, the hydraulic load, and the filling time; see Chapter 3). For the computation, observational data from 1976–81 were used and the extreme situation found in 1982 is illustrated.

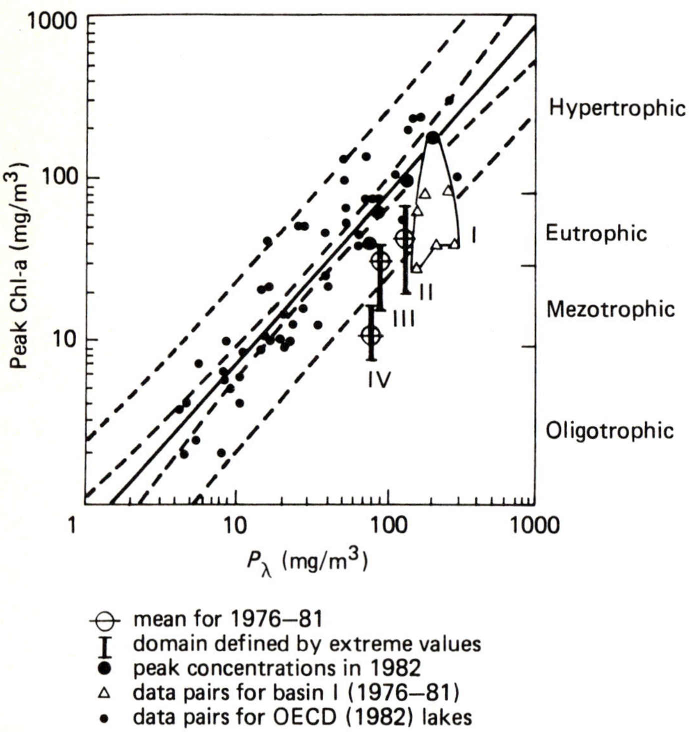

Figure 1.5. Comparison of Lake Balaton with other lakes.

As can be seen, Lake Balaton is not greatly different from other OECD lakes on the double logarithmic scale. The significant fluctuations in water quality and the large peak values of 1982 are apparent, although both are damped by the scale employed. For basin I data pairs for individual years are distinguished since the annual load is known with reasonable accuracy for this part of the lake. The (considerable) scatter is apparent, as is the fact that larger total P loads due to climatic influences do not necessarily produce higher Chl-a concentrations.

P played a limiting role in the eutrophication process until the end of the 1970s, but today N and light conditions are becoming more and more important (Chapter 8). This is a widely observed consequence of the development of eutrophication (see OECD 1982) since, through accumulation in the sediment and subsequent release, P is becoming overabundant in the water.

Observational data on Chl-a, biomass, and primary production reflect the fast dynamics of biological processes. Generally two algal peaks can be seen each year, in spring and around the end of summer (see Figure 1.3 and Chapter 3). In spring diatoms dominate, while in summer there is a mixed population dominated by (at present) blue-green algae (see Chapters 3 and 8). The seasonal changes in algal biomass are similar from year to year, but the actual dynamics depend largely on meteorological conditions that can cause significant deviations in observed peak values in subsequent years. The nutrient cycle generally is also influenced by consumer and decomposer organisms. For Lake Balaton, however, the importance of consumers is small: transport in the food chain is negligible in this direction and most of the organic material is decomposed by bacteria (Chapter 8).

As mentioned above, although P has played a decisive role in producing the present level of artificial eutrophication in the lake, it is also the only nutrient through which the process can be effectively controlled in practice. In other words, even if the ratio of N to other nutrients in the water is less than that in algal cells (i.e., if N is the limiting factor), a P deficiency can occur through P reduction, for which well developed, relatively simple and cheap technologies are available, making P the key element in eutrophication management. For these reasons the efforts were focused on P in the present study.

A relatively high proportion of the total P (30–35%) is found in the dissolved form in the lake. Since dissolved reactive P is constantly low (2–10 mg/m^3), a significant proportion of dissolved P consists of organic P and/or condensated polyphosphates. The temporal changes in the total P concentration reflect the strong influence of wind-induced resuspension, which affects the particulate P content. For this reason neither total P nor – due to its permanently low level – dissolved reactive P can be used to characterize the eutrophication of Lake Balaton.

Sediment and its Interaction with Water

About 90–95% of the P entering the lake accumulates in the top layer of sediment. A considerable proportion of this is not readily available for uptake by algae but, with the progress of eutrophication, it may be transformed to available forms depending on chemical and biological conditions (see Chapter 7). Thus the sediment is of paramount importance in eutrophication control.

Lake Balaton sediment is composed mainly of $CaCO_3$ and fine sand with a mean particle size of about 20–30 μm (larger along the southern shore). The organic material content is low (about 2%) although roughly one third of bacterial decomposition takes place in the sediment. The water content of the sediment is high, at 70–80%. The total P concentration is about 200–600 mg P/(g dry matter), while that of total N is larger by an order of magnitude (Tóth 1976). The total P decreases from north to south and the change in total N is the reverse. The vertical total P concentration does not vary significantly in the sediment (Tóth 1978). Data on the changes in ecologically important (mobilizable) nutrient fractions are not available.

Owing to the shallowness of the lake and wind action, oxygen conditions are ecologically favorable, and aerobic conditions can be observed even in the top layer of sediment. P is bound mainly to $CaCO_3$ and Fe (Chapter 7). Wind-induced resuspension and diffusion are important in sediment–water interactions. Resuspension depends on the turbulent kinetic energy available at the bottom and on wave motion, and can be measured by fluctuations in the suspended solids concentration, which can increase from a background value of about 5–10 mg/m^3 up to 100–200 mg/m^3 in open water, and 500 mg/m^3 in offshore areas (Hamvas 1967, Győrke 1978, Somlyódy 1980). Coupled to this process are changes in nutrient releases from the sediment (and light conditions). The annual net release of P (called the internal load) has recently been estimated to be of the same order of magnitude as the external load (see Chapters 3, 4, and 7).

1.4. Management Alternatives

There is now a wide range of possible eutrophication control methods (see OECD 1982). In the majority of cases the control of nutrient loads (primarily P) is the most effective technique. For large lakes such as Balaton, other methods (e.g., in-lake precipitation, biological ecosystem control, or dredging) are seldom applied, although dredging can be useful for small local areas of the lake.

As far as the control of the P load is concerned, a distinction can be made as to when and where the measures are taken. For instance, control of certain activities and processes can result in the reduction of nutrient emission into the environment (e.g., control of detergent and fertilizer application). If the emissions cannot be controlled, several protection measures can be initiated at various locations within the watershed, of which the most cost-effective combination of options should be selected. The most common techniques are as follows:

(1) Tertiary sewage treatment.
(2) Disposal of sewage residuals outside the watershed (assuming that these will not cause problems for the recipients; for a classic example of a successful application in Lake Washington see Edmondson 1972).
(3) Control through pre-reservoirs on tributaries.
(4) Other controls within the watershed (e.g., land-use practices, fertilizer application technologies).

Options (1) and (2) can be used to reduce point source pollution, and alternative (4) for nonpoint sources. Through pre-reservoirs, both point- and diffuse-sources can be controlled. In the case of Lake Balaton the first three alternatives can be considered as reasonable options; for further discussion see Chapters 4 and 5.

1.5. Data Basis and the Data Bank

The characterization of the lake–watershed system given above approximately reflects the data availability at the beginning of the study, but unfortunately rather little data were accessible on computer. A precondition for the research was the preparation of a data bank, which was then gradually extended. At present this incorporates (mainly for the period 1971–79) water quality data of the networks of VITUKI and the Central Transdanubian District Water Authority (9 and 16 locations, respectively, see Figure 1.2); water balance data (monthly means); meteorological observations (daily temperatures, global radiation, wind data); daily nutrient load observations from the West Transdanubian District Water Authority for two cross-sections of the Zala River (comprising weekly data on more than 20 chemical components, see Chapter 14); load data for other tributaries and wastewater discharges; and results of special observations performed within the frame of the study. In addition, historical data on the long-term development of the watershed (fertilizer applications, tourism, water uses, etc.), and information on sewage treatment plants in the region are available.

A preliminary analysis of the data has already revealed several difficulties. In many cases the temporal frequency and spatial density of observations are inadequate, and the data often reflect certain local influences or the consequences of incorrect averaging. Simultaneous observations for physical, chemical, biological, and hydrometeorological parameters (which would be of great importance from the viewpoint of understanding eutrophication) are not available, so that the significant role of data uncertainty has had to be recognized in the course of the study.

In this chapter we have given a brief overview of available knowledge of the lake and its region, on possible management alternatives, and on data availability. Next we discuss the approach developed and applied to analyze the eutrophication problem of Lake Balaton.

References

Baranyi, S. (1974) Study on the composition and exchange of Lake Balaton water. *Annual Report on the Research of VITUKI, 1971* (Budapest: VIZDOK) pp 141–159 (in Hungarian).

Baranyi, S. (1975) Hydrological characteristics of Lake Balaton. *Vizügyi Közlemények* 2:249–62 (in Hungarian).

Baranyi, S. (1979) Hydrological conditions of Lake Balaton, in S. Baranyi (Ed.) *Summary of Research Results Related to Lake Balaton* (Budapest: VIZDOK) pp 128–76 (in Hungarian).

Béll, B. and Takács, L. (Eds.) (1974) *The Climate of Lake Balaton* (Budapest: Hungarian Meteorological Service) Vol. XL, p 312 (in Hungarian).

Böszörményi, Z., Cseh, E., Felföldy, L., and Szabó, E. (1962) A note on the methodology of photosynthesis measurements in Lake Balaton using ^{14}C techniques. *Annal. Biol. Tihany* 29:39–63 (in Hungarian).

Edmondson, W.T. (1972) Nutrients and phytoplankton in Lake Washington, in G.E. Likens (Ed.) *Nutrients and Eutrophication: The Limiting Nutrient Controversy. Limnol. Oceanogr. Special Symp.* Vol. I, pp 172–93.

Felföldy, L. (1981) *Ecology of Waters: General Hydrobiology* (Budapest: Mezögazdasági Kiadó) p 289 (in Hungarian).

Györke, O. (1975) Studies of the factors affecting the morphological processes in shallow lakes by means of a hydraulic model. *Proc. 16th Congress of the IAHR* (Sao Paulo, Brazil: IAHR) Paper B4, Vol. 2, p 24.

Györke, O. (1978) Changes in suspended solids concentration in shallow lakes and reservoirs. *VITUKI Report No.* 7411/2-323 (Budapest: VITUKI) p 27 (in Hungarian).

Hamvas, F. (1967) A study on erosion in Lake Balaton. *Hidrológiai Közlöny* 47:483–91 (in Hungarian).

Herodek, S. and Tamás, G. (1973) The primary production of phytoplankton in Lake Balaton between April and September 1972. *Annal. Biol. Tihany* 40:207–18.

Herodek, S. and Tamás, G. (1978) The mass and production of phytoplankton, and the eutrophication of Lake Balaton, Part II: The Szigliget Basin in 1974–75. *Hidrológiai Közlöny* 58:384–91 (in Hungarian).

Hoffmann, I., Szilágyi, F., and Gitelsan, A.A. (1984) Application of automatic mobile systems in water quality investigations. *Vizügyi Közlemények* 66:126–33 (in Hungarian).

Horváth, V. and Kamarás, M. (1976) A real picture of the sedimentation problem of Lake Balaton. *Proc. Balaton Ankét* (Budapest: Hungarian Hydrological Society, MHT) p 17 (in Hungarian).

Muszkalay, L. (1973) *Characteristic Water Motions in Lake Balaton* (Budapest: VITUKI) p 91 (in Hungarian).

Muszkalay, L. (1979) Water movement in Lake Balaton, in S. Baranyi (Ed.) *Summary of Research Results for Lake Balaton* (Budapest: VIZDOK) pp 34–127 (in Hungarian).

OECD (1982) *Eutrophication of Waters: Monitoring, Assessment and Control* (Paris: Organization for Economic Cooperation and Development) p 154.

Somlyódy, L. (1980) Preliminary study on wind-induced interaction between water and sediment for Lake Balaton (Szemes Basin), in G. van Straten, S. Herodek, J. Fischer, and I. Kovács (Eds.) *Proc. 2nd Joint MTA/IIASA Task Force Meeting on Lake Balaton Modeling* (Veszprém: MTA VEAB) pp 26–49.

Tóth, L. (1974) Examination of the water quality of Lake Balaton. *VITUKI Report No.* 7782/4/8 (Budapest: VITUKI) p 48 (in Hungarian).

Tóth, L. (1976) Research on the water quality of Lake Balaton. *VITUKI Report No.* 7782/4/8 (Budapest: VITUKI) p 67 (in Hungarian).

Tóth, L. (1978) Determination of the carbon, nitrogen, phosphorus and mercury content of the sediment of Lake Balaton. *VITUKI Report No.* 7781/3/29 (Budapest: VITUKI) p 63 (in Hungarian).

Vörös, L. (1979) The change of quantity and composition of phytoplankton in Lake Balaton. *Proc. Annual Conference Hungarian Hydrologic Society*, MHT, Vol. III, A.16:1–17 (in Hungarian).

CHAPTER 2

The Approach

G. van Straten and L. Somlyódy

2.1. Introduction

In Chapter 1 we introduced the lake eutrophication problem and briefly outlined the many aspects that may play a role in the search for solutions. Obviously, the approach must cope with the characteristics of the problem, such as complexity, interdisciplinarity, and uncertainty. In addition, the approach must overcome the constant conflict between the need for scientific thoroughness and understanding, and the necessity to extract and provide information that can be employed at the policymaking level. Thus, the scientific problem arises as to how to interrelate and integrate system processes that essentially differ on temporal and spatial scales such that the final result not only provides insights, but can also be used as a tool for decision making. This is the theme of this chapter.

First, we outline in general terms the approach as it was adopted and further developed in our work on the Lake Balaton case study, based upon the principle of decomposition and aggregation (Somlyódy 1982a). The decomposition leads to submodels, which can be tackled with the usual modeling procedures. We then discuss the modeling methodology by briefly reviewing the steps in the model-building process, highlighting the role of uncertainty. We explain the principle of aggregation, where the only properties maintained are those that are important at a higher level of integration. Here, too, special attention is paid to the preservation of uncertainty in information.

Second, the features of the approach are illustrated on the basis of development of the lake eutrophication model (LEM), which plays a central role in the study as a whole. It is an account of aggregating complex hydrodynamic, physical, chemical, and biological phenomena (discussed in detail in Part II of this book) to yield a – hopefully – useful, integrated model. The purpose is illustrative, but it should also enhance the reader's understanding of the next two chapters.

2.2. Decomposition and Aggregation

The principle

The procedure starts with a reasonable decomposition of the system into smaller, more tractable units that are accessible for separate and detailed studies (e.g., laboratory and *in situ* experiments, mathematical submodeling based on independent *in situ* data, literature information, etc.), which form a hierarchical structure. We use the word "reasonable" intentionally because, in reality, processes cannot always be separated completely. For example, it is possible to study detailed hydrodynamic circulation models independently of water quality, but the same is not true for the in-lake water quality processes, because their appearance in the form of data will be influenced by prevailing circulation patterns. On the other hand, a "reasonable" separation is often still possible conceptually, because something is known beforehand about processes in water bodies, and a careful inspection of the data will frequently give clues as to which processes are important, irrespective of the disturbances and interrelations with others. It is true, however, that checks on the appropriateness of the decomposition of such interrelated subprocesses can only be made at a higher hierarchical level, unless physical isolation is also possible by performing specially designed experiments (e.g., bottle tests, enclosures, etc.).

The detailed subsystem studies are followed by an aggregation process in order to preserve and integrate only the essentials for higher levels of research in the hierarchical structure. One characteristic element of the aggregation process is the assessment of the relative importance of the subprocesses in view of the spatial and temporal detail in crucial data, with, of course, the objectives of the study in mind. In this way some unimportant processes can be eliminated *a priori* for consideration at higher hierarchical levels. Another characteristic procedure is to achieve aggregation by some form of averaging over space and time or by somehow combining groups of system variables (e.g., by putting different algal species into just one group), thus ruling out unnecessary detail. Sometimes aggregation can be performed by reducing the results of more complex detailed models to simple relationships between characteristic features, which can then be parameterized for use at a higher level.

As a result of the decomposition–aggregation process one can avoid the use of one large, fully coupled model, which is difficult to handle, and instead apply a sequence of correspondingly detailed and aggregated models. Only the highest-level aggregated models are coupled directly, resulting in a relatively simple model at the top of the model hierarchy where management issues are handled.

Application to the Balaton eutrophication problem

The practical application of the principle is best illustrated on the basis of Figure 2.1. The first decomposition that comes directly to mind is the distinction between lake and watershed, since the water quality problem arises

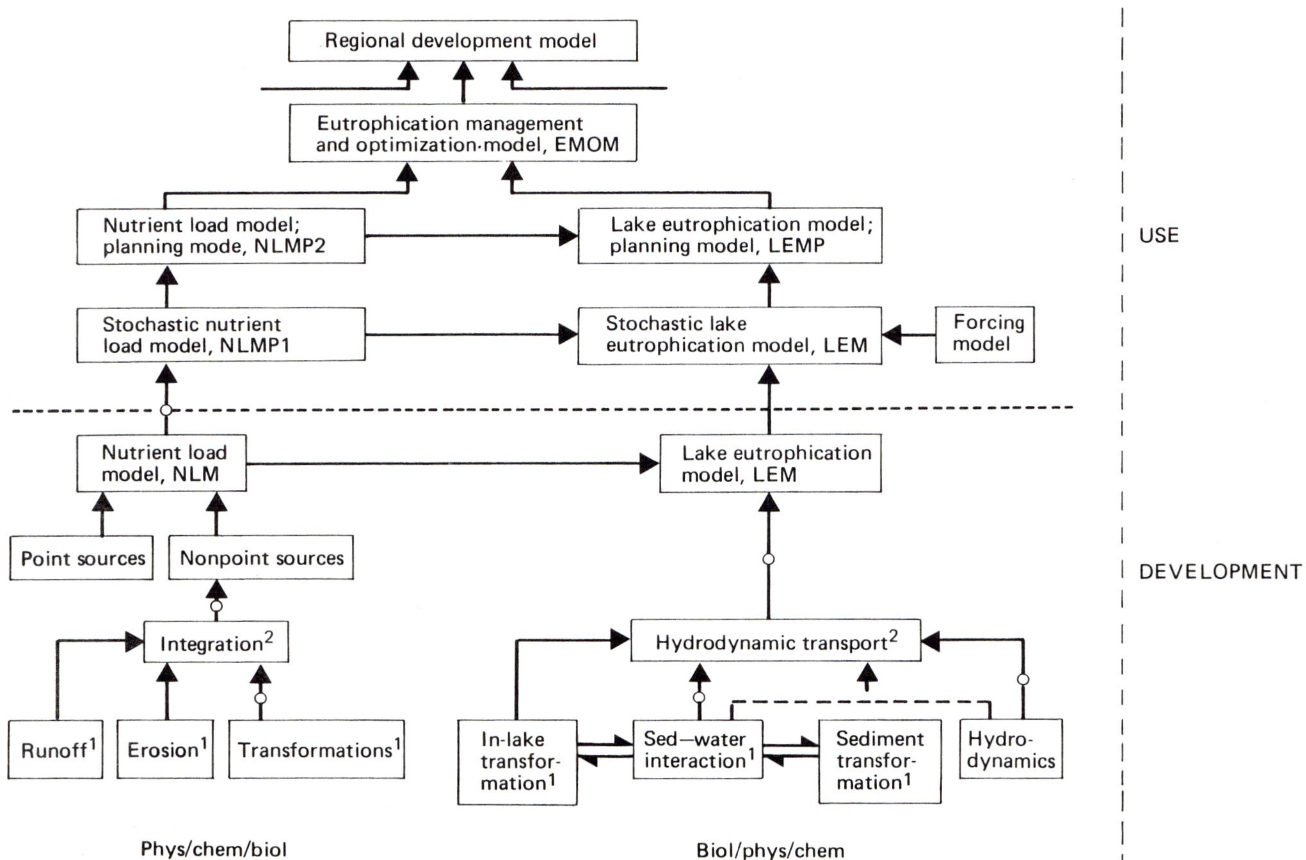

Figure 2.1. Application of the principle of decomposition and aggregation to the Lake Balaton shallow lake eutrophication problem. [1]Local subproblem models; [2]integration of local subprocess models; o points of aggregation.

in the lake, but the causes and almost all the control possibilities come from and must be applied to the watershed. Next, one may wish to separate the region and the lake into segments with more or less homogeneous properties. Here, judgment is important, because homogeneity will seldom be achieved. For example, in the lake local phenomena are affected by hydrodynamic transport and can thus not be separated in a strict sense. However, as was pointed out before, subsystems can still be identified for which "local" submodels can be developed. These are the mutually dependent in-lake transformation subsystem (biological, chemical; see Chapter 8), the sediment subsystem, and the sediment–water interaction subsystem (Chapter 7), which partly depends on the hydrodynamics. A similar procedure can be followed for the watershed, as pointed out in Figure 2.1.

Confining this illustrative discussion to the lake, the next step is to couple the local submodels by including transport terms. Here a first aggregation is made, because in the Balaton case interest focuses on the concentrations along the longitudinal axis of the lake, and cross-sectional details are ignored (see below). This allows the simplification of the complicated three-dimensional hydrodynamics (Chapter 9) into a one-dimensional advective–dispersive transport model (Chapter 10). This model can be simplified even further if one is prepared to accept some loss of spatial detail. For Balaton, the lake was divided into four basins, and the spatial data were averaged for each basin to allow for a comparison with model results. This is the lake eutrophication model (LEM), which is discussed in more detail in Chapter 3.

The LEM cannot be studied without knowing the forcing functions, that is, the external factors that drive the lake systems. Of these, nutrient loads are the most important, and they represent a factor through which water quality can be controlled in practice. Evaluation of historical nutrient input data is an important step in any eutrophication study, and can be quite time-consuming. Methods for evaluating the loads, separated into point and nonpoint sources, and the associated uncertainties are discussed in Chapter 6. Other forcing functions are related to meteorology: precipitation (which influences the water balance), wind speed and direction (which drive the hydrodynamics and mixing in the lake), and solar radiation and length of the photoperiod (as the energy source for photosynthesis). Temperature depends on all of these forcing functions (and thus can actually be modeled), but it is sufficient to handle it as an independent force by using actual data.

At this point it is vital to pay attention to an important distinction, also indicated in Figure 2.1. In the stages of model development, actual past data are used as forcing variables. However, in the next stage of the analysis, when applying the model to practical management problems (e.g., evaluation of water quality under changed load conditions) future scenarios should be employed. At this step forcing functions that reflect critical conditions (from the viewpoint of water quality) should be selected or should be considered as stochastic variables. Thus, these properties of the forcing functions also have to be modeled, as indicated in Figure 2.1; this is discussed further in Chapter 4.

We have now arrived at the management part of the approach, that is, at the stage where models are used to simulate the likely future effects of management actions (controls). First, the purpose of these controls is to change the nutrient loads and, depending on whether these actions are concentrated upon point or nonpoint sources, their variances as well. Aggregation is used here in the sense that groups of controls are combined because at this stage it is immaterial how the proposed reduction can be achieved most economically. The information resulting from this stochastic nutrient load model (NLMP1) is then used as an input to the (stochastic) lake eutrophication model, together with the generated stochastic forcing functions as mentioned above. At this point the structure of the LEM is not changed, only its mode of operation. Next, the output of the LEM, representing water quality variations over time for the four basins as a function of load levels and variability, is aggregated again to yield the LEM in planning mode (LEMP), as described in Chapter 4. This model simply outputs a maximum summer chlorophyll level and variance as a function of load level and variance and can, in fact, consist of a table or graphic plot. The input for LEMP comes from the nutrient loading model in planning mode (NLMP2), which specifies load level and variance as a function of actual concrete control actions in the watershed, but now on an annual basis. Here, each of the various options (sewage treatment, reservoirs, or diversion systems) is associated with costs.

Finally, this information is used in the eutrophication management optimization model (EMOM) in an optimization environment to find the most cost-effective alternative among the various control options. These final stages of the hierarchical modeling approach are discussed in detail in Chapter 4.

2.3. The Model-Building Procedure

Prior to discussing the development of the LEM it is useful to discuss briefly the modeling procedure (Eykhoff 1974, Beck 1983). Figure 2.2 summarizes some of the major activities (indicated by ellipses) and the various inputs needed for the activities plus the major products of each step (rectangles). Note that Figure 2.2 is itself a simplified "model" of the true modeling procedure. At the beginning of each analysis there is a problem that needs to be solved. Usually it is possible to identify by proper problem analysis which aspects are important. From this a selection can be made as to which parts of the total available basic scientific theory will have to be incorporated (in water quality modeling these fields are mainly hydrophysics, chemistry, and biology). Next, a conceptualization is made of how the system at hand operates: its basic processes and interrelations. In this step the objectives are important because they dictate the center of focus. This produces a conceptual base model, in the form of descriptions; no mathematical equations are involved yet. The translation of the conceptual base model into mathematical equations, which we call here model structure postulation (equivalent in some sense to the selection of model type), is dictated in practice by a number of constraints. The objectives play a part, especially with respect to the degree of detail needed. Also, the total time available for the study has a large

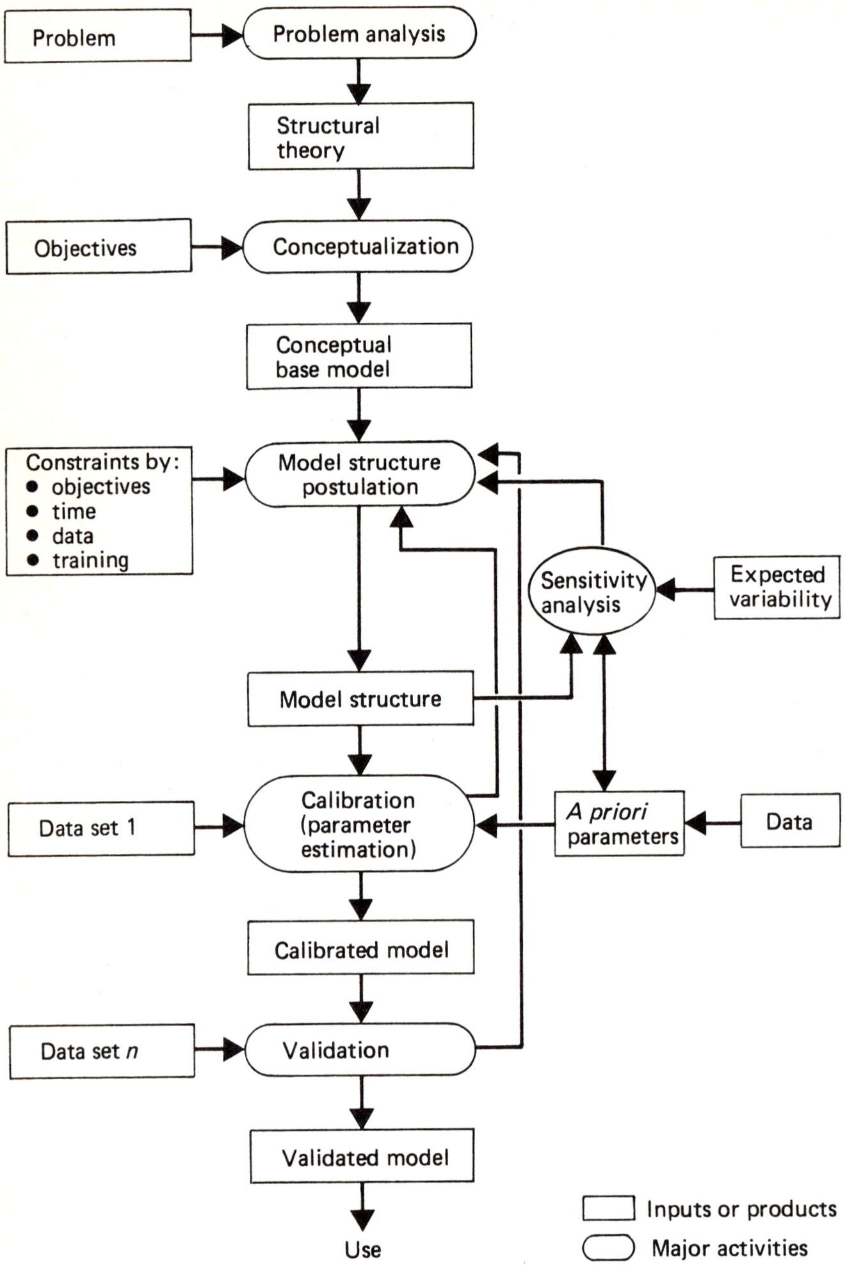

Figure 2.2. The procedure for model development.

influence on this step. Furthermore, the availability of data may guide the model structure choice, because inclusion of model segments for which no data are available is usually not a good practice unless sound theoretical knowledge is available. Finally, the expertise of the analyst or team of analysts usually has a large impact on the actual model postulation (see below).

The procedure outlined above leads to a model structure; this usually needs some computational implementation (not shown in the figure), a step that is not always obvious and may cause several difficulties, especially if the model consists of a set of partial differential equations. We will not deal with these problems in detail here; an example is given in Chapter 3. It is important to note that the model structure itself contains parameters, but does not yet contain values for them; these are obtained in the model calibration phase.

At first glance it may not be clear why parameter estimation is needed, so it is useful to elaborate this aspect further. There are various types of parameters: first, parameters may be well-established physical coefficients, such as the acceleration of gravity, or, for fresh water, the water density. Such parameters need not be estimated from field data; they can be considered as known *a priori*. However, not all parameters can be treated in this way. One of the major reasons is that simplifying the model structure in the process of model structure postulation not only affects the structure but also the parameters. To give an example from hydrobiology: the aggregation of numerous algal species (each with its own growth rate and light-optimum parameters) into one community group leads to new community parameters (community average growth rates and light-optimum parameters). The relationships between these new parameters and the original, individual ones (provided they are known) are not at all obvious. In fact, even the structure of the aggregated model may be affected. Thus, the most practical approach is to estimate such overall (garbage-bin) parameters by comparison with field data. It is immediately apparent from this example that such parameters can have only limited value; their validity is restricted to the given model of the system within the given environmental conditions. Of course, the fact that there may be aggregated parameters that have wider applicability cannot be ignored, but this is not often the case. However, by repeated application to a large number of similar situations we may succeed in specifying a reasonable parameter range. If so, then this is useful information for the next phases of the model-building process.

In the model calibration phase an attempt is made to estimate the parameters by comparing model results with field data. Many aspects are of interest here (see Eykhoff 1974), of which we discuss only a few. For instance, if parameter calibration is successful, one should not forget that the values cannot be considered as deterministic; rather, there are bound to be uncertainties, which arise from model uncertainties and from variance and error in the data, with respect to both observations on the models' state variables and to uncertainties in the forcing factors. A more detailed account of uncertainty in water quality can be found in Beck and van Straten (1983).

	Confident	*Nonconfident*
Sensitive	Fix or calibrate within narrow limits	Calibrate
Nonsensitive	Fix; occurrence in the model may enhance the model's predictive power	Occurrence may be questioned. If the model cannot be restructured: fix

Figure 2.3. Strategy to reduce the number of parameters to be calibrated in the calibration procedure.

Two aspects that can simplify the task of model calibration must be mentioned. First, sensitivity analysis may reveal those parameters to which the model is particularly sensitive. The calibration procedure can specifically be oriented toward these parameters and, because the model is sensitive, they can usually be estimated with reasonable precision. In contrast, it is unlikely that parameters that do not strongly influence the model outcome (i.e., to which the model is insensitive) can be estimated accurately. For such parameters very large confidence bounds appear, in which case it may be better to keep these parameters fixed from the beginning. Second, in some situations parameters can be estimated from isolated experiments, either in the field or with independent model segments. For example, algal parameters can be estimated independently from primary production experiments (e.g., van Straten and Herodek 1982), or settling and resuspension parameters can be estimated from time series data on suspended solids (e.g., Somlyódy 1982a). Parameter values obtained in this way can best be maintained constant, thus reducing the number of parameters to be calibrated. The strategy outlined above for selecting the parameters for model calibration is illustrated in Figure 2.3.

Both the calibration and sensitivity analysis may lead to a need to change the model structure. This loop, model structure postulation, calibration, and model update, is called system identification. If upon calibration no parameter values can be found such that the model output reasonably describes the data set, then the model must be modified. It may be that some process not considered important in the beginning actually was important. It may also be that certain aggregation assumptions appear to be unjustified. Although formal procedures exist, such as error sequence analysis (Eykhoff 1974, Draper and Smith 1966) it should be stressed that subjective elements are also important, especially in water quality. After all, judgment of whether the model captures the main features of the system depends strongly upon the objectives. So, some deviation might be acceptable, but then it would be correct to specify the remaining degree of misfit as uncertainty, which must be maintained in the later stages of model use.

Sensitivity analysis, too, may encourage model simplification, but here the modeler has more freedom of choice as to whether to simplify or not. For example, it may turn out that the model is insensitive to certain forcing functions; the modeler may then wish to simplify the model, but he or she can also intentionally decide not to do this, for example, if the insensitive sections are expected to become sensitive in certain situations in which model use is likely. This may also apply for the preservation of insensitive parameters, but additional confident information from other sources should then be available. The inclusion of nonconfident, insensitive parameters is debatable and should be avoided.

The final product of the identification phase is a model including its calibrated parameters. These then form the input to the validation phase, which is often a critical stage in modeling. The basic idea is to test the performance of the calibrated model against a different data set to that used for the calibration. In case of failure (again depending on subjective judgment related to the objectives) the model structure requires modification and the procedure starts again. Note that validation can be understood in several ways, each leading to different "grades" in validation (see also Thomann 1982). The simplest case is where the validation data set comes from the same system without essential changes in (environmental or managerial) conditions. We may call this "unchanged past" validation. A higher grade of validation is obtained if the model behaves well under changed conditions. Of course, this requires that changes have taken place in the past, and that sufficient data are available about the system's behavior in the past ("changed past" validation). If either condition is not fulfilled validation is essentially not possible, unless the model can be validated against data for another similar system or, perhaps, another spatial segment of the same system. Otherwise, the only remaining way is "the proof of the pudding is in the eating": the model is applied and its performance checked afterwards ("changed future" validation; Thomann calls this "model post-audit").

Ideally, the final product is a calibrated and validated model that can be used for the intended purpose. In practice, the completely ideal case is rarely achieved, if only for reasons of limited time and resources available for the study. Then, in the favorable situation that several, imperfect, models were developed for the same system, it is sometimes possible to compare the predictions of the various models. If no serious differences in overall behavior occur, then the predicted behavior can be accepted as the most likely.

2.4. Lake Eutrophication Model Development as an Illustration of the Approach

The aim of this section is to show how the principles of decomposition and aggregation can be applied in developing the lake eutrophication model. The example is based on experience gained during the Lake Balaton study.

Transport-oriented versus ecology-oriented models

Water quality models can be imagined as structures having three axes: time, space, and the degree of detail of the ecological component. The first axis defines whether the model is dynamic or steady state, the spatial resolution concerns the number of "boxes" or segments, while the third axis defines the number of state variables or compartments. In principle, the location of a model in this three-dimensional space should follow from the dynamics and relative importance of the various subprocesses in the entire system under study. Depending on the character of the problem, there can be any number of combinations, and this fact is well reflected by the large number of water quality models described in the literature.

The classification of water quality models in general is a huge task (see Cembrowicz *et al.* 1978), but simply by analyzing the degrees of detail along the spatial and ecological axes, most models can be broadly classified into two groups (Somlyódy 1982b):

(1) Transport-oriented water quality models, which emphasize the (hydrodynamical) transport equations and describe the fate of the water quality constituents in space and time. Consequently, the models take the form of partial differential equations. If chemical and biological transformations are taken into account these terms are usually strongly simplified, for example, by assuming simple first-order reaction kinetics.
(2) Ecology-oriented water quality models, which emphasize the chemical, biological, and ecological aspects, and in which water transport is simplified or ignored completely (e.g., by assuming complete mixing). These models consist of a set of ordinary differential equations.

A similar distinction has been made by Watanabe *et al.* (1983) who use the terms engineer-developed versus biologist-developed models. This terminology expresses clearly how preferences for one approach or the other are determined by professional background and training. There are, however, also more intrinsic explanations for the preservation of this distinction. We are concerned here with differences in the levels of theoretical knowledge available from the various disciplines, which should be integrated in a model. For example, hydrodynamics and transport phenomena have a quite solid theoretical background compared with that of the biology and chemistry of lakes or rivers. Consequently, hydrodynamic models contain only a limited number of calibration parameters, and measurement data have only limited or no effect upon the basic model structure. On the other hand, for the biological segments of a model, data (that is, *in situ* and laboratory observations; so-called measurement knowledge) are extremely important in setting up a model, in estimating its parameters, and in identifying the model structure. Thus, in this case there is a greater need for formal parameter estimation techniques – as explained in the previous section – and these are much better established for models with an ordinary differential equation structure than for one of partial differential equations. Or, in other words, spatial and transport detail is sacrificed in order "not to complicate the situation further".

In principle, there are no difficulties in developing, between these extremes, models that combine the transport- and biology-oriented approaches. One would then start by writing down transport equations for each biologically significant state variable, and add the proper interrelated biological and chemical transformation terms. Such a coupled transport–quality model would then consist of a set of nonlinear three-dimensional partial differential equations. Although examples of attempts in this direction exist, at the expense of enormous computing times (see Chen and Smith 1979 for deep lakes), it is debatable whether this is a feasible and useful approach, most notably because the calibration of such a model would be an almost insurmountable task. A general requirement for calibration is that the number of observations is larger than the number of state variables. Since such models typically contain a huge number of state variables, excessive requirements would have to be set on the precision and detail of the data collection programs, which could hardly be met in practice. Thus, there is a need to aggregate somehow the spatial detail in order to reduce the dimensionality and to approach better the ordinary differential equation structure in terms of the calibration needs. The next sections describe, on the basis of the Balaton case, how this goal can be achieved.

The basic water transport–quality equation

A full three-dimensional water quality–transport model in formal notation is given by a set of differential equations:

$$\frac{\partial c}{\partial t} = -u\frac{\partial c}{\partial x} - v\frac{\partial c}{\partial y} - w\frac{\partial c}{\partial z} + \frac{\partial}{\partial x}\left[\varepsilon_x \frac{\partial c}{\partial x}\right] \tag{2.1a}$$

$$+ \frac{\partial}{\partial y}\left[\varepsilon_y \frac{\partial c}{\partial y}\right] + \frac{\partial}{\partial z}\left[\varepsilon_z \frac{\partial c}{\partial z}\right] + \boldsymbol{r}(\boldsymbol{c}) \ ,$$

subject to initial conditions $\boldsymbol{c}(0, x, y, z) = \boldsymbol{g}(x, y, z)$ and boundary conditions for boundary points (x_b, y_b, z_b):

$$ac + b\frac{\partial c}{\partial j} = l(\boldsymbol{c}) \ , \tag{2.1b}$$

where

- $\boldsymbol{c}$ = n-dimensional mass concentration vector for the n state variables
- t = time
- x, y = horizontal direction coordinates
- z = vertical direction coordinate
- u, v, w = fluid velocity components in the x, y, and z directions, respectively

$\varepsilon_x, \varepsilon_y, \varepsilon_z$ = turbulent eddy diffusion coefficient for the directions x, y, z, respectively
$\mathbf{r}$ = n-dimensional vector of rates of change of the state variables due to biological or chemical reactions as a function of the concentration $\mathbf{c}$
g = formal function symbol
a = convective flow velocity at the boundary
b = diffusion coefficient at the boundary
j = direction perpendicular to the boundary
$\mathbf{l}$ = formal symbol for the load mass flux at the boundary for the n state variables, which may be a function of $\mathbf{c}$.

Equations (2.1) state that the rate of change in the concentration vector with time at a particular point in three-dimensional space equals the rate of change due to advective transport (first three terms), due to diffusive transport (next three terms), and due to biochemical reaction ($\mathbf{r}$). The final term is typically a complicated function of $\mathbf{c}$ and the environmental forcing variables. The effect of nutrient loads on the systems, and all other terms at the boundaries, such as sediment–water exchange, are expressed by the boundary conditions. For ease of notation we have chosen to write these in formal terms only. The equation merely states that the total load flux equals the sum of advective and diffusive mass fluxes at the boundaries. Note that $\mathbf{l}$ may be negative at some or all boundary points. It should be pointed out that all model coefficients are in principle functions of external variables, such as wind speed (which influences $u, v, w, \varepsilon_x, \varepsilon_y, \varepsilon_z$, and $\mathbf{l}$ through hydrodynamics), solar radiation, temperature, etc. [influencing the parameters hidden in $\mathbf{r}(\mathbf{c})$ and, again, $\mathbf{l}$].

It is obvious that in all practical cases equations (2.1) need to be simplified. It is perhaps good to note that even equations (2.1) are aggregated equations, because the representation of the turbulent diffusive fluxes as functions of gradients is a particular approximation of the more general Reynolds fluxes. As a consequence, $\varepsilon_x, \varepsilon_y, \varepsilon_z$ are in principle a function of time and space.

Aggregation

A logical start to the procedure of simplifying the basic equations (2.1) is to look at the geometry of the lake in question. For Lake Balaton, its elongated shape suggests that the longitudinal dimension is important. This is substantiated by the data, which show a remarkable longitudinal gradient. Thus, in aggregating we would like to preserve the longitudinal axis, but at the same time, the form of the lake suggests that it would be possible to aggregate the cross-sectional differences by spatial averaging of equations (2.1). There are three arguments to support this view. First, there are no data on the important state variables over the vertical and lateral directions (with the exception of infrequent chlorophyll data along the shore). So, even if a three-dimensional model could be successfully developed, it would not be possible to calibrate it. Second, the prevailing wind direction is perpendicular to the lake's horizontal axis, and, intuitively, this promotes a reasonable mixing in this direction. Of course, one may argue that this effect is counteracted by loads in the near-shore region, and it is certainly true that local water quality problems associated with these loads have been observed (e.g., hygiene problems related to bacteria from sewage). Also, it would be of interest to establish whether the shallow depth at the shore and hence higher average light conditions leads to algal growth that is more or less than average. These effects may all occur, but the preservation of the lateral dimension in order to describe these local problems would also demand that the hydrodynamical equations governing the velocities and diffusion coefficients in equations (2.1) could be satisfactorily solved for the near-shore region. Unfortunately, this is not quite the case to date, and this is another argument for removing the lateral dimension from equations (2.1).

Following this line of thought, our first step is, then, to average equations (2.1) over the cross section. Note that this process itself does not induce any assumptions: it can always be done. The new, one-dimensional advective dispersion equation is now

$$\frac{\partial C}{\partial t} = -U\frac{\partial C}{\partial x} + \frac{\partial}{\partial x}\left[D\frac{\partial C}{\partial x}\right] + R(C) + L_e + L_i(C) , \tag{2.2}$$

subject to the initial condition $C(0,x) = g(x)$ and the boundary conditions $D(\partial C/\partial x) = 0$ at $x = 0$ and $x = l$, where l is the length of the lake. Note that, after integration over the cross section, the boundary condition of equations (2.1) can be directly incorporated in equation (2.2). This is possible because the load at a certain point x can now be treated as being distributed over the cross section. A distinction is made between external loads L_e (associated with river inflow, sewage loads, and precipitation) and internal loads L_i, which are primarily associated with interface fluxes (such as exchange at the sediment–water boundary or at the water–air surface). The latter may depend upon the cross-sectional average concentration C. Both L_e and $L_i(C)$ are (possibly discontinuous) functions of x and t.

It is important to call attention to the fact that the reaction terms must now be understood as cross-sectional averages. Theoretically, this has consequences for the aggregated terms R: they are not the same as r if spatial nonlinearities occurred in the original formulation (e.g., in the depth-dependent primary production). These averaging errors are usually not recognized in the biologist-developed models. One may perhaps justify this by referring to the large uncertainties that already exist in the biological terms.

Of course, spatial averaging does have some important side effects. First, the averaging results in only one cross-sectional averaged flow velocity U, which depends on the full flow pattern. Thus, U is a parameter that must be derived from detailed *in situ* observations, or from averaging two- or three-dimensional models that have been calibrated from *in situ* observations. Second, cross-sectional averaging introduces nonzero averages of cross-products of the deviations from the average of velocity components and concentrations, and so an additional transport term in the longitudinal direction arises for which further assumptions have to be made. Analogously to the approximation of the Reynolds terms this is usually done by equating this flux to the product of the longitudinal concentration gradient and a proportionality constant, which is known as the dispersion coefficient, D. So, the dispersion coefficient is an artificial coefficient that encompasses the nonuniformities in flow velocity and concentration in the cross section. Its value must be obtained by derivation from calibrated hydrodynamic models. Only in special cases where the boundary fluxes are known can D be estimated directly from observed concentration profiles (e.g., in estuaries from salinity profiles). Both U and D are functions of space and time.

Thus, a significant step in the three- to one-dimensional aggregation is to find methods of evaluating U and D. Preferably, this should be performed using a full three-dimensional hydrodynamic model. Attempts made in the Balaton study show that this fails, essentially because of the difficult question as to the assumptions about the distribution of eddy viscosity over the vertical (Chapter 9 discusses this and the following steps in more detail). An essential compromise is to integrate the flow equations over the vertical, so again performing some aggregation, and to calibrate this model for the two basic parameters (bottom friction and drag coefficient). Data used for this were water levels and velocities at some strategic points. Thus, this two-dimensional model provides the vertically average velocities as a function of lateral and longitudinal dimension, dynamically, from which $U(x,t)$ can be computed. Theoretically, nonuniformities in the cross-sectional concentrations also influence the dispersion coefficient, although reasonable assumptions are possible here (for details see Chapter 9), and again $D(x,t)$ can be computed by an algorithm that uses the calculated nonuniformity of the cross-sectional flow velocity. It should be noted, however, that validation of the model that predicts the flow field was difficult for transverse winds, and could only be performed in a stochastic sense (Somlyódy 1983). This should warn the analyst not to place too much trust in the detailed properties of the results; rather, it should encourage a sensitivity analysis to identify the severity of these basic uncertainties.

Equation (2.2) is still highly dynamic. For example, flow may change rapidly within a time scale of hours due to seiche motion. The question now is whether this degree of detail is desirable and necessary. In principle, this must be studied by using the full coupled equation but, from inspection of the seiche excursion (on the order of 500 m) and from the fact that concentration gradients on this spatial scale are small, one can infer that seiche motion may be ignored. Again, this can be verified by comparison with a model where this degree of detail is preserved. In general, given the time scales of interest for the biological reactions (on the order of days), no large errors are made when U is replaced by hydrological throughflow only. In order to maintain consistency in temporal detail, D can also be averaged over a seiche period.

Next, one may wish to make further simplifications of the resulting water quality–transport model. For example, the dispersion coefficient is still a function of space and time, and this is difficult if the model is to be used in a planning phase. As shown in Chapter 10, no large loss of detail occurs if D is assumed to be time- and space-independent. The properties of the advective–dispersive model obtained by aggregating through integration of relevant information from an independent submodel (hydrodynamics) are interesting, as pointed out in Chapter 10. The most important achievement is that good spatial detail in the longitudinal direction is obtained. It should be noted that part of this detail originates from assumptions about the load distribution along the longitudinal axis of the lake. Errors in these assumptions will be reflected in the longitudinal distribution.

At this point, the analysis could stop. However, the question of parameter estimation for the biological part is not easy to solve with a transport equation of the type (2.2), although not impossible in principle. Again, data availability may be a limiting step here, so that further aggregation is useful, such as by looking for larger lake segments that are then treated as uniform. In this aggregation step hydrodynamics are simplified even further, resulting in a model with a series of interconnected, completely mixed cells or boxes, whereas the biological component is preserved. In Figure 2.1 this is represented by the transition of the coupled hydrodynamic transport model to the four-box lake eutrophication model. Clearly this again entails a loss of spatial detail but, on the other hand, it removes the detailed load distribution ambivalence to a large extent. The data problem is also less severe, although one must be careful here: spatial averaging requires that the data must be representative for the spatial average. In general, this condition is not fulfilled, and consequently one has to take an aggregation error in the data into account.

Chapter 10 discusses in detail a framework for ensuring the maximum similarity between box models, resulting from this further aggregation and the advective–dispersive equation. The concentrations after averaging the predictions for the coupled model over the basins turn out to be almost equal to those of the four-box model, both in level and as far as the dynamics are concerned. So, the result of the aggregation operation is that the representation of Lake Balaton by four ideally mixed basins of various sizes, originally chosen on intuitive grounds and suggested by Figure 1.1, is quite appropriate.

With this conclusion we close this illustrative example of application of the decomposition–aggregation principle. The reader will encounter this principle again in Chapter 4 for the higher hierarchical levels of Figure 2.1.

References

Beck, M.B. (1983) A procedure for modeling, in G.T. Orlob (Ed.) *Mathematical Modeling of Water Quality: Streams, Lakes, and Reservoirs*. Wiley–IIASA International Series on Applied Systems Analysis, Vol. 12 (Chichester, UK: Wiley) pp 11–41.

Beck, M.B. and van Straten, G. (Eds.) (1983) *Uncertainty and Forecasting of Water Quality* (Berlin: Springer).

Cembrowicz, R.G., Hahn, H.H., Plate, E.J., and Schultz, G.A. (1978) Aspects of present hydrological and water quality modelling. *Ecological Modelling* 5:39–66.

Chen, C.W. and Smith, D.J. (1979) Preliminary insights into a three-dimensional ecological–hydrodynamic model, in D. Scavia and A. Robertson (Eds.) *Perspectives on Lake Ecosystem Modeling* (Ann Arbor, MI: Ann Arbor Science Publishers) pp 249–80.

Draper, N.R. and Smith, H. (1966) *Applied Regression Analysis* (New York: Wiley).

Eykhoff, P. (1974) *System Identification – Parameter and State Estimation* (Chichester, UK: Wiley).

Somlyódy, L. (1982a) Modelling a complex environmental system: The Lake Balaton study. *Mathematical Modelling* 3:481–502.

Somlyódy, L. (1982b) Water-quality modelling: A comparison of transport-oriented and ecology-oriented approaches. *Ecological Modelling* 17:183–207.

Somlyódy, L. (1983) Input data uncertainty and parameter sensitivity in a lake hydrodynamic model, in M.B. Beck and G. van Straten (Eds.) *Uncertainty and Forecasting of Water Quality* (Berlin: Springer) pp 129–56.

van Straten, G. and Herodek, S. (1982) Estimating algal parameters from vertical primary production profiles. *Ecological Modelling* 15: 287–311.

Thomann, R.V. (1982) Verification of water quality models. *J. Env. Eng. Division, ASCE* 108(EE5):923–40.

Watanabe, M., Harleman, D.R.F., and Vasiliev, O.F. (1983) Two- and three-dimensional mathematical models for lakes and reservoirs, in G.T. Orlob (Ed.) *Mathematical Modeling of Water Quality: Streams, Lakes, and Reservoirs.* Wiley–IIASA International Series on Applied Systems Analysis, Vol. 12 (Chichester, UK: Wiley) pp 274–336.

CHAPTER 3

Lake Eutrophication Models

G. van Straten

3.1. Introduction

The subject of this chapter is the relation between nutrient input and response of lakes, particularly the biological response. In the last decade, the complexity of physical, chemical, and biological processes in lakes has prompted the development of modeling as an instrument for providing a simplified, and hopefully more understandable, picture of the system. Modeling must be understood in a broad sense as a means of organizing and representing available information in a systematic and useful way. As was pointed out in Chapter 2, models can serve as an aid to improve our understanding, and to guide us in further research. This is important, but in view of the subject of this book the usefulness of models for judging the effectiveness of management options for lake eutrophication abatement should receive special attention. Thus, the aim of this chapter is to explore not only what we have gained from modeling exercises thus far, and what are the strong and weak spots, but also what we can conclude about the expected effects of lake eutrophication abatement programs. This discussion is based upon the experiences gained from our work on the Lake Balaton case study, but the conclusions should also be valid for other lake eutrophication problems as well.

Loosely speaking, in lake eutrophication modeling three broad classes may be distinguished:

(1) Statistical models, i.e., models based upon correlations between a limited number of aggregated variables, collected for a large number of lakes. A typical output is an annual mean or summer peak algal concentration as a function of annual nutrient load. Time information is absent.

(2) Dynamical models, i.e., models based upon mass balance considerations, mostly in the form of differential equations, with time as an independent variable. The output is the evolution of algal concentrations and other variables throughout the year, as a function of dynamically varying environmental and load conditions.

(3) Steady state models, i.e., models equally based upon the mass balance concept, but in which only equilibrium, steady state, or a series of steady states are considered. Thus time is eliminated, or is maintained only in a rudimentary form as a discrete sequence series. The output is in the form of predictions of maximum possible algal concentrations (i.e., the algal bloom potential, not necessarily true values) as a function of the prevailing environmental conditions, which are assumed to be constant during the selected period of time.

3.2. Statistical Modeling

In the early 1970s the OECD initiated a study to investigate the problem of lake eutrophication on a global scale. All over the world scientists volunteered to collect and supply data about the lake system of their interest. Such data were generally aggregated values, e.g., annual average total P concentration, annual average chlorophyll, summer average chlorophyll, maximum summer chlorophyll, etc., together with nutrient load information and lake geomorphological and hydrological data such as average residence times and surface hydraulic loads. In addition, the investigators were asked to classify their lake(s) as oligotrophic or eutrophic. The information was first used to elaborate a method for classifying a lake as eutrophic or oligotrophic when hydraulic and areal P loads were given (see Jones and Lee 1982), but since the distinction between them is based upon subjective judgment, and consequently is rather uncertain, this method is only indicative and is of little interest for management purposes.

More useful for management is the attempt that was made to relate some of the observed lake variables to P loads. After numerous trials the promoter of the OECD study, R.A. Vollenweider, proposed the following empirical model for the relation between annual average total P concentration and load, with hydraulic load and "filling time" as parameters (Vollenweider and Kerekes 1980, 1982):

$$P_\lambda = \frac{L_s}{q_s} \frac{1}{1 + \sqrt{\tau_w}} , \tag{3.1}$$

where

P_λ = annual average total P concentration (mg/m^3)
L_s = areal annual total P load (mg/m^2yr)
q_s = hydraulic load = inflow per unit surface area (m/yr)
τ_w = filling time (residence time) = volume/inflow (yr).

Conceptually, this equation can be understood by comparison with a mass balance analysis. If the lake is ideally mixed, and if it is assumed that settling occurs in proportion to the total P concentration, a dynamic P balance reads

$$H \frac{dP(t)}{dt} = L_s(t) - v_s P(t) - q_s P(t) \ , \tag{3.2}$$

where

H = average depth = volume/surface area (m)
$P(t)$ = actual P concentration (a function of time) (mg/m^3)
v_s = settling velocity (m/yr).

By averaging this equation over a year the left-hand terms become equal to zero, provided that the lake is in a steady cycle from year to year. Under the additional assumption that v_s and q_s do not vary too strongly throughout the year, the final result derived from equation (3.2) is

$$P_\lambda = \frac{L_s}{q_s} \frac{1}{1 + v_s/q_s} \ . \tag{3.3}$$

Comparing this mass balance result with equation (3.1) we see that the ratio of P settling velocity and hydraulic load in any lake is empirically proportional to the square root of the filling time. It should be stressed that the Vollenweider model is empirical; there is no obvious physical explanation for the proportionality observed.

Figure 3.1 shows the line of best fit of the OECD lakes when the annual average total P concentration [the left-hand side of equation (3.1)] is plotted against the right-hand side, sometimes referred to as the "normalized annual total P load" (Jones and Lee 1982). Also shown are the ranges observed around this line of best fit. The 45° line is the line that would have occurred had equation (3.1) been exact. We may see that this "model" line is well within the boundaries of the empirical best fit, but since the plot is double logarithmic the deviations in concentration terms are considerable.

Some simple calculations, based on available data, allow the positioning of the four Balaton basins in the plot of Figure 3.1 (shown as dots), and it can be seen that all four fall well within the OECD study boundaries. This seems to suggest that Lake Balaton is not an exception in this respect, as is sometimes believed given the exceptionally large Ca content and its associated P binding capacity (cf. Park 1978). On the other hand, the wide range, masked somewhat by the logarithmic scale, makes such conclusions less relevant, the more so since the OECD line was obtained for an ensemble of lakes, most of which were deep. In fact, when only shallow lakes and reservoirs were incorporated a slightly better correlation could be obtained using the model

$$P_\lambda = \frac{L_s}{q_s} \frac{1}{1 + 2\sqrt{\tau_w}} \tag{3.4}$$

(Clasen 1980). However, for lakes where τ_w is relatively short, such as Lake Balaton, no large differences are obtained, and the improvement is only marginal in the light of the uncertainty bounds.

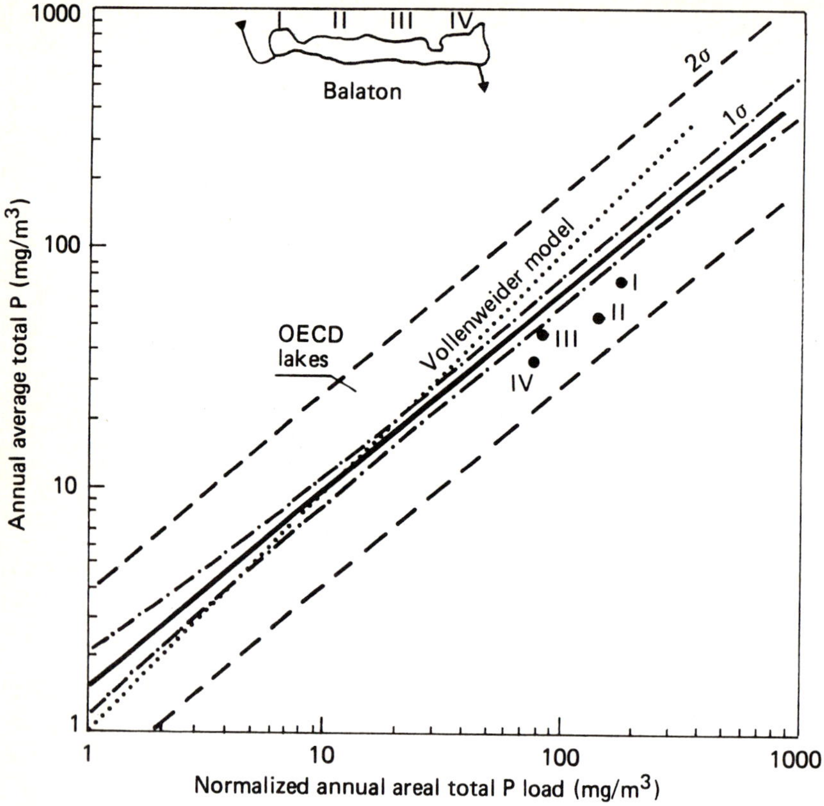

Figure 3.1. Empirical relationship between annual average total P concentration and normalized load for OECD lakes. —— OECD best fit; —·—·— 1σ and – – – 2σ boundaries; .... Vollenweider model line [equation (3.1)]; • the four Balaton basins.

Since Lake Balaton fits within the framework of the OECD studies, it can be concluded that load reductions are likely to lead to a reduced annual total P concentration in the lake, although the uncertainties are large. Perhaps a more direct impression of the extent of these uncertainties can be obtained by plotting the annual areal load against the annual average total P for a particular lake on a linear scale. This is done in Figure 3.2 for the Keszthely basin. The conclusion is obvious: all we can say is that the trend is clear, but more quantitative statements cannot be made. In lake eutrophication management the total P load is only an indirect indicator of water quality; one would also be interested in quantities like biomass or chlorophyll-a (Chl-a) as indicators of possibly objectionable algal blooms. This aspect has been dealt with within the OECD study, and in general a strong correlation was found between maximum Chl-a or summer average Chl-a and total P concentration. Consequently, similar plots can be made to express the dependence of these

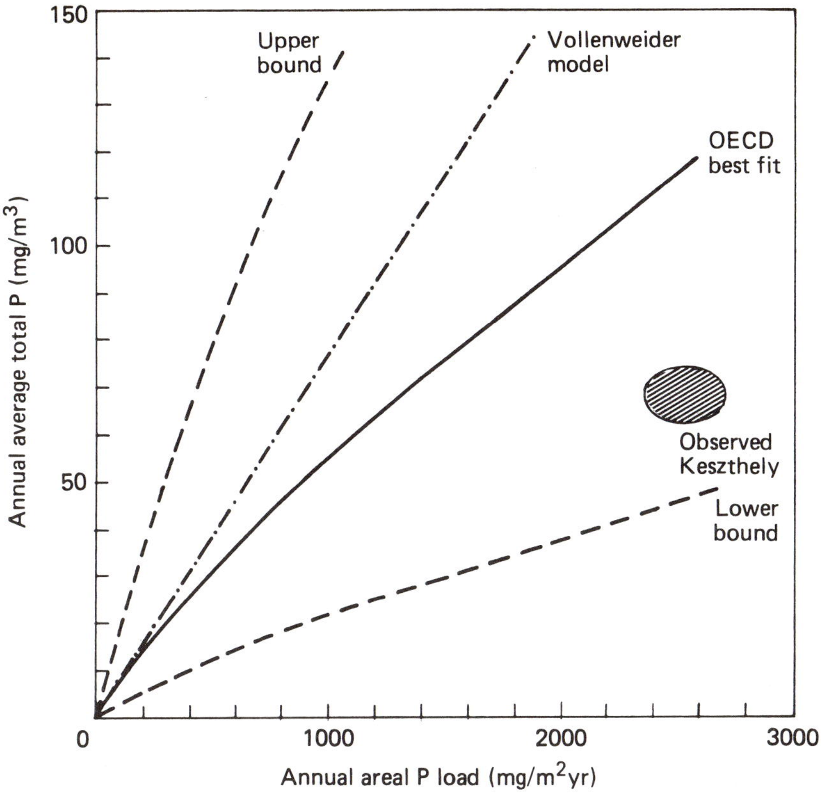

Figure 3.2. Relation between annual average total P and areal annual load predicted by OECD for the Keszthely basin, on a linear scale.

quantities on total P loads, such as that shown in Figure 1.5. Again, the trend is that load reductions will lead to lower Chl-a concentrations, but, again, the uncertainties are large.

Finally, Figure 3.3 shows the trends suggested by the OECD results for the four Balaton basins on a linear scale. Note that the load is expressed here per unit basin volume. The differences in response slopes result from the different hydraulic loads and residence times for the four basins. An impression of the implicit variability is provided by plotting the actual, uncontrolled position for the Keszthely basin for a number of consecutive years. The lines drawn indicate the average direction of change predicted by the OECD best fit curve, but it is clear that the actual behavior does not follow that prediction, at least not on a year-by-year basis. One of the reasons for this is that meteorological fluctuations have quite a strong influence on the actually observed maximum Chl-a concentrations, as shown in Chapter 4, so that the statistical approach cannot be used on a year-by-year basis.

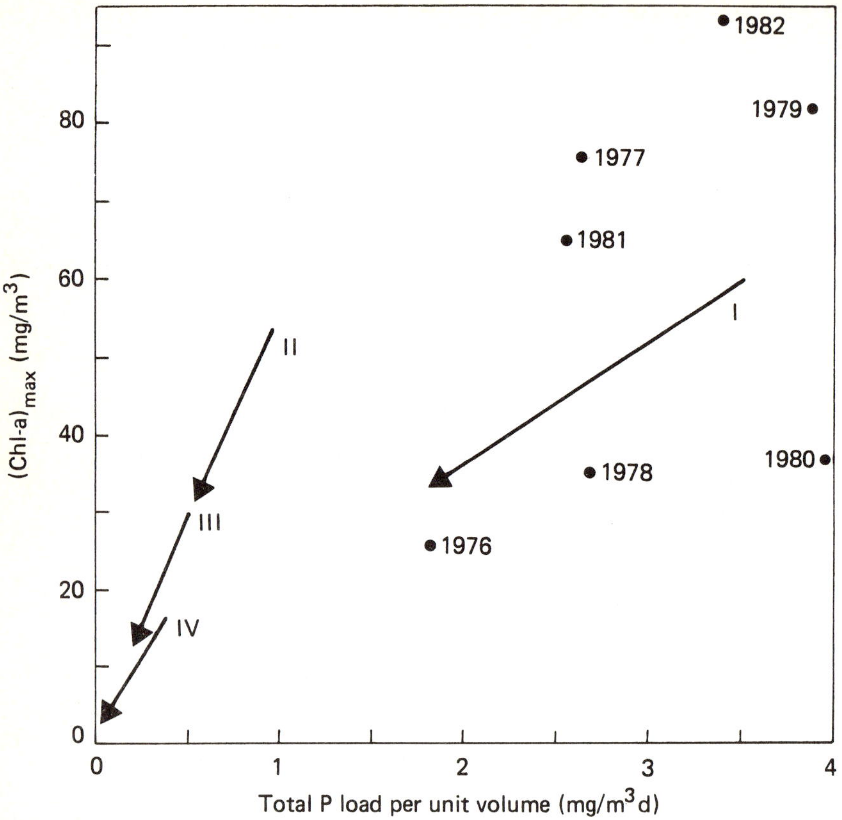

Figure 3.3. Trend in reaction of $(Chl\text{-}a)_{max}$ to P load reductions according to OECD empirical relations. Historical variability for Keszthely basin is also shown.

The discussion in this section leads to the following conclusions:

(1) The statistical approach is a tool for rough and preliminary analyses of long-term effects of management options. If the management is straightforward, with clear-cut solutions without much balancing among various alternatives – usually the case in the initial stages of implementation – the Vollenweider plot is useful for judging management effectiveness.

(2) The statistical approach leads to the conclusion that the long-term response of lake water Chl-a concentrations is practically linear with the load per unit volume.

(3) The statistical approach must not be used to predict the behavior of a particular lake on a year-by-year basis.

(4) If management requires more sophisticated cost-effectiveness information (e.g., in the allocation of resources for lake restoration), then the statistical approach has just too large an uncertainty to be usefully employed.

(5) If spatial and temporal details of water quality are important (e.g., if there are choices as to where measures should be implemented along a lake), and if differentiation in management actions is considered (e.g., not only load reductions, but also dredging, or changing the ratio of available to unavailable nutrient forms), then the statistical approach simply cannot be applied at all.

In all situations where more accurate or more detailed answers are needed, other methods must be explored. It has frequently been claimed that dynamic models could provide such answers, and so we examine this in greater detail in Section 3.3.

3.3. Dynamic Modeling

In contrast with the statistical approach, dynamic models are based on dynamic mass balance equations which represent in mathematical form a conceptual picture of the system. These models have advantages over the statistical approach in that they allow for more temporal and spatial detail, they generally consider a number of relevant variables (such as phytoplankton, orthophosphate, detritus, zooplankton, etc.) rather than just one or two, and they are potentially more accurate once properly calibrated. The increased spatial and temporal resolution may be desirable if the results are to be used for management purposes, especially when it comes to the description of peak events that cause most of the nuisance. The disadvantages of the dynamic modeling approach stem partly from the same items constituting its advantages: the larger detail must be bought with large, sometimes exhaustive, data requirements. Moreover, accuracy is frequently affected in a negative way because the mechanisms of some of the processes incorporated in the models may not be known to a satisfactory degree. Despite these shortcomings dynamic models have proved successful in organizing and disseminating results of scientific research, and have been helpful as guidelines for further *in situ* investigations. And, as will be shown below, specific conclusions of management interest can be drawn, even with the limitations outlined above.

Brief overview of dynamic modeling

The literature on dynamic lake water quality modeling is vast. An almost classic example is the study by Di Toro *et al.* (1974) on modeling the phytoplankton dynamics in the San Joaquin delta. Many other examples exist, such as the fine account of marine system modeling by Kremer and Nixon (1978). Overviews and discussions of new developments can be found in Canale (1976), Scavia and Robertson (1979), and Orlob (1983).

In Chapter 2 we discussed the various terms that constitute a dynamic lake eutrophication model as an example of decomposition and integration. For the sake of convenience we restate the various terms in the mass balance equation from a slightly different point of view: (*a*) hydrological and hydrodynamic transport terms; and (*b*) terms related to physical, biological, and chemical transformations. The latter can be separated into terms related to processes within the water body, and terms related to processes at the interfaces of the water body (loads, exchanges at the bottom or surface).

Transport Terms

In the literature there is little debate on the inclusion of hydrological throughflow. This term determines the flushing rate of a water body, which can constitute a significant loss factor for algal biomass and nutrients. Much more discussion arises as to how hydrodynamic phenomena should be incorporated. The most important internal transport mechanisms are wind-induced mixing and heat-generated buoyancy, although the latter can usually be ignored for shallow lakes. Frequently, in the literature, attempts are made to circumvent the difficult internal transport and mixing problem, such as by assuming that the system is ideally mixed. If there are no strong systematic spatial gradients in the lake and if a lake-wide average is sufficient for management, then the use of a fully mixed model is quite appropriate, as long as model results are compared with lake-averaged data rather than single spot points (e.g., Thomann *et al.* 1975). This simplification works pretty well because the nonlinearity of algal growth is fairly weak; that is, surplus growth in areas of high concentration is roughly balanced by less than average growth elsewhere. Of course, a consequence of the fully mixed assumption is that spatial detail is lost. If strong gradients exist – which can be discovered by careful examination of the data – a more thorough analysis is needed, as is pointed out in Chapter 2 and discussed in more detail in Chapter 10.

Transformation Terms

The core of most eutrophication models comprises the chemical and biological terms. Figure 3.4 presents in a sketchy form the most important cycles of C and nutrients in a segment of a water body. Typically, C assimilation by photosynthesis is accompanied by nutrient uptake in some form or another. The nutrients are then recycled by the sequence mortality, hydrolysis, and mineralization, so that most become available for uptake in the next cycle. During each loop a certain amount of material is transferred to the sediment where it is stored for shorter or longer periods of time.

Now, even this complex figure is still a strong simplification of reality. There is, however, nothing disturbing about this: in fact, simplification and schematization are the quintessence of model building (Imboden 1982). The choices made in each individual modeling effort on how to translate the generally accepted principles outlined in Figure 3.4 into a practical model are, or should be, greatly influenced by the specific goal in mind. So, a large number of model variants exist, especially with respect to the number of dependent variables considered ("state variables", sometimes called "components") (e.g., Scavia and Robertson 1979, Canale 1976), and with respect to the mathematical formulation of the subprocesses (for reviews see Swartzman and Bentley

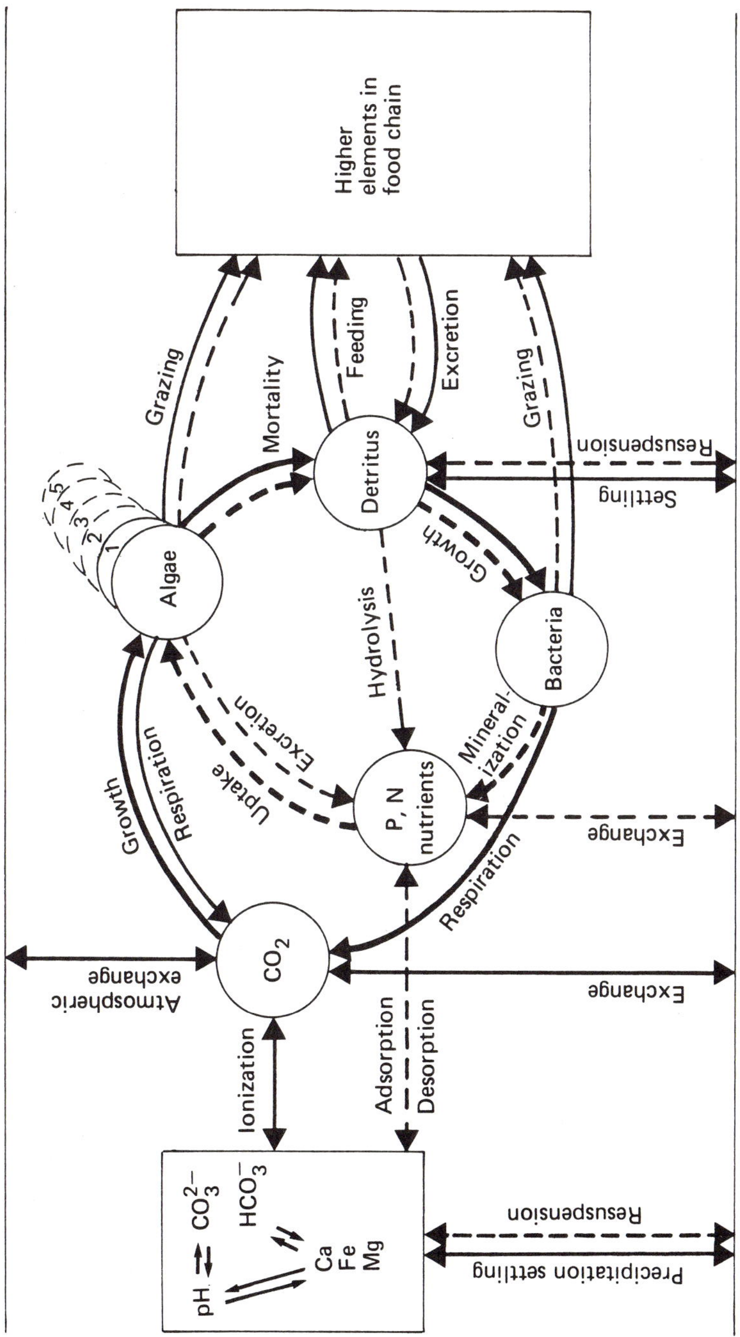

Figure 3.4. Schematic diagram of main C (——) and nutrient (– – –) cycles of interest for eutrophication. Thicker lines indicate major cycles.

1979, Jørgensen 1983). Differences in the number of state variables between models arise from more or less subjective judgments on the degree of detail desired, or required, to yield sufficient realism; in practice, the choice is frequently dictated by data availability. Examples of decisions influencing the number of state variables are: inclusion of only a single nutrient (mostly P) or of several nutrients (e.g., N, Si); consideration of zooplankton and of higher trophic levels, or not; detailed chemical component description (e.g., orthophosphate, condensed P, particulate organic P, dissolved organic P, etc.) versus simplification (e.g., only dissolved available P and phytoplankton P); and inclusion of a single or a few algal species versus multispecies modeling. The inclusion of more than one algal species can be particularly important for management if the aim is to avoid conditions promoting blooms of obnoxious species such as blue-green algae. Examples of differences in mathematical formulation are: mortality as a first- or higher-order process; multiple nutrient limitation by a multiplicative versus a minimum-rule formula; growth rate light dependence by a Steele or Smith relation, etc. An overview can be found in Orlob (1983).

An important fundamental distinction between models is related to the problem of how measures of biomass (biomass weight, algal counts, Chl-a) are related to nutrient content. Most models use the so-called "constant cell stoichiometry", i.e., the ratio of biomass to nutrient content is a fixed constant that does not vary over time. In these models no distinction is made between algal growth and nutrient uptake: they are strictly coupled so that the phytoplankton concentration can be expressed in terms of nutrients directly, leading to a reduction in the state variables needed. However, this advantage of simplicity is bought at the expense of a less realistic description, because the nutrient content of the algae is known to vary quite considerably. Apart from the dynamic implications this fact hampers the interpretation of constant cell stoichiometry models, since comparison with field data, such as Chl-a, requires a conversion from calculated phytoplankton nutrients with a conversion factor that is unknown, and, moreover, variable in time. Representations that overcome this difficulty, and that are probably also more correct dynamically, are known as "variable cell stoichiometry" or "internal cell quota" models (Bierman *et al.* 1973, Bierman 1976), in which algal growth and nutrient uptake rates are treated as separate processes. Consequently, more state variables and system parameters have to be introduced because biomass and internal nutrients must be monitored simultaneously. Thus these models tend to be more complex than constant stoichiometry models. It should be noted, however, that this disadvantage can be mitigated somewhat, as was demonstrated in an analysis by Di Toro (1980), because the relatively fast response of the internal cell nutrient concentration to external concentration changes allows for a fair computational simplification. An example of the application of an internal cell quota model to the Keszthely basin of Lake Balaton and a comparison with fixed cell quota models are presented in Chapter 13.

Most eutrophication models concentrate on the biological cycle, and much less attention is paid to the chemical component, even though chemical reactions can have a significant effect on the eutrophication process of a water body. An important example is the regulation of the pH of natural waters, mainly by the (bi)carbonate equilibria and dissolution or precipitation of carbonates such as $CaCO_3$. The biological cycle itself influences these reactions because of fixation or liberation of CO_2 (by photosynthesis and mineralization, respectively), and so is responsible for the differences in pH between, e.g., a productive water body (pH 8–10) and a consumptive sediment (pH < 7). In addition to pH influences on the life cycle itself, there are also indirect effects such as phosphate nutrient coprecipitation with biogenic lime, which may be a significant factor in the total nutrient loss to the sediment, especially in summer. Another phosphate removal mechanism may be associated with Fe and Ca compound precipitates formed when river water comes into contact with lake water of a higher pH. In this light it is somewhat strange to observe that most eutrophication models described in the literature ignore this field, or treat the matter in an extremely simplified way. Examples of recent developments that do include chemistry explicitly are Di Toro (1976), de Rooij (1980), and Lum *et al.* (1981).

Chemical subprocesses are usually modeled by assuming that chemical reactions are faster than biological reactions, although there are certainly some doubts about this (see for example the uptake and coprecipitation experiments described in Chapter 7). But once the assumption has been made, the problem reduces to a calculation of the chemical equilibrium for each species, which is a fairly straightforward procedure. The redistribution at the end of each time step involves the solution of a set of simultaneous equations, but with today's numerical library routines this does not constitute a problem. Problems arise in slow transformations of chemical components, in adsorption-type reactions, and in complexing reactions, most notably with organic compounds partly of biological nature. In this field rapid progress can be expected in conjunction with the increasing attention being paid to chemical aspects of the acid rain problem.

Interface Terms

An important class of terms in this category is formed by the external loads on the system. The external loads constitute the link between watershed and water body, and considerable efforts are generally needed to evaluate these terms (this is illustrated in detail in Chapter 6). It should be stressed here that load information is needed irrespective of the choice of the lake eutrophication modeling approach, whether statistical, dynamic, or steady state.

Another class of interface terms is associated with exchange processes, e.g. of O_2 and CO_2 at the water surface, and the exchange of P compounds with the sediment. The latter process is extremely complex (see Sly 1982). Physicochemical processes such as adsorption or chemisorption on Ca and Fe compounds are affected by O_2 conditions and pH, which in turn depend on the biological processes in the water column, as well as the transport and transformation processes within the sediment (see Chapter 7).

In most lake eutrophication models the sediment, if represented at all, is extremely simplified. There is little doubt that ignorance or oversimplification of the sediment is largely responsible for the unsatisfactory performance of dynamic models for long-term predictions, especially for shallow lakes. The reason is that the immediate internal source related to the sediment can be estimated on the basis of detailed mass balance data, but this gives no information on the long-term behavior under changed management conditions. As long as fundamental research in this field is lacking there is not much hope for improvement. On the other hand, with proper analysis based on the limited information now available, some useful and relevant statements can still be made regarding the role of the internal sediment pool in lake restoration, as is shown in Section 3.4.

Application to Lake Balaton

In the early stages of the IIASA Lake Balaton case study an attempt was made to apply one of the Manhattan models (Thomann *et al.* 1975) to Lake Balaton, but several difficulties were encountered that soon led to this approach being abandoned. For example, sediment exchange, which is crucial in shallow lakes, was missing from this model, because it had been developed for Lake Ontario, a deep stratified lake. The zooplankton cycle, which is important in Lake Ontario, did not seem to play a major role in Lake Balaton. On the other hand, for Lake Balaton more data on various P fractions were available than for Ontario, and they seemed to be of major significance. Altogether, the necessary structural adaptations were so large that a fresh start was considered to be the best option.

Eventually, three phytoplankton dynamics models were developed in the frame of the IIASA case study, all of the constant stoichiometry type: BALSECT (Balaton sector model; see Leonov and Vasiliev 1981, Leonov 1982), BEM (Balaton eutrophication modelers' group model; see Chapter 12), and SIMBAL (simple Balaton model; see Chapter 11). Here, first we give a brief outline of the major features, assumptions, and possible shortcomings of the models, and then we review the results. Finally, we attempt to draw conclusions about the lessons learned from these exercises, first with respect to Balaton, and second with respect to the state of lake eutrophication modeling in general.

Structure, Features, and Drawbacks of Lake Balaton Eutrophication Models

All three models use a separation of Lake Balaton into four basins, which are assumed to be ideally mixed. Spatial data within the selected basin boundaries have been averaged. These averages will contain errors, because it cannot be expected that the measurement points are distributed such that a truly basin-wide average can be computed. This has to be accepted as just one of many factors of uncertainty. It is clear that it would have been unjustified to treat Lake Balaton as just one single, ideally mixed reactor, because the gradients along the longitudinal axis are considerable, and persistent. However, the particular choice of four basins has not been made on the basis of a proper analysis as described above, but rather on historical and geometric grounds. Fortunately, it turned out *a posteriori* that the initial and subjective choice of four boxes to represent the various basins was quite appropriate from a hydrodynamics point of view, as demonstrated in Chapter 10. Of course, there is some loss of spatial detail: in the Keszthely region in particular the Chl-a concentration at the outer shore of the basin is likely to be higher than the fully mixed basin average. This is not a problem as such, but it can be of interest for management, because the public perception of lake quality is largely determined by the experience of bathers (see Chapter 5). At the same time, it implies a warning not to judge the results of abatement programs on the basis of nonrepresentative, single-point field measurements.

Figure 3.5 gives a schematic overview of the biological and chemical structures of the three models. BALSECT is a P cycle model. P is taken up from the dissolved inorganic P pool to form phytoplankton P, which then moves into detritus P by mortality and into dissolved organic P by direct excretion. Dissolved organic P is mineralized by uptake by bacteria and subsequent excretion, whereas part of the bacterial P reenters the detritus P pool by mortality. Of the three models, BALSECT has the most detailed description of the various organic P components. In contrast, SIMBAL assumes that these three components can be simplified into just one, and thus assumes that the internal transformations are either relatively fast or insignificant. Of course, a problem arises in this case because settling in SIMBAL works upon the total detritus pool, and although a correction on the settling rate is made to account for the proportion of dissolved material, the dynamics of this presentation are not exactly the same as in BALSECT. A similar problem arises in BEM, where an "organic material" pool is recognized, containing both particulate and dissolved material. BEM is not a P cycle model, but a mixed biomass–P–N model; however, because of the constant stoichiometry a conversion is possible to yield a P cycle model. The most remarkable difference with respect to BALSECT and SIMBAL is the inclusion of nutrient N as a state variable. Thus, possible N limitation can be covered with BEM.

Another difference between the models lies in the degree of detail with respect to phytoplankton. BEM is the most complete, covering summer phytoplankton, autumn phytoplankton (mainly blue–green algae), and winter/spring phytoplankton (mostly diatoms). SIMBAL considers communities dominated by warm-water species ("summer phytoplankton", most notably *Ceratium hirundinella*) and those dominated by cold-water species ("winter phytoplankton"). The distinction between the two groups was made on the basis of field data on primary production, from which a fairly adequate estimate of the necessary additional parameters could be made (van Straten and Herodek 1982), so that no detail was incorporated in the model that could not be supported by experimental evidence. This is also a good example of isolated parameter estimation explained in Chapter 2. BALSECT has the most simple algal compartment. Succession over the season is largely covered by

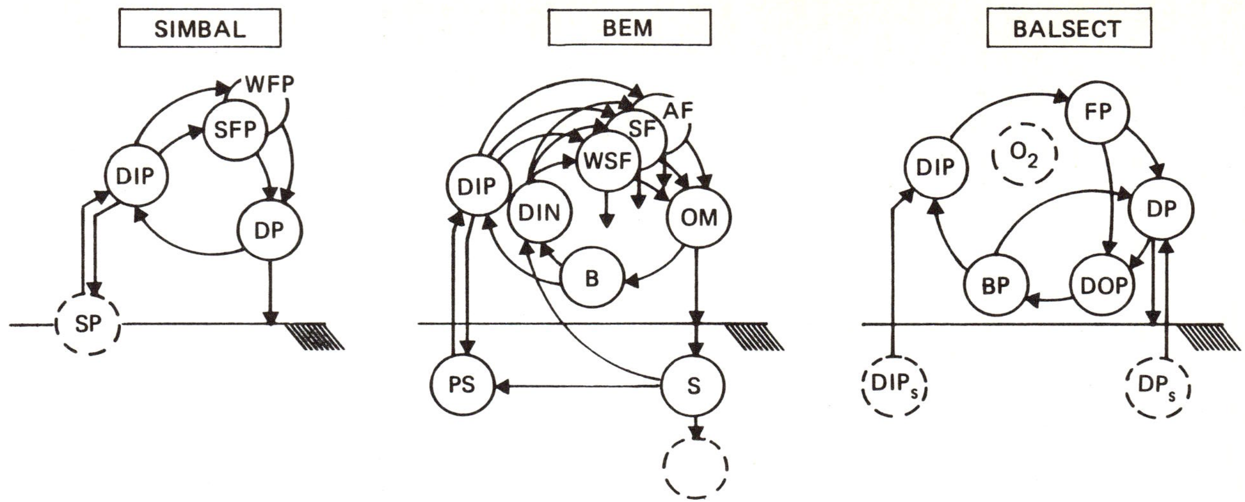

P cycle model
(van Straten, Chapter 11)

WFP: Cold water phytoplankton P
SFP: Warm water phytoplankton P
DP: Detritus P
DIP: Dissolved inorganic P

SP: Adsorbed exchangeable P

Mixed biomass–nutrient model
(Kutas and Herodek, Chapter 12)

AF: Autumn phytoplankton biomass
SF: Summer phytoplankton biomass
WSF: Winter and spring phytoplankton
OM: Organic material (detritus)
B: Bacteria biomass
DIN: Dissolved inorganic N
DIP: Dissolved inorganic P

PS: Exchangeable P in sediment
S: Organic material in sediment

P cycle model
(Leonov 1982)

FP: Phytoplankton P
DP: Particulate detritus P
DOP: Dissolved organic P
BP: Bacteria P
DIP: Dissolved inorganic P

DP_s: Detritus P in the sediment
DIP_s: Dissolved inorganic P in sediment

Figure 3.5. Comparison of the structures of Balaton phytoplankton dynamics models.

introducing a two-peak temperature function for algal growth, a construction initially employed in SIMBAL but later abandoned. It should be noted that a comparison of BALSECT results with field data was made by comparing the sum of the calculated algal P and nonliving particulate P with measured particulate organic P. This procedure, although correct as such, may mask the seasonal variability in the ratio between algae and detritus.

The models also show remarkable differences in the way that sediment–water exchange is treated. Each model accounts for settling of detritus material. In BALSECT this is modeled as two counteracting processes: settling and resuspension. Resuspension is governed by wind action, and its parameters are derived from a time series analysis of suspended solids data (Somlyódy 1982). This is another example of isolated calibration of portions of the model. However, the need for such detail can be questioned. The authors experimented with a much more complex version of SIMBAL and found that resuspension by wind in conjunction with an assumed desorption process is indeed able to describe rapid fluctuations in algal concentration. At the same time it became clear that the thickness of the mixed sediment layer and the transport processes within the sediment were of paramount importance, and the model proved to be very sensitive to these unknown quantities. Without further experimental evidence on a short time scale, calibration of such a detailed model is hopeless, and the modeling has to await results of detailed sediment research before further progress can be made.

Fast resuspension dynamics were ignored in BEM and SIMBAL, where only net settling is considered. Consequently, the apparent settling velocity assumed in these models must be much lower than in BALSECT. Dissolved P exchange is incorporated in all the models, and is modeled as a diffusive or dynamic adsorptive process. A difference, however, is whether the sediment dissolved inorganic P is assumed constant, and thus forms just a parameter in the model (SIMBAL) or whether it is modeled dynamically (BEM). Dynamic modeling is, of course, preferable in principle, because long-term effects of load changes can only be predicted with a proper bookkeeping of the sediment P household. However, the present state of knowledge about sediments makes it almost impossible to attach sufficient confidence to the results.

The differences outlined above are structural differences, but there are also variations in the mathematical treatment of the various transformation processes (see Chapters 11 and 12 for details). The most remarkable difference can be seen in the description of phytoplankton mortality. In both SIMBAL and BEM mortality is modeled simply as first order in algae, whereas BALSECT assumes that the mortality rate is proportional to the ratio of the biomass to the net specific growth rate. Thus, mortality becomes a fairly complicated second-order process in phytoplankton, where a high growth rate due to favorable light and temperature conditions, for example, leads to amplified growth because of a simultaneous relatively low mortality rate. There is much to say in favor of such a mortality behavior in the model because, while growth by cell division can be modeled effectively as a first-order process, a first-order model for algal mortality might be questioned. For example, it may

Table 3.1. Number of state variables and parameters in the three models.

	BEM	BALSECT	SIMBAL
Water body			
State variables	6+1[b]	5	4
Parameters:			
basic rate constants	13	11	8
nutrient limitation factors	3+3[b]	1	2
light limitation factors[a]	5	3	4
temperature coefficients	12	12	8
Sediment			
State variables	2	2	(1)[c]
Parameters:			
coefficients	4	2	3
temperature coefficients	1+1[b]	–	1

[a]Including self-shading coefficients and base extinction.
[b]For BEM model, + indicates extra needs for the inclusion of N.
[c]Considered constant.

be that algal mortality depends on the age distribution, in which case mortality will occur after growth has ceased, a dynamic delay that cannot be properly described by the usual first-order kinetics. In this light it is somewhat strange to see that the expected stronger dynamics is not present in the BALSECT results.

Other differences in mathematical formulation can be found in the description of the temperature dependences of the various rate functions. It turns out, in fact, that temperature functions have a decisive influence upon the dynamic behavior of the models. At the same time very little experimental information is available about these temperature dependences, and this is an essential weakness of all the models. It should be noted that temperature relationships cannot simply be taken from laboratory experiments, because model processes are aggregated, and so the temperature dependence of, say, phytoplankton growth will probably be a very complex composite of the temperature dependences of individual species in the population. The same holds for the dependence of the algal community as a whole on light. Experimental evidence from *in situ* observations is therefore of much value (cf. van Straten and Herodek 1982). Table 3.1 summarizes the number of state variables for the three models, as well as the number of parameters that must be specified. The demanding role of the temperature functions is very apparent from this table.

At this point we would like to stress that the final versions of the models were quite different from those first formulated when the study began. This reflects the iterative process of model building where changes are made on the basis of observed deficiencies. From a scientific point of view this is a logical procedure, since new insights and new measurements call for updating, so maybe even further versions would have been developed had the study not come to an end. At the same time this illustrates that the ultimate model simply does not exist. So, if time is a precious resource, as is often the case when management decisions have to be taken, we are forced to draw conclusions on the basis of the actual state of affairs, i.e., on the basis of results, obtained within a finite amount of time, reflecting the present state of ignorance, and shaped by the skills and subjective judgment of the modeler. However, we may be more confident if the results of different approaches, though individually imperfect, point in the same direction. So, let us now examine the results.

Model Performance

Despite the considerable progress made in eutrophication modeling there are very few examples of a full model calibration and validation against an independent data set for the same water body. Apart from the fact that models are necessarily simplifications of reality, another reason certainly also lies in the quality of available field data, due to both analytical and sampling problems. In the case of Lake Balaton the situation is similar. Although each of the three models has been calibrated more or less for one particular year, quite strong deviations from reality can still be observed. Figure 3.6 gives a comparison of simulations by SIMBAL and BEM for 1977. As pointed out above, in each case a conversion was needed from the units of the model state variables (like biomass or phytoplankton P) into Chl-a. This causes difficulties because the ratio may shift throughout the year. Furthermore, for example in BEM, emphasis was sometimes put upon a calibration of biomass, with less attention to a good fit for the other state variables. In BALSECT (not shown) all P fractions had the same weights, but the calibration only included total organic P, so that a comparison with Chl-a can only be made if the algae/detritus ratio is assumed to be constant. As a result, the algae level predicted by BALSECT was generally too low. Also, BEM generally predicts lower phytoplankton levels than have been observed. SIMBAL was calibrated using a Monte Carlo technique in order to represent roughly the observed behavior. The main aim of this procedure was not calibration but hypothesis testing. Only by postulating an additional adsorption–desorption process acting upon orthophosphate could feasible parameters be found (see Chapter 11). It should be noted that analyses of primary production data from vertical profiles (van Straten and Herodek 1982) revealed that the maximum specific growth rate parameter had to be considerably larger than is usually assumed in other eutrophication studies. Consequently, the mortality rate had to be larger in order to maintain the fit. But calibration was possible with both a "low" and a "high" assumption on these parameters, which shows that there are dangers in calibration in cases where parameters are closely related in their effects on model behavior. A separation on the basis of field data alone is not possible, and only independent experiments can resolve this problem.

Despite problems in the calibration phase, attempts were made to run each of the models for a sequence of years, maintaining the calibration parameter set. In general, the results of this validation are rather disappointing. Table 3.2 summarizes the results for Chl-a for the Keszthely and Szigliget basins for the years 1976–78 for SIMBAL (where calibration was based on

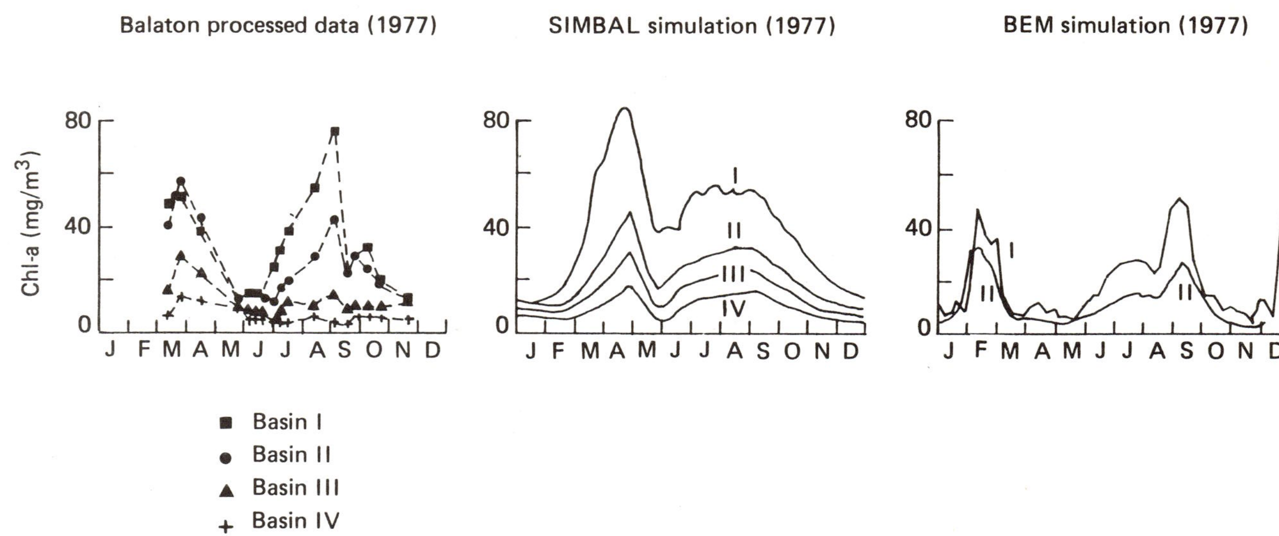

Figure 3.6. Phytoplankton simulation with SIMBAL and BEM for 1977 (basins I, Keszthely; II, Szigliget; III, Szemes; IV, Tihany/Siófok). Processed Chl-a data are given for comparison on the left. Conversion from original model variables: SIMBAL phytoplankton P/Chl-a = 2.0; BEM biomass/Chl-a = 17.8.

Table 3.2. Comparison of validation attempts for BEM and SIMBAL. Averages of three peak Chl-a values during spring and summer (mg/m^3).

	Szigliget basin			Keszthely basin		
	Data	SIMBAL[a]	BEM[b]	Data	SIMBAL[a]	BEM[b]
Spring 1976	8	16	15	19	32	31
Summer 1976	17	23	33	26	32	39
Spring 1977	48	30	27	46	52	39
Summer 1977	34	27	22	56	40	49
Spring 1978	20[c]	24	–	28[c]	44	–
Summer 1978	22	32	–	29	54	–

[a]Growth, mortality, and mineralization rates twice those in Chapter 11.
[b]Taken from Figures 12.2 and 12.3.
[c]Based on two observations only.

1977), and for the years 1976–77 for BEM (calibration based on a one basin model for 1977). For ease of presentation averages of the three largest values in spring (January–May) and summer (June–October) were taken as the basis for comparison. Various observations can be made. First, it is quite remarkable that the peak Chl-a levels do not even coincide for the calibration year 1977. One of the reasons is that the models were also checked against P fractions, and in order to maintain a reasonable fit for these some mismatch for Chl-a was unavoidable. Upon validation, even more serious deviations occur. For example, the trend that 1976 Chl-a tends to be lower than 1977 is followed by SIMBAL, but not by BEM for Szigliget in the summer, whereas SIMBAL fails to predict a lower summer peak for 1978. In both 1976 and 1978 the models considerably overpredict Chl-a levels. It seems that the actual annual variations are larger than those in the models. A similar tendency can be observed for the dynamics within a year. SIMBAL persistently gives a fairly flat chlorophyll curve in summer [cf. Figure 11.3(*d*)], whereas the actual data are much more peaky. Although this is better in BEM it is still not really good (cf. Figures 12.2 and 12.3). As is shown in Chapter 13, some improvement in short-term dynamics can be obtained with variable cell quota models. But for this type of model a convincing validation is also lacking.

The results obtained during the Balaton study are by no means exceptional in the field of eutrophication modeling. In fact, it is difficult to find an example of a really satisfactory validation for shallow lakes in the literature, although several attempts have been made (Di Toro *et al.* 1977, Jørgensen *et al.* 1978). Obviously, there are gaps in our knowledge that need further research. At this point, it is only speculative to give a list of items of which further investigation would probably lead to improvements, but it is worth considering at least a few candidates. First, we believe that the chemistry, including sediment adsorption and desorption mechanisms, together with biogenic lime coprecipitation, play important roles in shallow lakes. It may very well be that the ratio between loads of adsorptive material (Ca and Mg compounds for Lake Balaton) and P loads is decisive in the long-term behavior of the lake. In this context it is worth noting that recent investigations suggest that the low Chl-a concentration in the very warm summer of 1978 might be attributed to a reduced P availability as a consequence of coprecipitation of P with larger than usual amounts of $CaCO_3$ precipitate, formed in that year because of the exceptionally high evaporation. A second factor that might influence the short-term dynamics of algal growth is that mortality processes have been relatively underexposed to the attention of modelers, as mentioned before. More systematic research in this field may lead to substantial improvements. Finally, perhaps somewhat more detail is needed on the problem of algal species succession, although some attempts have been made, particularly in respect to blue–green algae (Bierman 1976). In general, blue–green algae have special properties (N fixation, buoyancy) and so do not seem to behave according to our usual structural equations. For Balaton, this factor is important for the Keszthely and Szigliget basins, under present conditions.

So, there is much to be improved. But the question now is: can we use such apparently deficient models for management? Of course, there are hazards in doing so, because a lack of fit usually means a lack of understanding. If we were asked to make a detailed prediction about future algal dynamics we would certainly not be able to do so, even if there were a way to eliminate the stochastics in loads and meteorology. However, since the models on a rough, annual average basis follow the phytoplankton levels from year to year – admittedly with some uncertainty – and since this means that the models respond reasonably well to different loads and weather conditions, we may still use them to investigate how at least the *models* respond to load reductions. Figure 3.7 shows the results for SIMBAL, where the annual average phytoplankton P is given as a function of volumetric "available" P load for each basin. The peak level is usually some 50–100% larger. The stochastic aspects are discussed in more detail in Chapter 4. Also shown are results for two of the models used for the study presented in Chapter 13, namely model I (basically equivalent to SIMBAL) and model III, based on variable cell stoichiometrics (summer peak levels; Keszthely basin only). Similar plots were obtained for BEM and BALSECT, although a direct comparison is not possible because of the different load reduction scenarios used and differences in the definition of the lake response.

Whatever the situation, the load reduction behaviors of models show remarkable similarities. First, despite quite strong nonlinearities in the models, the response curve is almost linear. This suggests a fairly simple relationship for management purposes, and this behavior is also in line with empirical predictions based upon the OECD study (Figure 3.3). Second, in contrast with the OECD results, the models seem to agree that a full suppression of algal blooms is not achievable, not even at zero loads, if this were ever possible. There are specific reasons for this, which we discuss in Section 3.4.

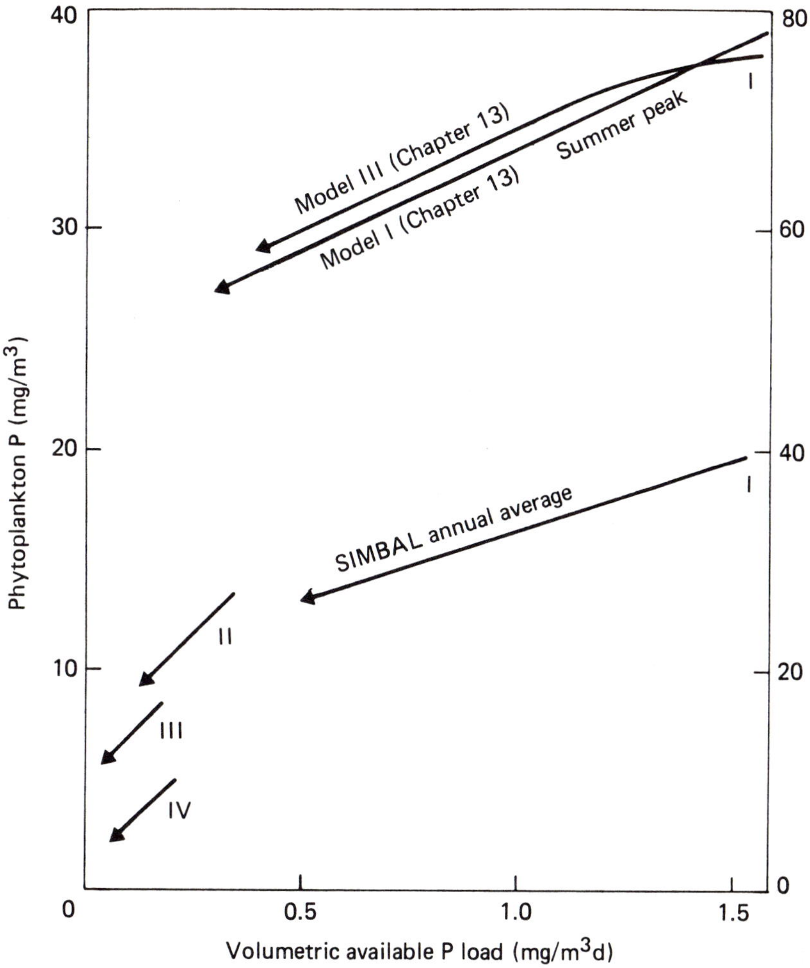

Figure 3.7. Phytoplankton concentration response to P load reductions according to SIMBAL (annual average values, all basins) and to models I and III from Chapter 13 (summer peak values, basin I only).

Conclusions

To conclude this section on dynamic modeling we summarize briefly what we have learned from our attempts to develop dynamic lake eutrophication models for Lake Balaton, and perhaps also about lake eutrophication in general:

(1) Dynamic models are helpful in analyzing the major processes of a lake, to form and test hypotheses, and to elucidate weak spots in our knowledge. For shallow lakes one such weak spot is the sediment.

(2) It is extremely difficult to calibrate and validate a dynamic lake eutrophication model properly. In the case of Lake Balaton we have essentially failed to do this. Data inadequacy, lack of essential knowledge, difficulties in communication between model builders and field workers, have all contributed to this failure. Whether the problem can be solved, in principle, we do not know. It is a fact, however, that in the field of eutrophication modeling there exists hardly any example of a modeling attempt for which the validation has been really successful.

(3) Despite problems in the validation phase, to the best of our knowledge models that have already been developed show a remarkable linearity in their response to loads when examining an aggregated water quality indicator, such as annual maximum Chl-a concentrations. So, by ignoring the dynamic details, an answer is obtained that can still be used for management purposes.

(4) The uncertainty ranges of dynamic eutrophication models to date are not smaller than those of statistical models. However, when further developed, dynamic models have the potential to yield more detailed and more precise answers, particularly in their capability to specify separately the contributions of meteorology and of controllable and uncontrollable loads, which will be a great advantage.

We postpone further conclusions until after a discussion of the results from the steady state modeling approach.

3.4. Steady State Modeling

The modeling technique in this category is still based on dynamic model representations of the system, but rather than calculating the full dynamics in time a simplification is sought by analyzing steady state conditions only. In general, this is done by setting the time derivatives equal to zero.

In 1975 Lorenzen and Mitchell analyzed the steady state condition for light-limited algal growth in a closed reservoir, in order to investigate the effect of mixing depth upon the maximum algal standing crop that could be reached under specified solar irradiation levels and a given natural extinction coefficient of the water. The maximum standing crop is obtained if the shading effects over the mixed depth of the algae themselves reduce growth to a level such that depth and daily average growth are balanced by depth and daily average mortality. The same principle was the basis for a series of models developed originally by the Rand Corporation, but a new element was the inclusion of a number of different algal species. The idea here was that nature would strive for a biomass composition that would lead to the maximum possible biomass under prevailing environmental conditions. It turned out that this maximization principle could be cast in a linear programming framework to calculate the optimal community composition (Bigelow *et al.* 1977, Loucks 1983). The effect of nutrient availability is accounted for by considering the total nutrient mass as a boundary condition; i.e., nutrient limitation here is the standing crop limitation, and dynamic rate limitations are not incorporated.

A further extrapolation of this modeling concept is the BLOOM-II model developed within the frame of the Policy Analysis for Water Management in the Netherlands study (PAWN, Los *et al.* 1982). In this approach the time element is reintroduced by calculating the potential for algal bloom from week to week assuming constant environmental conditions. Thus, a series of steady states is computed where each is fully independent of the previously calculated situation. Of course, it is recognized that the model provides an upper limit for algae only, because there is no guarantee that a steady state is actually reached within each weekly period. The model contains 10 algal species. Possible crop limitation is incorporated for three nutrients (P, N, Si), but nutrients are not modeled independently; the input of an actually measured time series of nutrient concentrations is required instead. The model was applied to a number of (shallow) Dutch lakes with varying degrees of success.

When reviewing the ideas above it can be said that the principal innovation is in the concept of predicting the possible algal composition under certain environmental conditions; in particular, the ability to detect situations where blooms of blue–green algae can be expected is of practical value. But this is limited by the fact that only those algal species for which growth, light, and mortality parameters have been introduced can be used in the model's prediction. For each new species these data have to be established, either from the literature or from specially designed laboratory experiments. On the other hand, one may perhaps hope that these data are fairly universal, because they are coupled to named species; this is not the case when groups of algae are considered.

In the case of the sequence of steady state calculations, the question may arise as to what advantages this method has over a full dynamic simulation. In fact, the claim that data requirements are less is only correct because nutrient concentrations are not modeled but are treated as known model inputs. If this was done similarly in a dynamic modeling exercise, the only additional requirement would be that Monod coefficients (i.e., the concentration for which the growth rate is half the unlimited rate) be known. This small disadvantage would be largely compensated by the elimination of the need to discuss the difficult question in the steady state approach of whether steady state is really reached all the time.

From the management point of view, the requirement of the approach outlined above for nutrient concentration data as inputs is rather disappointing. The nutrient concentration within a lake is the result of loads and in-lake processes such as algal growth, and consequently the resultant algal concentration will be determined by the load level and the removal rate generated by the algae themselves. The approaches discussed do not specify how the observed nutrient levels relate to these counteracting forces, and consequently a direct relationship between load and algal level cannot be made.

Since, in this book, eutrophication management is our major concern, it is precisely the relationship between load and maximum potential algal standing crop that we are largely interested in. By returning to the original equations it turns out that the requirement of known nutrient levels is by no means necessary. We discuss this form of a more complete steady state analysis below. The main objective is to analyze important properties of the models without the need to consider the full dynamics, and to see whether conclusions from this can be drawn with respect to either modeling or management.

To illustrate the method consider the following simplified structure:

$$\frac{dA}{dt} = k_g f_I \frac{P}{P_k + P} A - k_d A - q_v A \tag{3.5a}$$

$$\frac{dD}{dt} = k_d A - k_m D - k_s D - q_v D + L_D \tag{3.5b}$$

$$\frac{dP}{dt} = k_g f_I \frac{P}{P_k + P} A + k_m D - xPf_I \frac{P}{P_k + P} A + L_{\text{int}} + L_P \tag{3.5c}$$

where

A = phytoplankton P
D = detritus P
P = dissolved inorganic P
k_g = maximum specific growth rate coefficient for algae
k_d = algal mortality rate coefficient
k_m = detritus hydrolysis and mineralization rate coefficient
k_s = detritus net settling rate coefficient
q_v = reciprocal residence time = flow rate/volume
x = biogenic lime coprecipitation factor
f_I = light attenuation factor for algal growth, depth, and day averaged
$= G(I)/(\varepsilon_0 + \alpha A)H$, (3.5d)
where $G(I)$ is a function dependent upon optimal light intensity for growth, day length, and global radiation (cf. van Straten 1979); ε_0 is the extinction coefficient of water without algae; α is the self-shading coefficient; and H is the depth
P_k = half saturation constant
L_D = external detritus P load
L_P = external dissolved inorganic P load
L_{int} = internal load

This structure reflects the basic properties of the SIMBAL model, and, with inclusion of bacterial effects in the detritus mineralization rate, those of BEM as well. For the present analysis not more than one algal species and not more than one nutrient (P) are considered. An analysis of more complete structures is possible, but is not needed here.

Just as before, we can now analyze the steady state behavior of this system by setting time derivatives equal to zero. Zero derivatives are also a necessary condition for an extremum to be reached, which is what we are usually interested in for management purposes. It is important to note that the parameters in equations (3.5) depend on environmental conditions such as temperature and light. Hence, a different set of environmental conditions will lead to a different extremum, and whether this is a minimum or a maximum will be determined by the time dependence of the parameters. Generally, one expects algae to reach a minimum in winter and a maximum in summer.

For the analysis various possibilities still exist:

(1) We can set to zero the derivative of the algae only. There is no ***a priori*** requirement for the time derivatives of orthophosphate or detritus to be zero at the same time. What we obtain is a condition for a local extreme (minimum or maximum) of the algal concentration (provided that the second derivative changes sign at this point).

(2) We can set to zero the derivatives for all state variables simultaneously and also assume that the input functions (temperature, light, loads) remain constant for a sufficient period of time. In this case an equilibrium condition is obtained that is valid for the prevailing input situation.

(3) The equilibrium condition (2) is also a local extremum condition. However, in order to know the global extreme equilibrium condition, we have to scan through the various possible forcing combinations in order to find out which environmental conditions will cause the highest (or lowest, in the case of a minimum) algal bloom potential.

Let us first examine a local extremum condition for the algae. Reaching an extremum of phytoplankton requires that the time derivative is zero, i.e., a (necessary, but not sufficient) extremum condition is:

$$k_g f_I \frac{P}{P_k + P} = k_d + q_v \ , \tag{3.6}$$

where P denotes the dissolved inorganic P level at the algal extreme. It is more useful to reshape this equation in the form

$$P = gP_k \ , \tag{3.7a}$$

with

$$g = \frac{k_d + q_v}{k_g f_I - (k_d + q_v)} \ . \tag{3.7b}$$

Since f_I depends on the algal level, we cannot yet solve this equation in order to find the extremum algal level A'; some additional assumptions have to be made. However, two situations exist for which a simplified solution is possible with little difficulty:

(1) The lake is not P limited (i.e., it is light limited). In this case $P/(P_k + P) \simeq 1$ and A' is obtained by solving equation (3.6) directly, using equation (3.5d). This is essentially the method of Lorenzen and Mitchell (1975), but expanded with the effects of hydrological throughflow.

(2) The lake is P limited and turbid, i.e., $\alpha A \ll \varepsilon_0$ in equation (3.5d). In this case f_I is not a function of A.

The latter situation is of particular interest for Lake Balaton: for P limited turbid lakes equations (3.7) specify that the dissolved inorganic P level is directly proportional to the Monod constant, with a proportionality factor given by the ratio of loss by mortality and flushing, and the maximum potential non-P-limited net growth. The inorganic P level is independent of the load: any additional load will be taken up by algae and distributed over algae and detritus such that condition (3.6) is fulfilled again. The result presented by equations (3.7) also explains why the orthophosphate level is so remarkably constant in Lake Balaton during the summer. At this point it is important to make two remarks. First, the parameters in equations (3.7) are functions of external variables, most notably temperature, and – with respect to f_I – light, which are time dependent. Consequently, the parameters in equations (3.7) are also time dependent. Since generally larger orthophosphate values are observed in winter, the factor g must be larger in winter than in summer, provided P_k remains the same. Consequently, the temperature dependences of the growth and mortality rates must be such that this condition is fulfilled. This fact is helpful in selecting and checking the temperature functions when building the model. A second remark is related to the hypothesis of orthophosphate adsorption or desorption. Seemingly, if desorption takes place, this would tend to increase the orthophosphate level, but at the same time it improves the situation for algal growth, and algae will grow until conditions (3.7) are met again. Clearly, this can only occur at higher algal levels. So, adsorption and desorption simply act as an internal sink and source, affecting the algal levels but in the end not those of orthophosphate. Depending upon the time scales of the sorption and uptake processes orthophosphate fluctuations will be dampened, so that it is difficult to tell from inspection of orthophosphate time series alone whether sorption takes place or not.

Although the results of the local extremum analysis provide a good insight into the behavior of the model (and, hopefully, of the lake), the goal of calculating the extremum algal level can only be achieved for light-limited lakes. For P-limited lakes, more assumptions are needed, and the next logical step is to perform an equilibrium analysis. Of course, for equilibrium, equations (3.7) still hold.

Equation (3.5b) provides an expression of the equilibrium detritus level D^* as a function of the equilibrium algal level A^*:

$$D^* = \frac{k_\mathrm{d}}{k_\mathrm{m} + k_\mathrm{s} + q_\mathrm{v}} A^* + \frac{L_\mathrm{D}}{k_\mathrm{m} + k_\mathrm{s} + q_\mathrm{v}} \; . \tag{3.8}$$

Equation (3.8) states that the detritus level increases linearly with the particulate P load, whereas detritus P is coupled to algal P by the ratio of algal mortality and total detritus loss rate.

Finally, equations (3.7) and (3.8) can both be substituted into equation (3.5c) to obtain the equilibrium algal P level:

$$A^* = \frac{\left[\dfrac{k_m}{k_m + k_s + q_v} L_D + L_P + L_{int}\right]}{\left[(k_d + q_v)(1 + xgP) - \dfrac{k_m k_d}{k_m + k_s + q_v}\right]} \tag{3.9}$$

In order to elucidate the significance of equation (3.9) a further simplification is made by setting $q_v = 0$. This is not unreasonable for most lakes; for example for the Keszthely basin $q_v = 0.001\,d^{-1}$, which is at least one order of magnitude less than k_d and $k_m + k_s$. Thus, ignoring q_v in equation (3.9) and introducing the shorthand notation

$$\sigma = \frac{k_s}{k_m + k_s} , \tag{3.10}$$

we finally obtain

$$A^* = \frac{(1 - \sigma) L_D + L_P + L_{int}}{k_d (xgP_k + \sigma)} . \tag{3.11}$$

This important result states that for nutrient-limited, turbid lakes the equilibrium algal level is practically proportional to the total P load, including both external *and* internal loads. The slope of the load–algal peak line is governed by only a few parameters: algal mortality rate and the sum of losses due to biogenic lime coprecipitation and settling, the latter expressed as the fraction of the total detritus loss rate (settling plus mineralization) that is due to settling alone (σ). Thus, equation (3.11) provides an elegant explanation for the remarkable linearity of the model's load responses. In addition, now, it is clear why the predicted algal level does not drop to zero if the external load ($L_D + L_P$) is totally removed: there remains the internal load, which is, from the model's point of view, just like any other load.

It is interesting to elaborate this internal load aspect a little further, for which equation (3.11) is an excellent basis. Let us suppose that the observed lake position in a Chl-a–load plot is given by the dot in Figure 3.8. Furthermore, one may assume that a dynamic calibration yields values for the parameters k_d, x, and σ such that the load response corresponds with the solid line (similar to the SIMBAL result of Figure 3.7). Then, this calibration has led automatically to an estimate of the internal load; in other words, the internal load can approximately be derived from in-lake data. However, the estimate is not very precise; as can be seen in the plot different estimates for the mortality rate and/or settling/mineralization rates, and thus different slopes, lead to quite different values for the internal loads. Turning the argument around, we can also say that information on the internal load from sediment research would also be helpful in reducing the calibration uncertainty of the in-lake model segment. Although the internal load cannot be estimated precisely, its value for Lake Balaton must be of the same order of magnitude as the external load.

In terms of the developments discussed thus far, two questions remain that require some consideration. First, one may ask whether a static

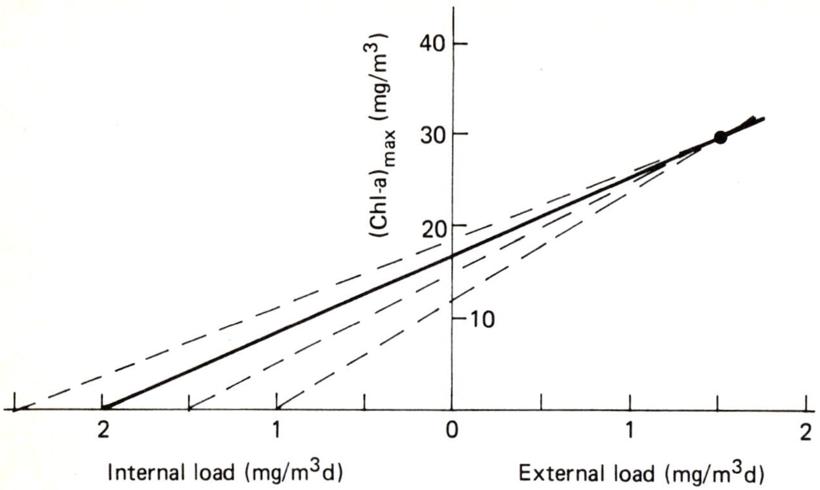

Figure 3.8. Estimation of internal load by in-lake parameter estimation and the effect of parameter uncertainty on load uncertainty.

condition is ever really reached in practice. It is possible to study this problem by perturbation analysis (cf. van Straten 1981). In general, for Lake Balaton, load response times are of the order of a few weeks to more than a month. This means that load changes are followed by slow responses only, and consequently it takes time to reach equilibrium. However, if there are periods with relatively small load changes, such as is frequently the case in summer, reaching equilibrium is determined mainly by environmental factors such as temperature and light. For these factors responses are faster, and minima and maxima follow each other in a rapid sequence. As was pointed out before, the problem is that it is not known at which time the most favorable conditions occur for algal blooms. On the other hand, it is known that algal peaks frequently occur in August in Balaton, so scanning the temperature and light conditions that prevail around that time should yield a fair upper limit to the truly expected algal maximum.

The second question is whether or not a load reduction will affect the internal load. Of course, such a question can only be answered when appropriate knowledge about the sediment is available. Here we are interested in investigating the implications of an (assumed) feedback mechanism within the sediment. The starting point is Figure 3.8, which holds if no feedback occurs. Now, let us assume a relationship between external and internal loads. Such a hypothetical relationship (probably determined by the ratio of adsorbents load and P load) is depicted in the bottom right of Figure 3.9. Then, any external load reduction entails a change in internal load (which may, however, be delayed, as discussed in Chapter 7; these dynamics are ignored here). If this new internal load is projected on the internal load axis, a new lake response line may be drawn. As long as algal mortality, mineralization, and settling rate coefficients are not affected this line will be

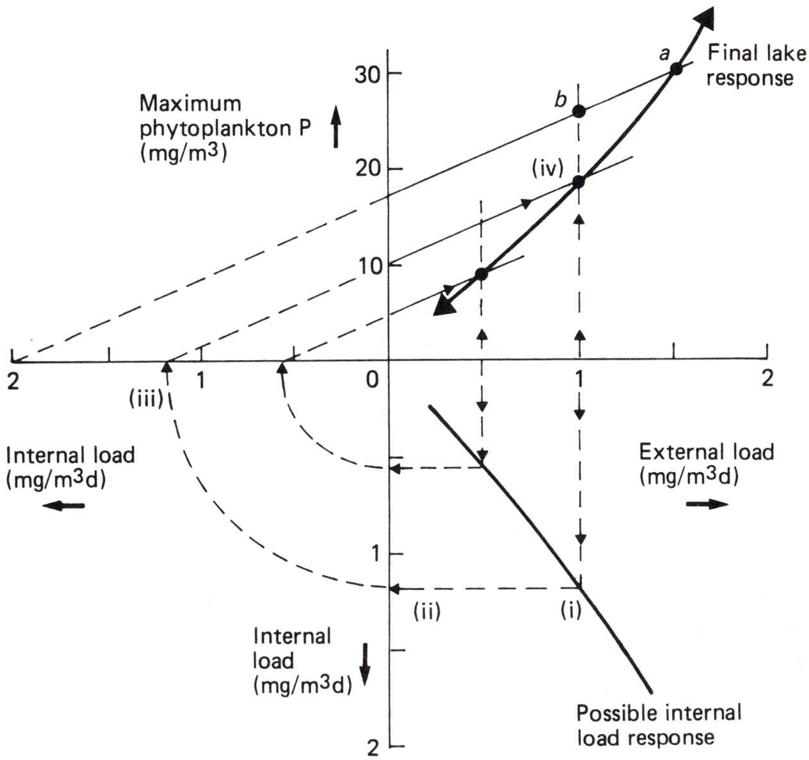

Figure 3.9. Phytoplankton response to external load reductions with (hypothetical) feedback on internal load. A load reduction at point *a* initially brings the system to point *b* along the immediate response line. The final, long-term response is found by reading the new equilibrium internal load from the bottom right sector [points (i)–(ii)], projecting onto the horizontal internal load axis [point (iii)], and drawing the new immediate response line parallel to the original one. The final, long-term response is at the intersection of this line with the reduced external load [point (iv)].

parallel to the original response. The new algal peak is now lower than it was originally due to the effect of the feedback. Finally, when the external load reaches zero, the algal peaks drop to almost zero too. This outcome is more in agreement with the empirical findings of the OECD study than with the original Balaton models without long-term feedback.

The results of the analysis outlined above are very important for management. In the light of Figure 3.9 the load response of the dynamic models may be viewed as the immediate response, i.e., the response that can be expected in the year of implementation of load reduction measures. It should be noted, however, that this response may be masked considerably by annual fluctuations in meteorological factors (cf. Figure 3.3 and discussion). The immediate response is not very large and may discourage the implementation of abatement programs. However, Figure 3.9 suggests that this is not justified: due to the feedback mechanism within the sediment the internal load will also drop, and sooner or later algal peak levels will decrease further. Thus, the prospects for management are better than suggested by the present version of the dynamic models, although the time delay is uncertain (see Chapter 7). The results also indicate that inclusion of long-term sediment dynamics in the models is absolutely necessary for the achievement of better long-term predictions. The lack of appropriate sediment knowledge is an essential weakness of eutrophication models for shallow lakes.

3.5. Perspectives of Eutrophication Modeling

The experience gained from the Lake Balaton study in applying models of various levels of complexity to the eutrophication problem calls for a discussion of the perspectives of eutrophication modeling. As far as models based on mass balance considerations are concerned, two extreme philosophies exist.

The first school assumes that a good representation of the system as a whole is best obtained by putting together mathematical formulations for a large number of ecological subprocesses. This approach strives for completeness, and the resultant model is believed to be, in a sense, universal in that it can be applied to a large variety of situations. For example, the model CLEAN (Park *et al.* 1974), later expanded to CLEANER (Park 1979), now comprises no less than 40 state variables. The constituent subprocess formulations and associated parameters (several hundred) are derived from laboratory experiments or other independent investigations. The claim is that in this fashion one should be able to predict the future, even for large deviations from the present situation, because, after all, "everything" is included in the model. Although there is no proof that this is true, the idea is appealing in principle. However, in practice, serious problems can arise because despite careful experimentation it is impossible to know the subprocesses and process parameters without error. Consequently, the complete model must reflect these uncertainties, but because the number of inevitable subprocess errors is so large, the effect on model performance as a whole is practically unpredictable. As a result the final answers remain unreliable and uncertain. It should be noted that the usual escape of calibrating the model against field data is impracticable here because with so many variables it is unlikely that appropriate field data will be available. To some extent BALSECT is a member of this school.

The counterpart of the modeling outlined above starts from the field data and a limited amount of *a priori* knowledge of those processes that are believed to be dominant in the behavior of the system. So, models tend to be simple initially, and are updated only if deviations between observations and model outputs are considered unsatisfactorily large in the light of the objectives of the study. Thus, these type of models are "tailor-made" rather than "universal". From deviations new hypotheses are generated and only those improvements are allowed to enter the model that find support in the field data. In this approach parameter estimation (or calibration) and structure identification are closely related. An example of this procedure can be found in Beck (1982). To some extent SIMBAL is a member of this school.

Frequently, measurement data are not accurate or complete enough to allow for a fully deterministic description. In such cases it is better to recognize this as an explicit stochastic factor and to carry the fundamental uncertainties along in the generation of hypotheses (Hornberger and Spear 1980) or in the predictions (Fedra *et al.* 1981, van Straten, Chapter 11). The relatively simple model structure also allows for a simplified analysis, such as an extremum analysis (e.g., Verhagen 1976; Section 3.4). Both the transparency of the structure and the explicit treatment of uncertainties constitute the basic advantages of this approach. But there is also a fundamental disadvantage, namely that predictive power is often present only for excursions not too far from the present situation, and this limits the practical applicability of these models.

Reviewing the status of model building for the eutrophication problem, one seems to end up with the dilemma of choosing between universal models that cannot be calibrated, and therefore perhaps correctly predict the future but with impracticably large uncertainties, and simpler models that can be calibrated more or less, but with which structural changes in response to strong management actions cannot be predicted (Beck 1981, Beck and van Straten 1983). The results of the Balaton study have demonstrated that both complex and simple dynamic models are far from well developed. Yet, for management purposes where only aggregated results are demanded, simple models are preferable because they reflect the major phenomena while bypassing the fine detail. Improvements in predictive power can only come from the inclusion of broadly observed ecological phenomena, for which the judgment of experienced limnologists is indispensable.

3.6. Conclusions

From the experience gained during the Lake Balaton eutrophication study the following final conclusions emerge:

(1) Empirical relationships such as the Vollenweider/OECD plots are useful to underline the close relationship between external P loads and algal blooms. However, the accuracy for application to a specific lake is not always sufficient.

(2) Dynamic modeling provides a powerful tool, in principle. The development of models has led to significant increases in understanding and has served as a guide to direct further scientific research. However, in their present state, the models are not yet adequate for detailed prediction, mainly because of the lack of appropriate sediment dynamics and the lack of fundamental limnological knowledge.

(3) Simple load–peak relationships by steady state analysis of dynamic models have proved helpful in detecting the shortcomings of these models. The sediment behavior determines the transient lake response between the present and future load situations. But in any case, in the long term, P load reductions are more beneficial than expected.

(4) The application of various approaches to the lake eutrophication problem is one way of partly overcoming the deficiencies of each individual approach, and allows for a fair judgment of the associated uncertainties at the level of aggregation required for management.

References

Beck, M.B. (1981) Hard or soft environmental systems? *Ecological Modelling* 11:233–51.

Beck, M.B. (1982) Identifying models of environmental system's behaviour. *Mathematical Modelling* 3:467–80.

Beck, M.B. and van Straten, G. (Eds.) (1983) *Uncertainty and Forecasting of Water Quality* (Berlin: Springer).

Bierman, V.J., Jr. (1976) Mathematical model of selective enhancement of blue–green algae by nutrient enrichment, in R. Canale (Ed.) *Modeling Biochemical Processes in Aquatic Ecosystems* (Ann Arbor, MI: Ann Arbor Science Publishers) pp 1–32.

Bierman, V.J., Jr., Verhoff, F.H., Poulson, T.L., and Tenney, M.W. (1973) Multinutrient dynamic models of algal growth and species competition in eutrophic lakes, in E.J. Middlebrooks, D.H. Falkenborg, and T.E. Maloney (Eds.) *Modeling the Eutrophication Process* (Logan, Utah: Utah Water Research Laboratory, Utah State University) pp 89–109.

Bigelow, J.H., Bolten, J.G., and De Haven, J.C. (1977) *Protecting an Estuary from Floods – A Policy Analysis of the Oosterschelde: Vol. IV, Assessment of Algae Blooms, A Potential Ecological Disturbance*. Research Report R-2121/4-Neth (Santa Monica, CA: Rand Corporation).

Canale, R. (1976) *Modeling Biochemical Processes in Aquatic Ecosystems* (Ann Arbor, MI: Ann Arbor Science Publishers).

Clasen, J. (1980) *Regional Project Shallow Lakes and Reservoirs. Final Report.* OECD Report (Medmenham, UK: Water Research Centre).

Di Toro, D.M. (1976) Combining chemical equilibrium and phytoplankton models – a general methodology, in R. Canale (Ed.) *Modeling Biochemical Processes in Aquatic Ecosystems* (Ann Arbor, MI: Ann Arbor Science Publishers) pp 233–56.

Di Toro, D.M. (1980) Applicability of cellular equilibrium and Monod theory to phytoplankton growth kinetics. *Ecological Modelling* 8:201–18.

Di Toro, D.M., O'Connor, D.J., and Thomann, R.V. (1974) A dynamic model of the phytoplankton population in the Sacramento–San Joaquin Delta, in *Nonequilibrium Systems in Natural Water Chemistry*, Advances in Chemistry Series No. 106 (American Chemical Society), pp 131–80.

Di Toro, D.M., Thomann, R.V., O'Connor, D.J., and Mancini, J.L. (1977) Estuarine phytoplankton biomass models – verification analysis and preliminary applications, in E.D. Goldberg (Ed.) *The Sea: Ideas and Observations on Progress in the Study of the Sea* (New York: Wiley).

Fedra, K., Beck, M.B., and van Straten, G. (1981) Uncertainty and arbitrariness in ecosystems modelling: a lake modelling example. *Ecological Modelling* 13:87–110.

Hornberger, G.M. and Spear, R.C. (1980) Eutrophication in Peel Inlet, I: Problem defining behaviour and a mathematical model for the phosphorus scenario. *Water Research* 14:29–42.

Imboden, D.M. (1982) Modellvorstellungen über den Phosphor-Kreislauf in stehenden Gewässern. *Z. Wasser und Abwasser Forschung* 15(3):89–95 (in German).

Jones, R.A. and Lee, G.F. (1982) Recent advances in assessing impact of phosphorus loads on eutrophication-related water quality. *Water Research* 16:503–15.

Jørgensen, S.E. (1983) Modeling the ecological processes, in G.T. Orlob (Ed.) *Mathematical Modeling of Water Quality: Streams, Lakes and Reservoirs*. Wiley–IIASA International Series on Applied Systems Analysis, Vol. 12 (Chichester, UK: Wiley) pp 116–49.

Jørgensen, S.E., Mejer, H., and Friis, M. (1978) Examination of a lake model. *Ecological Modelling* 4:253–78.

Kremer, J.N. and Nixon, S.W. (1978) *A Coastal Marine Ecosystem, Simulation and Analysis*. Ecological Studies Series, Vol. 24 (Berlin: Springer).

Leonov, A.V. (1982) *Transformations and Turnover of Phosphorus Compounds in the Lake Balaton Ecosystem, 1976–1978*. Working Paper WP-82-27 (Laxenburg, Austria: International Institute for Applied Systems Analysis).

Leonov, A.V. and Vasiliev, O. (1981) Simulation and analysis of phosphorus transformations and phytoplankton dynamics in relation to the eutrophication problem of Lake Balaton, in D.M. Dubois (Ed.) *Progress in Ecological Engineering and Management by Mathematical Modelling* (Liege, Belgium: Editions CEBEDOC) pp 627–56.

Lorenzen, M.W. and Mitchell, R. (1975) An evaluation of artificial destratification for control of algal blooms. *J. Am. Water Works Assoc.* 67(5):373–6.

Los, F.J., de Rooij, N.M., Smits, J.G.C., and Bigelow, J.H. (1982) *Policy Analysis of Water Management for the Netherlands, Vol. VI: Design of Eutrophication Control Strategies*. RAND Note N-1500/6-NETH, Rand Corporation/Delft Hydraulics Laboratory (Santa Monica, CA: Rand Corporation).

Loucks, P. (1983) Models for management applications, in G.T. Orlob (Ed.) *Mathematical Modelling of Water Quality: Streams, Lakes and Reservoirs*. Wiley–IIASA International Series on Applied Systems Analysis, Vol. 12 (Chichester, UK: Wiley) pp 468–502.

Lum, L.W.K., Armstrong, N.E., and Fruh, E.G. (1981) Incorporation of chemical equilibrium, physico-chemical reactions and the simulation of inorganic qualities in a reservoir ecological model. *Ecological Modelling* 14:95–124.

Orlob, G.T. (Ed.) (1983) *Mathematical Modeling of Water Quality: Streams, Lakes and Reservoirs*. Wiley–IIASA International Series on Applied Systems Analysis, Vol. 12 (Chichester, UK: Wiley).

Park, R.A. (1978) *A Model for Simulating Lake Ecosystems*. Report No. 3 (Troy, NY: Center for Ecological Modeling, Rensselaer Polytechnic Institute).

Park, R.A. (1979) The aquatic ecosystem model MS. CLEANER, in S.E. Jørgensen (Ed.) *State-of-the-art in Ecological Modelling* (Copenhagen: Int. Society of Ecological Modeling) pp 579–602.

Park, R.A. *et al.* (1974) A generalized model for simulating lake ecosystems. *Simulation* 23(2):33–50.

de Rooij, N.M. (1980) A chemical model to describe nutrient dynamics in lakes, in J. Barcia and L.R. Mur (Eds.) *Developments in Hydrobiology, Vol. 2* (The Hague: Junk) pp 139–49.

Scavia, D. and Robertson, A. (1979) *Perspectives on Lake Ecosystem Modeling* (Ann Arbor, MI: Ann Arbor Science Publishers).

Sly, P.G. (Ed.) (1982) Sediment/freshwater interaction, in *Developments in Hydrobiology, Vol. 9* (The Hague: Junk). Reprinted from *Hydrobiology* (1982) Vol. 91/92.

Somlyódy, L. (1982) Modelling a complex environmental system: The Lake Balaton study. *Mathematical Modelling* 3:481–502.

van Straten, G. (1979) A note on the effect of light penetration fluctuations on algal growth in Lake Balaton, in G. van Straten, S. Herodek, J. Fischer, and I. Kovacs (Eds.) *Proc. 2nd Joint MTA/IIASA Task Force Meeting on Lake Balaton Modeling, August 1979* (Veszprém: Veszprém Academy Press) pp 176–94.

van Straten, G. (1981) Aspects of nutrient loading in relation to lake eutrophication modelling, in J.H.A.M. Steenvoorden and W. Rast (Eds.) *Impact of Non-Point Sources on Water Quality in Watersheds and Lakes.* UNESCO Man and the Biosphere Programme, Project No. 5. Workshop Proceedings, May 1980 (Amsterdam: UNESCO) pp 17–32.

van Straten, G. and Herodek, S. (1982) Estimation of algal growth parameters from vertical primary production profiles. *Ecological Modelling* 15:287–311.

Swartzman, G.L. and Bentley, R. (1979) A review and comparison of plankton simulation models. *International Society of Ecological Modelling J.* 1:30–81.

Thomann, R.V., Di Toro, D.M., Winfield, R.P., and O'Connor, D.J. (1975) *Mathematical Modeling in Lake Ontario, 1: Model Development and Verification.* EPA Ecological Research Series, EPA-660/3-75-005 (Washington, DC: Environmental Protection Agency).

Verhagen, J.H.G. (1976) The construction of models of simulation: a guideline for investigations into and control of water quality in reservoirs. *Hydrobiological Bulletin* 10(3):172–8.

Vollenweider, R.A. and Kerekes, J.J. (1980) The loading concept as basis for controlling eutrophication. Philosophy and preliminary results of the OECD programme on eutrophication. *Progress in Water Technology, Norway* 12:5–38.

Vollenweider, R.A. and Kerekes, J.J. (1982) *Background and Summary Results of the OECD Cooperative Programme on Eutrophication.* OECD Report (Paris: OECD).

CHAPTER 4

Eutrophication Management Models

L. Somlyódy

4.1. Introduction

In Chapter 3 the relationship between nutrient loads and lake water quality was discussed. The results obtained should now enable us to determine – at least approximately – the load reductions necessary to achieve the required level of water quality (e.g., for the lake to be shifted from a hypertrophic to a eutrophic state). We still do not know, however, how such load reductions can be achieved in practice. What control alternatives are available? How should the lake eutrophication model (LEM) be used in a planning mode? What are the costs, benefits, and constraints associated with such controls? What are the most important trade-offs (e.g., between agriculture, tourism, and environmental objectives)? In other words, we still have to consider the technical and economic implications of various control methods that can be taken within the watershed. A certain level of water quality can be achieved by various management practices. When considering alternative policies the aim is to find the "best" strategy and for this purpose optimization techniques serve as useful tools. Accordingly, most management models involve optimization of some sort, and this is described in this chapter.

Referring to Chapter 2, a management model should incorporate the aggregated information obtained from various "microscopic" subprocesses in order to make reliable "macroscopic" decisions on water quality management. The aim in this chapter is to establish a eutrophication management model based on the decomposition–aggregation principle, and to apply it to Lake Balaton. With such a model, important new questions can be answered, such as how sensitive is the model to various watershed and in-lake processes? And how do uncertainties propagate and appear at the decision-making level? The model should also involve the influence of subjective factors that are inherent to the policymaking procedures within which such models are used. Again, we need to know how sensitive model performance is to these factors, especially in comparison with the influence of scientific subprocesses and uncertainties.

The primary objective of this study was not only to solve methodological questions, but also to contribute to the solution of the water quality problem of Lake Balaton. Meanwhile, a methodology was developed that was felt to be applicable to a wider range of lake eutrophication problems. During the study several alternative management models were formulated (as well as those for hydrodynamics and other in-lake processes; see Chapters 3 and 9–13). Again, the comparison with conclusions of various other models is of interest.

The chapter is organized as follows. In Section 4.2 the management or control of a lake–watershed system is discussed in general terms, and the type of control appropriate to Lake Balaton decided upon. In section 4.3 is outlined the structure of the analysis, which stems from the principle of decomposition and aggregation, and the use of the LEM in a planning mode. A brief discussion is also given of the generation of forcing functions as synthetic time series based on historical data. The major outcome of Section 4.4 is an aggregated, stochastic load response model derived from a Monte Carlo-type use of one of the lake eutrophication models (see Chapters 3 and 11) under generated forcing functions. The model is linear, in accordance with the findings of Chapter 3, and describes changes in the annual peak chlorophyll-a (Chl-a) concentrations – a proper indicator for eutrophication management – as a function of the average annual biologically available P load and its stochastic variability. In Section 4.5 various protection measures are discussed and a stochastic nutrient load model is derived, which describes the (stochastic) load as a function of control variables. Then in Section 4.6, two alternative eutrophication management (optimization) models are described. The two models differ from each other with respect to formulating the objective functions and result in linear programming and stochastic programming with recourse, respectively. Results of the short-term control strategy for Lake Balaton over the coming years are then given in Section 4.7. Subsequently, four other models and their conclusions are discussed briefly, and these are compared with the findings of the approach discussed in detail.

4.2. Management of a Lake–Watershed System

First we discuss the management of a lake–watershed system from the "engineering" viewpoint of artificial eutrophication. We assume that the major processes, appropriate control methods, and their related costs and constraints, etc., can be qualitatively identified, but first we must ascertain whether there is sufficient information and data with which to elaborate an appropriate quantitative description. Second, the influences of socioeconomic factors on eutrophication management are considered. The conclusion of this two-step discussion is that eutrophication control – as a cause–effect problem of the lake and its region – in a broad, regional sense is hampered by several scientific and institutional shortcomings. Thus, in the majority of cases (including Lake Balaton) the action space of management – in which reliable, quantitative evaluation is possible – should be significantly narrowed.

Management based on a coupled description of the lake and its watershed

In Chapter 3 lake eutrophication models (LEMs) of various levels of complexity were outlined: empirical models, dynamic simulation models, etc. For the development of such models the nutrient load, as one of the forcing functions of the LEM, should be known with reasonable accuracy. This information can be obtained from either (*a*) historical, *in situ* observations, or (*b*) watershed nutrient load models (NLMs).

At this stage of the analysis loads should be spatially aggregated, as dictated by the objectives of the study and the corresponding segmentation of the lake (e.g., four basins for the Lake Balaton models discussed in Chapter 3, or the 40 elements of the coupled hydrodynamic–water quality model outlined in Chapter 10). Option (*a*) is quite straightforward and widely employed; the main precondition for applying this method is that nutrients reaching the lake should be adequately monitored (see Chapter 6).

The objective of option (*b*) is not only to derive the aggregated loads of the lake segments, but also to gain insights into watershed processes (both natural and artificial) and how these eventually influence the resultant pollution of the lake (that is on the emission–transmission–immission chain). The structure of such models, often called chemical transport models, is quite similar to that of the lake water quality model illustrated in Figure 2.1. They incorporate sources (or emissions) distributed throughout the watershed (e.g., sewage discharges, fertilizers) and describe the "fate" of nutrients through transformation and transport processes (biochemical changes, runoff, erosion, infiltration, river transport, etc.). Again, similar to LEMs, various approaches can be developed (simple or complex, static or dynamic) and at this stage the models are used in a descriptive fashion (for details of the many different models, see the review of Haith 1982). It is sufficient to note here that none of the complex NLMs has yet been successfully calibrated and validated for "large" watersheds and simple, static models based on the concept of export coefficients are usually used (the situation is similar to that discussed in Chapter 3 for LEMs). In order to develop a eutrophication management model, not a descriptive, but a planning-type watershed nutrient load model (NLMP) containing control variables should be developed (see Haith 1982) and coupled to the LEM (Figure 4.1). Watershed control methods include cropping management, and control of fertilizers, sewage, feedlot effluent, erosion, and runoff. One class of these actions influences the emissions, while the other represents removal of nutrients in the course of their transport within the catchment. The wide range of control alternatives allows – at least in principle – consideration of the entire watershed as the action space for eutrophication management. In practice, however, not only should the shortcomings of the descriptive watershed model be recognized, but also the fact that the costs and benefits of many control measures can rarely be evaluated well.

As indicated above, a management model is defined as an optimization model of some sort. This means that the planning-type watershed model

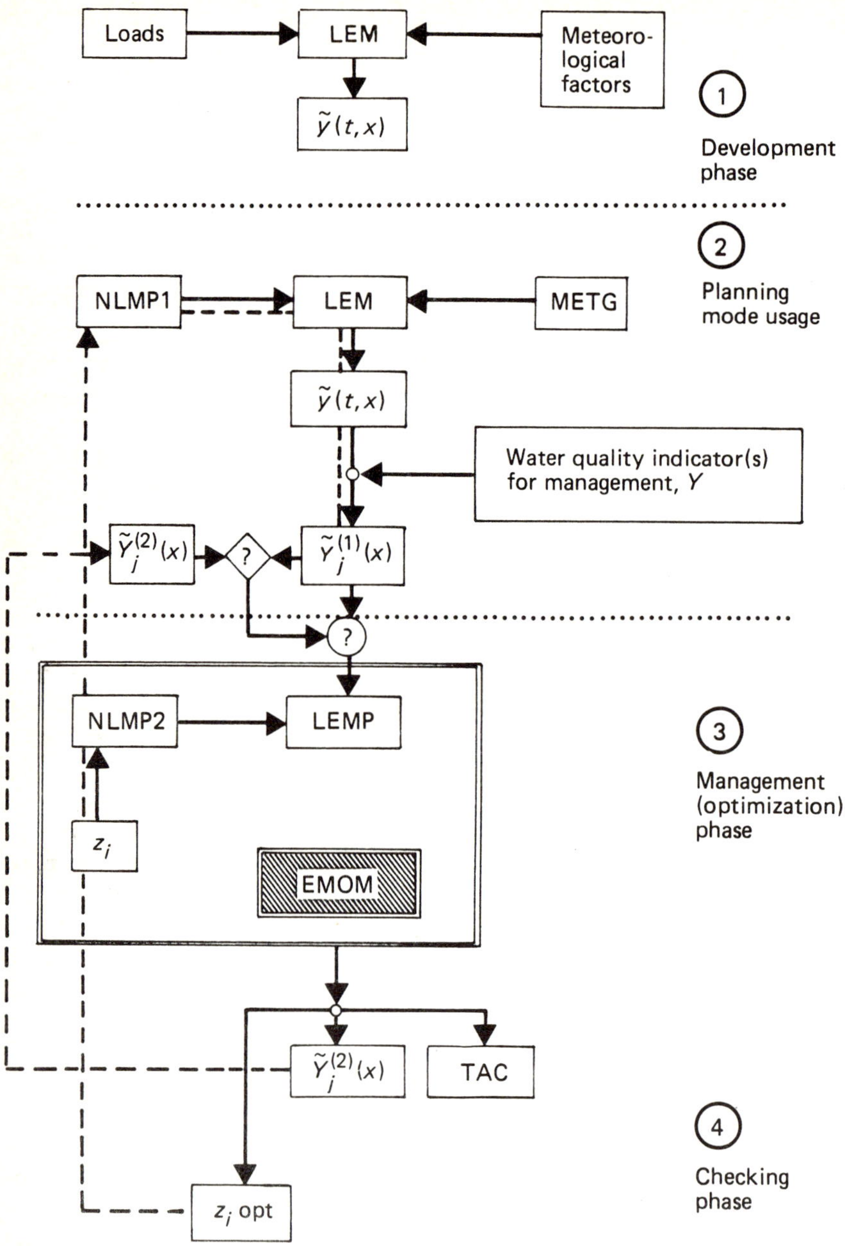

Figure 4.1. Structure of the analysis. LEM, lake eutrophication model; NLM, nutrient loading model; EMOM, eutrophication management optimization model; METG, meteorology generator; c_j state variables of LEM; Y_j water quality indicators; z_i control variables; TAC, total annual cost.

should be coupled to the LEM and then incorporated in the optimization. This procedure leads to further difficulties; the final model can be extremely complex and the solution is often hampered by methodological problems. In principle, simulation by changing the decision variables (systematically, randomly, sequentially, or by trial and error) can be used, not for establishing the optimal solution, but rather to evaluate alternative management plans. However, the procedure can be computationally very expensive and tedious, and may lead to a solution that is still far from the best possible (Loucks *et al.* 1981).

All these difficulties explain why in the field of eutrophication the management model is most often replaced by the LEM or by the watershed model NLMP. In the first case, the LEM is used to determine the load reductions required to improve water quality, but a systematic economic evaluation of alternative plans is rarely performed. In the second case, the NLMP is put in an optimization framework; and e.g., the external load is minimized under budgetary constraints.

Although changes in the ambient lake water quality should be an important element in decision making, no lake is given in the literature (to our knowledge) for which a proper description of in-lake processes has been accounted for in the management model (an example of a different problem can be found in Spofford *et al.* 1976). Therefore, it is our intention here to describe the development of such a management model for Lake Balaton that incorporates in-lake and watershed processes in a simplified way and combines the advantages of simulation and optimization.

Conflicts and trade-offs

The Lake Balaton region is characterized by intensive agricultural production and recreational use. Consequently, solutions to eutrophication control should be looked for in the spheres of agriculture, tourism, and environmental protection. For example, increases in fertilizer use lead to higher agricultural output, but also to water quality deterioration, which then reduces income from tourism. Similarly, increasing tourism results in declining income due to the feedbacks associated with the growing sewage loads. The issue is even broader, since deteriorating water quality can lead, in extreme cases, to the devaluation of the immediate surroundings of lakes, which can have serious social implications. In sum, eutrophication management cannot be separated from the socioeconomic management of the region, and perhaps even the country, and properly formulated, it should form part of a much broader area of control. This, however, is beyond the scope of this book.

In practice, the task of a lake manager is to utilize the available budget as effectively as possible, even though this may be marginal compared with the total funds allocated to the region. Consequently the manager cannot consider socioeconomic side effects; this has to be done at a higher level in the institutional hierarchy.

Eutrophication management for Lake Balaton and its watershed

From the above discussion the following conclusions on factors influencing model-building strategy can be drawn:

(1) Owing to practical reasons eutrophication management is understood in a narrow sense (as an engineering economics problem).
(2) Only those processes and related protection measures should be considered for which appropriate information is available.
(3) The planning-type watershed nutrient load model (NLMP) and the lake eutrophication model (LEM) used for management should be matched in terms of their complexity.

Owing to the current critical level of water quality in Lake Balaton our aim is to establish a short-term "optimal" control strategy for the next few years. For that purpose the only major alternatives available are the introduction of tertiary sewage treatment at existing plants and the building of pre-reservoirs on tributaries before they enter the lake. Through these two options some of the load, point, and surface nonpoint sources (mainly of agricultural origin) can be controlled. Accordingly, the NLMP is relatively simple, which is – in light of the findings of Chapter 3 – in compliance with the conclusion drawn in (3) above.

The control strategy that will be derived is obviously "defensive", since it aims to answer the question of how the lake can be protected from pollution already discharged into the environment, but it does not consider the socioeconomic activities that lead to the emissions.

4.3. Structure of the Analysis

The procedure adopted is illustrated in Figure 4.1 (see also Figure 2.1 and Somlyódy 1983); and it consists of four stages:

Phase 1. This is the development phase of the dynamic LEM which has two sets of inputs: controllable inputs (mainly artificial nutrient loads), and uncontrollable inputs (meteorological factors, as discussed in Chapter 3).

Phase 2. The purpose here is to apply the LEM to solve a management problem. In order to do so two important steps should be taken:

(1) A decision has to be made as to the kinds of loads and meteorological inputs to be used in planning scenarios (e.g., in analyses of most river water quality problems, low flow conditions are assumed). In principle, the "critical" scenario can be deterministic (large loads, high temperatures, favorable light conditions), but in practice such a combination of environmental factors is rare because of the irregular annual dynamics of these forcing variables and the decisive role of their correlative properties. We failed to find such "design" environmental conditions for Lake Balaton and the situation is probably the same for many other shallow water bodies. Thus for future planning the inputs should be

considered as stochastic functions and generated in a random fashion (NLMP1 and METG).

(2) The nutrient load model (NLM) – independently of its complexity – should incorporate control variables (NLMP1) for deriving the loads of individual lake basins. In this step, aggregated control variables can be used to express the combined effects of management actions at various locations in the watershed on the expectation and variance of the total basin load (e.g., control variables for direct sewage sources can be replaced by a single variable acting on the sum of the sewage discharges).

The LEM can then be run systematically under stochastic inputs with different control variable vectors, resulting in the temporal and spatial changes in various ecological components, j, as stochastic variables, $\tilde{y}_j(t, x)$ [in contrast to $y_j(t, x)$ of the development phase, which is deterministic and "uncontrolled"]. These changes are of interest in developing understanding, but for decision making much simpler indicators, Y, reflecting the global behavior of the system can be used. For Lake Balaton, perhaps the only suitable parameter is the Chl-a concentration; this is frequently employed in the literature to define the trophic state of water bodies. The chlorophyll content affects the color of water and thus can be used as a measure of water quality and, perhaps more importantly, of the recreational value of the lake.

Eventually, we selected the annual peak Chl-a concentration as the major indicator; taking into account the cumulative nature of the eutrophication process and the relatively slow response to control measures (of the same order as flushing time) the annual dynamics are of secondary importance. The situation is similar in many other lakes since a relatively close correlation can usually be observed between the annual average and maximal Chl-a concentrations (OECD 1982). The selection of $Y = (\text{Chl-a})_{\text{max}}$ as the main management indicator means that time can be disregarded (see Chapter 3) and the dynamic lake model can be used in an off-line fashion for the subsequent optimization.

Phase 3. There are various ways in which the LEM can be involved in the optimization, and a final conclusion requires careful analysis. For example, the $\tilde{Y}_j(x)$ indicators obtained from systematic computer experiments can be stored as functions of the corresponding control vectors. These form a "surface" on which the "optimal" solution is looked for later. The other possibility followed here is to parameterize the results obtained under various control vectors and to arrive at an analytical expression (called LEMP; see Figure 4.1). As indicated in Figure 4.1, the success of this procedure is not *a priori* obvious and depends mainly on the complexity and major features of the system.

LEMP is an aggregated version of the LEM that can be used directly for planning purposes. It describes approximately the indicator as a function of the nutrient load and its stochastic variability, and accounts also for the random effect of meteorological factors. As shown in Chapter 3, the lake Chl-a response is practically linear with the load; thus LEMP has a simple algebraic structure that is easy to handle later on. LEMP is interconnected with a nutrient load model, NLMP2, including control variables z_i (where i refers to the ith element of the control vector, $1 \leq i \leq I$). Compared with NLMP1, the model used here shows three significant differences: (*a*) it covers only that portion of the load that is thought to be controllable based on available, realistic measures; (*b*) it contains more details on the watershed (location of control measures, and pollutant sources, etc.); and (*c*) it is more aggregated with regard to time (see below).

The coupled NLMP2–LEMP models are then put through an optimization procedure, which results (depending on the formulation of the eutrophication management optimization model, EMOM), for example, in corresponding values of "optimal"[1] control variables, z_i, indicators, $Y_j^{(2)}(x)$, and total annual costs, TAC.

Phase 4. In the course of this procedure various simplifications and aggregations are made without a quantitative knowledge of the associated errors. Accordingly, the last step in the analysis is checking; that is, the LEM can be run with the "optimal" load scenario, as indicated in Figure 4.1 by the dashed line, and the $\tilde{Y}_j^{(1)}(x)$ "accurate" and $\tilde{Y}_j^{(2)}(x)$ approximate solutions compared. An unacceptable agreement between the two results means that the extraction of the LEMP was not satisfactorily performed, so that a new derivation is necessary. In the following this procedure is described step-by-step for Lake Balaton.

4.4. Derivation of the Aggregated Lake Eutrophication Model, LEMP

At this stage of the analysis it is assumed that we already have a dynamic LEM of acceptable quality (Chapter 3). From the models discussed previously, SIMBAL (simple Balaton model) was selected, but all the important conclusions reached with SIMBAL were also tested with other models.

Input data averaging

The computer experiments revealed that LEMP is not sensitive to day-to-day changes in forcing variables, and that monthly mean values can be satisfactorily used for loads, temperature, and global radiation. This is an important practical finding since the generation of forcing functions for shorter averaging periods would be hampered by lack of data and information. The same conclusion can be drawn for the P load from an analytical perturbation analysis on a simplified lake P cycle model (van Straten 1981).

In possession of the "time scale" for the forcing functions we consider next the development of NLMP1 and the meteorology generator, METG (Figure 4.1). The emphasis is placed primarily on the random generation of forcing

[1] i.e., feasible or quasi-optimal solutions.

functions. In relation to the structure of NLMP1, the major load components distinguished in Chapter 6 – tributary loads (dissolved and particulate), and direct and indirect sewage loads – are supplied by control variables covering the controllable portion of the load[2] (further details are given in Section 4.5, in which we derive NLMP2).

Input generation in a random fashion

Nutrient Loads (NLMP1)

Daily observations on the Zala River for the period 1976–79 (see Chapters 6 and 14) and 25-year records of the stream flow rate, Q (Rákóczi and Varsa 1976) were used to generate monthly mean P loads, $\bar{L}_m$ (where the bar indicates mean and m is the index of months). A simple linear regression expression between $\bar{L}_m$ and $\bar{Q}_m$ was found acceptable (Chapter 6). The statistical properties of $\bar{Q}_m$, which reflect the stochastic influence of the hydrological regime, were given by various gamma and log-normal distributions (Rákóczi and Varsa 1976), while the residual error of the regression equation $\bar{L}_m = f(\bar{Q}_m)$ was approached by a normal distribution.

Again using the Zala River data set, a Monte Carlo analysis was performed (Somlyódy 1984, and Chapter 6) in order to study uncertainties created by infrequent sampling (one or two samples per month for most tributaries). The related error was described by three parameter gamma distributions fitted to the results of Monte Carlo simulations. The load generator developed for the Zala River, covering 50% of the Lake Balaton catchment, was fairly accurate for the most heavily polluted area of the lake, the Keszthely basin (although records were too short). An example for the biologically available P load, L_{BAP}, is given in Figure 4.2 for the Keszthely basin (the figure contains the volumetric load $L^*_{BAP} = L_{BAP}/v$, where v is the basin volume).

The load generator was subsequently extended to other tributaries and finally the extrapolated sampling error term was added[3] (this was negligible for basin I). The nutrient load model, NLMP1, derived in this manner can be regarded as merely an approximation for basins II–IV since distributions typical for the Zala watershed had to be used due to the lack of detailed data on other catchment areas. Bearing in mind, however, that the volumes and residence times of water in these basins are much larger and consequently the influence of the random changes on water quality is expected to be smaller, the model can be considered realistic for present purposes. As noted before, the model was operated on a monthly basis. In the management model, however, annual averages can be used satisfactorily (see below), so the model was further aggregated for calculating annual means (NLMP2; see Figure 4.1).

[2] Certainly the total load can also be modified in the course of computer runs, irrespective of the technical realization of load reductions.

[3] An uncertainty term related to direct sewage discharges (overloading of treatment plants in summer) was also added, but its influence later proved insignificant.

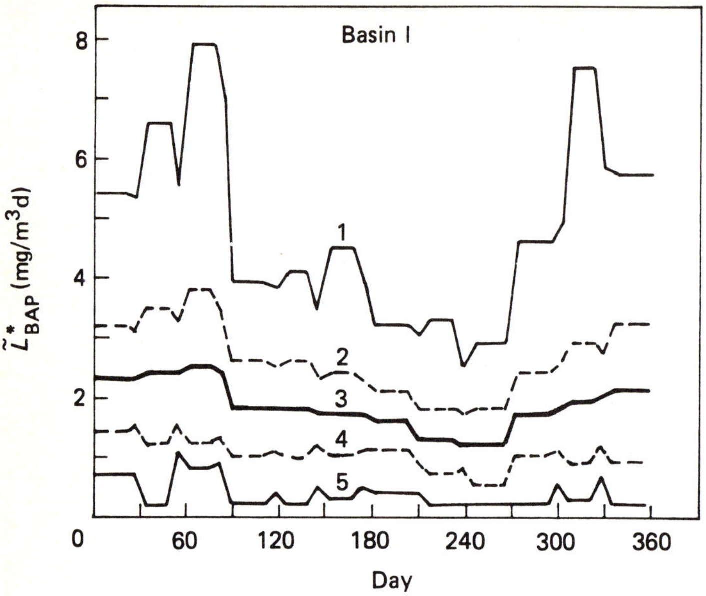

Figure 4.2. Generation of the volumetric biologically available P load. 3, mean; 2 and 4, ± standard deviation; 1 and 5, extremes.

Temperature and Global Radiation (METG)

Based on a ten-year historical record a simple autoregressive model was developed to calculate daily water temperatures (the duration of ice-cover was generated independently). While daily temperature changes are moderate, global radiation is characterized by large fluctuations. For this reason radiation data were *a priori* aggregated to weekly averages. Since cross-correlation is important here, radiation was related to weekly mean temperatures and eventually a model incorporating temperatures in two subsequent weeks was derived. Finally, daily temperatures and weekly global radiation were aggregated to monthly means for inputs to the LEM, as shown in Figure 4.3 (for further details see Somlyódy 1984).

Use of the LEM with stochastic inputs

Shallow lakes are strongly affected by random fluctuations in environmental factors, the quantification of which is of great importance for effective management. For this reason, the input generators outlined above were coupled to the LEM, which was then used in a Monte Carlo fashion. Figure 4.4 shows the influence of (uncontrollable) climatic factors on Chl-a concentrations (a conversion factor of 2 from phytoplankton P to Chl-a was adopted; see Chapter 3). Compared with simulation results obtained with SIMBAL for 1977 (Figure 3.6), changes in the average trajectory are less pronounced because

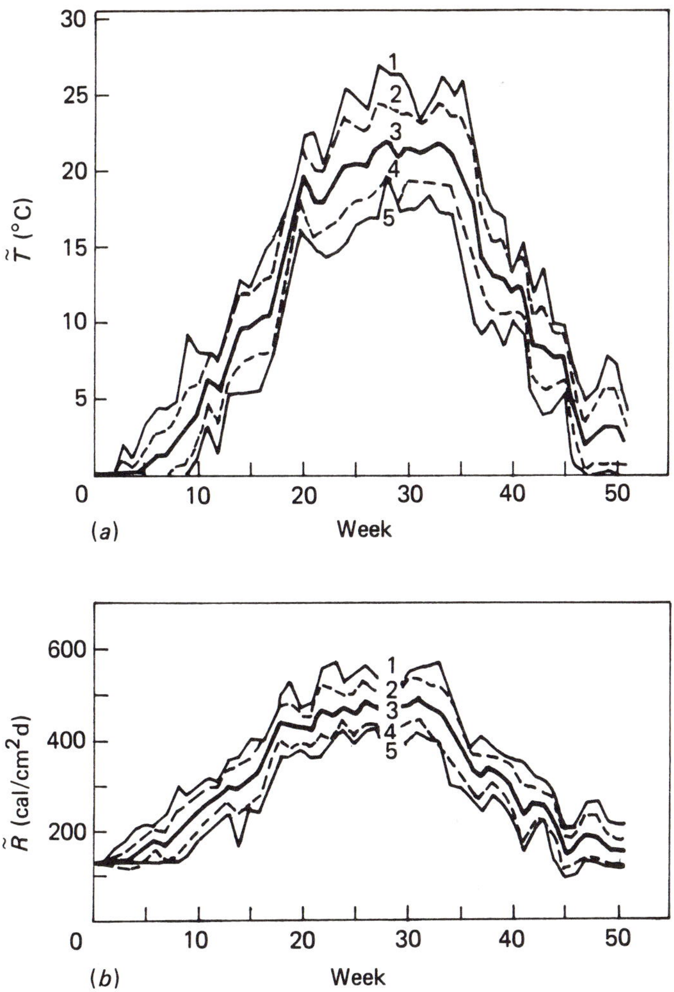

Figure 4.3. Generation of (*a*) water temperature, $\tilde{T}$, and (*b*) global radiation data, $\tilde{R}$. 3, mean; 2 and 4, ± standard deviation; 1 and 5, extremes (all from 1000 Monte Carlo simulations).

the mean trajectory of Monte Carlo simulations is determined primarily by the average input scenario, which in fact hardly ever occurs in nature. It is apparent from Figure 4.4 that the annual peak Chl-a concentration $(\text{Chl-a})_{max}$ can vary between 30 and 90 mg/m^3 (± 40% around the mean) depending solely on climatic factors. Such strikingly wide fluctuations can appear to mask the effect of considerable load reductions. The sensitivity of other lake basins was found to be smaller: the coefficients of variation of $(\text{Chl-a})_{max}$ for basins II–IV are 10, 6, and 5% (13% for basin I); upper extremes were 50, 30, and 20 mg/m^3, respectively.

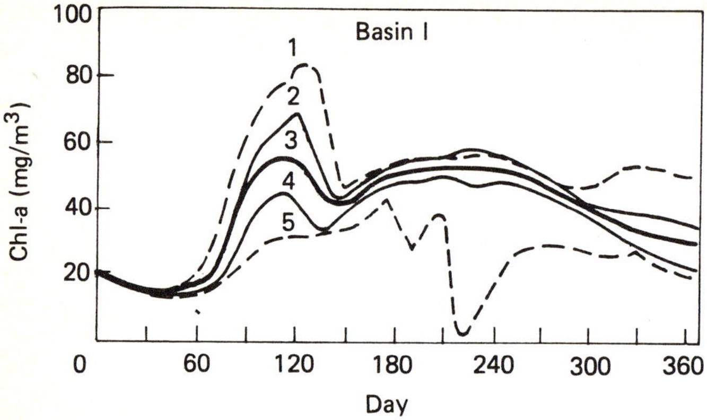

Figure 4.4. Influence of uncontrollable factors (e.g., climatic) on water quality. 3, mean; 2 and 4, ± standard deviation; 1 and 5, extremes (from 200 Monte Carlo simulations).

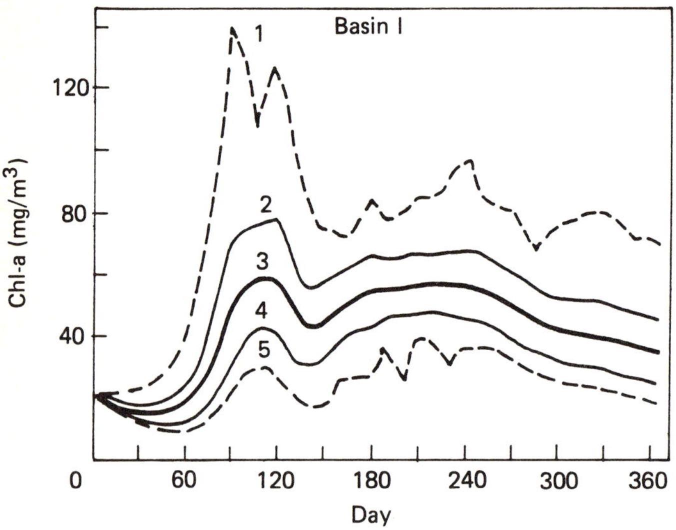

Figure 4.5. Influence of natural and controllable factors on water quality. 3, mean; 2 and 4, ± standard deviation; 1 and 5, extremes (from 200 Monte Carlo simulations).

If stochastic load changes are also taken into account, the result is as shown in Figure 4.5. The coefficient of variation is now about 20% and the extreme value of $(\text{Chl-a})_{max}$ can reach 140–150 mg/m^3; upper extremes for subsequent basins are 60, 35, and 25 mg/m^3, respectively. While empirical distributions of $(\text{Chl-a})_{max}$ are approximately symmetrical in the first case (Figure 4.4), here typically skewed distributions are produced, partially reflected by Figure 4.5.

Aggregated Lake Eutrophication Model (LEMP): Deterministic Version

So far we have considered the stochastic influence of forcing functions under the "original" load conditions, but the dynamic LEM was also run under systematically changed loads. Subsequently $(\text{Chl-a})_{max}$ was plotted against annual average values of L^*_{BAP}, and the linear relationship obtained is shown in Figure 4.6 (see also Chapter 3). The linearity is surprising at first glance only: the same trend is also reflected by empirical models (OECD 1982). It is perhaps a little disappointing that the practical essence of sophisticated dynamic models is so simple; but as the "extremum" analysis given in the previous chapter shows, the linear load response may be quite widely valid for nutrient-limited, turbid lakes (at least at our present level of understanding). All the other models developed for Lake Balaton show this feature (see Chapters 12 and 13), as do those for Lake Erie (Lam and Somlyódy 1983).

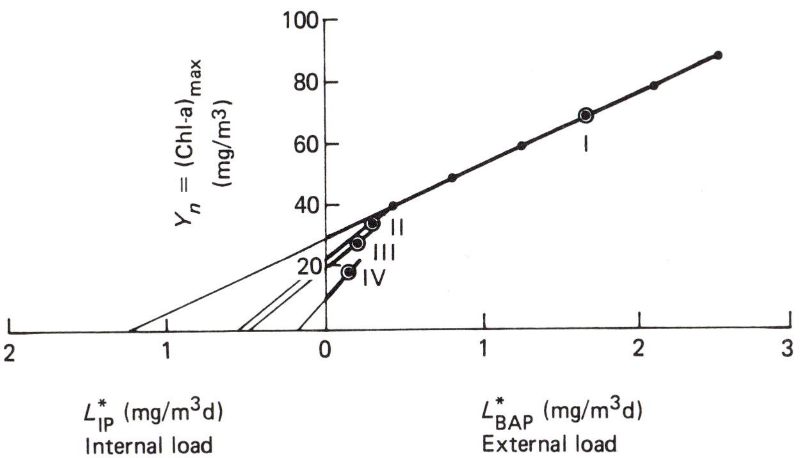

Figure 4.6. Aggregated lake eutrophication model: deterministic version.

It is apparent from Figure 4.6 that $(\text{Chl-a})_{max}$ is linearly related to the sum of external and internal volumetric loads (L^*_{BAP} and L^*_{IP}), and thus $(\text{Chl-a})_{max}$ does not necessarily approach zero (or a relatively small value) if $L^*_{BAP} \to 0$. From Figure 4.6 the annual average internal load can be estimated at about 600–800 kg/d (1.2 mg/m^{3}d), which is of the same order as the external load. This value is in agreement with the estimate given in Chapter 7 on the basis of experimental results.

The particular model used here, SIMBAL, incorporates only implicitly the influence of sediment, but the long-term behavior of sediment (enrichment and renewal) was not modeled[4] due to the lack of relevant data and

[4] Formally, it is easy to include the "memory effect" of sediment (see Chapter 12), but calibration can be based solely on intuition, at the present level of understanding.

experience (see Chapters 3 and 7). Consequently, Figure 4.6 does not give information on the progress of eutrophication or how future "equilibria" are reached. The plot illustrates the short-term (deterministic) response of the lake. Since external and internal loads are coupled (a reduction in external load generates a time-lagged reduction in the internal load; see Chapter 7) the long-term improvement is greater than suggested by Figure 4.6 (e.g., according to the plot basin I would remain eutrophic even if external loads were cut completely). The important conclusion from Figure 4.6 is that the dynamic LEM can be replaced by an aggregated, linear model version (LEMP) at this level of analysis (Somlyódy 1983):

$$Y = Y_0 + \mathbf{A}(L^*_{BAP} - L^*_{BAP_0}) = Y_0 - \mathbf{A}\Delta L^*_{BAP} \quad , \tag{4.1}$$

where Y is a vector representing $(\text{Chl-a})_{max}$ (mg/m^3), L^*_{BAP} is a vector representing the mean volumetric biologically available P load ($\text{mg/m}^3\text{d}$) for the four basins, 0 indicates the present (or nominal) state, and $\Delta L^*_{BAP} = L^*_{BAP_0} - L^*_{BAP}$. The elements of matrix **A** are reciprocals of lumped reaction rates [d]. The main diagonal comprises primarily the effects of biological and biochemical processes (with their forcing functions), while the other elements refer to those of interbasin exchange (due to hydrological throughflow and mixing).

For Lake Balaton, only the neighboring basins were coupled (in a unidirectional way) so that matrix **A** takes the specific form

$$\mathbf{A} = \begin{bmatrix} a_{11} & 0 & 0 & 0 \\ a_{21} & a_{22} & 0 & 0 \\ 0 & a_{32} & a_{33} & 0 \\ 0 & 0 & a_{34} & a_{44} \end{bmatrix} \tag{4.2}$$

indicating that any management actions performed on the eastern subwatersheds will have no effect on water quality of the western basins. Values of $a_{11}, \ldots, a_{44}$ (slopes of load response lines of Figure 4.6) range between 20 and 35, while $a_{21}, \ldots, a_{34}$ are an order of magnitude smaller.

It should be mentioned that the global influence of sediment renewal can be easily included in equation (4.1). If the equilibrium relation L_{IP} (L_{BAP}) is linearized in the vicinity of the nominal point (see Chapter 3) and the corresponding tangents are $k_1, \ldots, k_4$ (>0), the long-term response can be obtained formally using a modified matrix

$$\mathbf{A}^* = \begin{bmatrix} (1+k_1)a_{11} & 0 & 0 & 0 \\ (1+k_1)a_{21} & (1+k_2)a_{22} & 0 & 0 \\ 0 & (1+k_2)a_{32} & (1+k_3)a_{33} & 0 \\ 0 & 0 & (1+k_3)a_{34} & (1+k_4)a_{44} \end{bmatrix} . \tag{4.3}$$

Again, the problem is that the k_n values are unknown and only loose statements can be given on the time required to reach equilibrium (see Chapters 3 and 7). Nevertheless, it is worth mentioning that the derived simplified model LEMP still preserves the influence of all the important subprocesses in an aggregated way.

Aggregated Lake Eutrophication Model (LEMP): Stochastic Version

Monte Carlo simulations were also performed under various reduced load conditions using the "lumped" control variables of NLMP1. In order to separate the influence of controllable and uncontrollable factors on water quality indicators, two runs were done in each case: (*a*) with the climatic generator only, and (*b*) with simultaneous generation of nuturient loads and climatic factors. Results for the Keszthely basin are given in Figure 4.7 (pre-reservoir control).

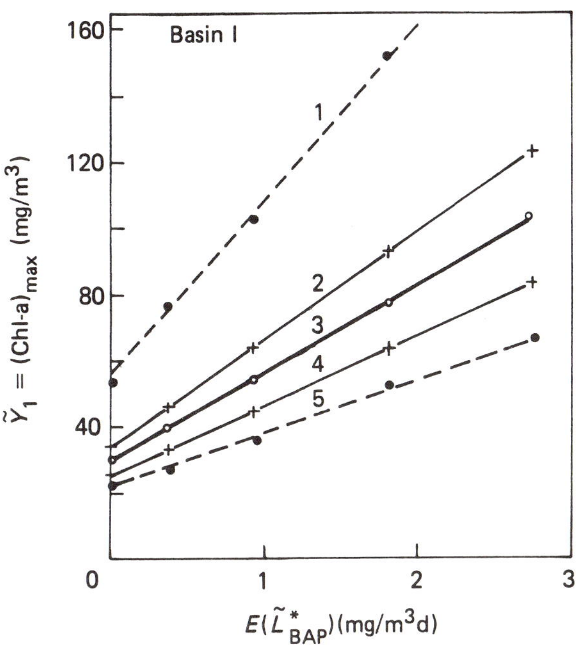

Figure 4.7. Aggregated lake eutrophication model: stochastic version. 3, mean; 2 and 4, ± standard deviation; 1 and 5, extremes; E, operator of expectation.

As can be seen from the figure, linearity is held as before, not only for the mean, but also for statistical properties (standard deviations and extreme values). When controlling sewage only, the expected $(\text{Chl-a})_{max}$ values change similarly, but the variance obviously remains more or less unchanged since most of the stochastic features and uncertainties of loads are associated with tributaries. The sensitivity of the model (and the lake) to variations in weather conditions is strikingly large. By also taking into account the "slow" recovery of sediment, this means that after reducing the external loads, the lake approaches a new equilibrium via considerable fluctuations: managers and public alike should not expect spectacular improvements in water quality.

A careful analysis of the results of Monte Carlo experiments shows that the influence of natural and controllable factors can be separated approximately. The expression is as follows (Somlyódy 1983):

$$\tilde{Y} = Y_0 + \tilde{Y}_w - (\mathbf{A} + \mathbf{b}\tilde{Y}_w)\Delta\tilde{L}^*_{BAP} , \tag{4.4}$$

where $\mathbf{A}$ and $\mathbf{b}$ are derived from model simulations, and the tildes indicate stochastic variables. $\tilde{Y}_w$ represents random changes caused by climatic factors (see Figure 4.4), and $\Delta\tilde{L}^*_{BAP}$ is the difference between the expectation of the "nominal" load and the controlled stochastic load (annual means; the nominal state is again indicated by 0): $\Delta\tilde{L}^*_{BAP} = E(L^*_{BAP_0}) - \tilde{L}^*_{BAP}$ (the operator E refers to the expectation of a stochastic variable). This definition means that if there is no control, equation (4.4) gives uncertainties in $\tilde{Y}$ in the "nominal" situation. $\Delta\tilde{L}_{BAP}$ can be expressed in detail as a function of the control variables; this is done in Section 4.5.

According to equation (4.4) water quality varies randomly for three reasons: (*a*) random changes in temperature and solar radiation (elements of $\tilde{Y}_w$ are given by gamma distributions); (*b*) stochastic changes and uncertainties in the loads, as discussed above; and (*c*) the combined effect of climatic and load factors. It is noted that the character of b_n is similar to that of a_{nn}, that is $\tilde{Y}_{wn}$ has a minimum if the sum of the external and internal loads is zero. With equation (4.4) the aggregated planning-type lake eutrophication model, LEMP, is obtained. This describes the (Chl-a)$_{max}$ concentrations of the four basins as a function of annual mean L_{BAP} loads, and also accounts for various stochastic features and uncertainties of the system. This model is used in subsequent analyses after deriving the corresponding nutrient load model, NLMP2.

4.5. Nutrient Load Model, NLMP2

Now we consider the controllable portion of L_{BAP} (the time scale is one year, as required for LEMP). The basic control options and variables are (Figure 4.8):

(1) Direct sewage control (tertiary treatment), z_{DS}, which may be thought of as a removal coefficient ranging between 0 and 1 (if 0, no action is taken).

(2) Indirect sewage control, z_{IS}, which has features similar to those of z_{DS}.

(3) Pre-reservoir systems[5] established on tributaries before they enter the lake, assumed to consist of two parts (see Figure 4.8 and Chapter 14): the removal of particulate P through sedimentation, z_{r1}, and the removal of dissolved P (benthic eutrophication in reed lakes, sorption, etc.), z_{r2} ($0 \leq z_r \leq 1$, and again, if $z_r = 0$, no action is taken).

[5] Such systems may be multipurpose; in the present context they are considered as tools for water quality control only.

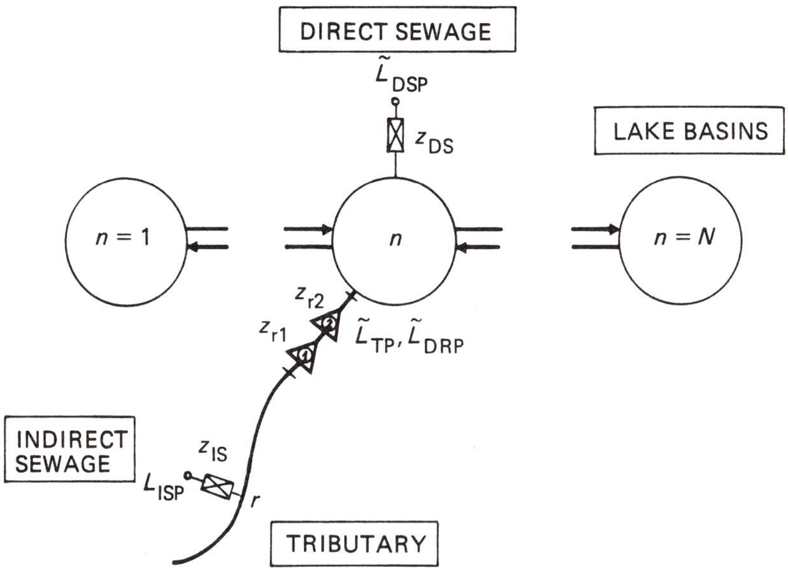

Figure 4.8. Development of the nutrient load model NLMP2. $\tilde{L}_{TP}$, total P load of tributaries; $\tilde{L}_{DRP}$, dissolved reactive P load of tributaries; L_{DSP}, direct sewage P load (considered deterministic); L_{ISP}, indirect sewage P load (considered deterministic).

Now consider the simple situation given in Figure 4.8 for the nth basin of the lake. The original, uncontrolled $\tilde{L}_{BAP}$, can be expressed as follows:

$$\tilde{L}_{BAP_{n0}} = \tilde{L}_{DRP} + \alpha(\tilde{L}_{TP} - \tilde{L}_{DRP}) + L_{DSP} + \tilde{L}_{NC} \;, \tag{4.5}$$

where

$\tilde{L}_{DRP}$ = dissolved reactive P load of tributaries
$\tilde{L}_{TP}$ = total P load of tributaries
L_{DSP} = direct sewage P load
$\tilde{L}_{NC}$ = portion of the load that is beyond the control space considered here ("uncontrollable" load, e.g., atmospheric pollution)
α = availability ratio of particulate P ($\alpha \simeq 0.2$).

$\tilde{L}_{DRP}$ and L_{DSP} are assumed to represent the P fractions that are wholly available for algal growth.

The controlled load of the nth basin is

$$\tilde{L}_{BAP_n} = (1 - z_{r2})\left[\tilde{L}_{DRP} - (1 - r) z_{IS} L_{ISP}\right] + \alpha(1 - z_{r1})(\tilde{L}_{TP} - \tilde{L}_{DRP}) + (1 - z_{DS}) L_{DSP} + \tilde{L}_{NC} \;, \tag{4.6}$$

where r is the retention coefficient, and L_{ISP} is the indirect sewage load. The expression $(1 - r)$ replaces a river P transport model and defines the portion of P that reaches the lake from an indirect sewage discharge at a given point on the tributary ($r = 0$ means no retention). It is apparent from equation (4.6) that the tributary load (due to point and nonpoint source effects) can be controlled by tertiary treatment and/or pre-reservoirs. The latter influence both the expectation and variance of L_{BAP_n}, while P precipitation (or tertiary treatment) influences only the expectation; thus there is an obvious trade-off between these two alternatives. Equation (4.6) is nonlinear on the control variables. For z_r we used the conventional definition (inflow minus outflow divided by the inflow), but for the sake of simplicity it was assumed that the upstream and downstream loads of the intended reservoirs would be equal in the uncontrolled case.

With equations (4.5) and (4.6) we can now derive the changes in $\tilde{L}_{BAP}$ for each basin in the aggregated lake model [see equation (4.4)] described above $[\Delta\tilde{L}_{BAP_n} = E(\tilde{L}_{BAP_{n0}}) - \tilde{L}_{BAP_n}$; from which for equation (4.4) $\Delta\tilde{L}^*_{BAP_n} = \tilde{L}_{BAP_n}/V_n$, where V_n is the basin volume]:

$$\begin{aligned}
\Delta\tilde{L}_{BAP_n} = {} & z_{r2}\left[E(\tilde{L}_{DRP}) - (1-r)z_{IS}L_{ISP}\right] && \text{(Term 1)} \\
& + (z_{r2}-1)\left[\tilde{L}_{DRP} - E(\tilde{L}_{DRP})\right] && \text{(Term 2)} \\
& + \alpha\left\{\left[(z_{r1}-1)\tilde{L}_{TP} + E(\tilde{L}_{TP})\right] - \left[(z_{r1}-1)L_{DRP} + E(\tilde{L}_{DRP})\right]\right\} && \text{(Term 3)} \\
& + (1-r)z_{IS}L_{ISP} && \text{(Term 4)} \\
& + z_{DS}L_{DSP} \; . && \text{(Term 5)}
\end{aligned} \tag{4.7}$$

Terms (1) and (4) express reduction in the expectation of the river's dissolved reactive P load, $\tilde{L}_{DRP}$; term (2) represents the effect on the fluctuations of this load; term (3) gives the modification in the particulate P load of the river; while term (5) shows the influence of direct sewage control. It is noted that the reduced indirect sewage load term (4) is subtracted from the river's uncontrolled $\tilde{L}_{DRP}$ load prior to the reservoir term (1) because obviously this part cannot be removed twice. If we set all the zs to zero in equation (4.7) fluctuations in the original, uncontrolled load are obtained, the expectation of which is zero.

For a more general situation when the nth lake basin is fed by I_1 direct sewage discharges $(1 \le i_1 \le I_1)$ and I_2 tributaries $(1 \le i_2 \le I_2)$, each with $I_3^{i_2}$ indirect sewage discharges $(1 \le i_3 \le I_3^{i_2})$, equation (4.7) can be generalized as follows:

$$\Delta\tilde{L}_{\mathrm{BAP}_n} = \sum_{i_2=1}^{I_2} \left\{ z_{\mathrm{r2}}^{i_2} \left[E(\tilde{L}_{\mathrm{DRP}}^{i_2}) - \sum_{i_3=1}^{I_3^{i_2}} (1 - r^{i_2 i_3}) z_{\mathrm{IS}}^{i_2 i_3} L_{\mathrm{ISP}}^{i_2 i_3} \right] + (z_{\mathrm{r2}}^{i_2} - 1) \left[\tilde{L}_{\mathrm{DRP}}^{i_2} - E(\tilde{L}_{\mathrm{DRP}}^{i_2}) \right] \right\}$$
$$+ \alpha \left\{ \left[(z_{\mathrm{r1}}^{i_2} - 1) \tilde{L}_{\mathrm{TP}}^{i_2} + E(\tilde{L}_{\mathrm{TP}}^{i_2}) \right] - \left[(z_{\mathrm{r1}}^{i_2} - 1) \tilde{L}_{\mathrm{DRP}}^{i_2} + E(L_{\mathrm{DRP}}^{i_2}) \right] + \sum_{i_3=1}^{I_3^{i_2}} (1 - r^{i_2 i_3}) z_{\mathrm{IS}}^{i_2 i_3} L_{\mathrm{ISP}}^{i_2 i_3} \right\} + \sum_{i_1=1}^{I_1} z_{\mathrm{DS}}^{i_1} L_{\mathrm{DSP}}^{i_1} \quad (4.8)$$

We have now arrived at the desired model, NLMP2. Equation (4.8) operates on an annual basis, incorporates control variables, and accounts for various uncertainties and stochastic effects. Equations (4.4) and (4.8) together serve as a coupled NLMP2–LEMP model (see Figure 4.1), which forms a direct relationship between water quality and control variables.

Below we return to the nonlinearity caused by the product terms $z_{\mathrm{r2}} \cdot z_{\mathrm{IS}}$ [see equations (4.7) and (4.8)], which can lead to difficulties in the optimization procedure, especially if someone wants to use one of the well developed and readily available linear programming (LP) techniques. This problem can be resolved by introducing a new variable $z^* = z_{\mathrm{r2}} \cdot z_{\mathrm{IS}}$, expressed as a linear function of z_{r2} and z_{IS}, which is then included in the constraint equations. In this case the optimization requires an additional iteration (see Loucks *et al.* 1981).

Although this procedure was used in the LP version of the management model outlined later, linearization is possible by modifying the definition of z_{r2}; namely, at our limited present understanding, we can only estimate the P removal per unit of reed-lake surface area. This implies the assumption of a constant efficiency, which is independent of the inflow concentration (or load). Under this approximation, however, z_{r2} can be defined in terms of the original, uncontrolled reservoir load (z_{r2}^*), which is not influenced by indirect sewage control. The price for such an elimination of nonlinearity is marginal:

(1) An upper limit should be specified for z_{r2}^* by the condition

$$z_{\mathrm{r2}}^* E(\tilde{L}_{\mathrm{DRP}}) + r z_{\mathrm{IS}} L_{\mathrm{ISP}} \leq E(\tilde{L}_{\mathrm{DRP}}) \;, \quad (4.9)$$

which simply states (in terms of expectation) that no more nutrients can be removed other than those that originally reached the lake via a particular tributary.

(2) A new variable, z_{r2}^{**}, should be introduced ($z_{r2}^{*} \leq z_{r2}^{**}$), which is essentially related to the same reed lake in order to express that the impact of the reservoir on the fluctuation [see equation (4.8), term (2)] is not restricted by the physical condition given by equation (4.9).

Through these steps iteration can be avoided, and this is a major practical advantage.

4.6. Formulation of the Eutrophication Management Optimization Model, EMOM

A large variety of alternative models can be developed to solve a management problem. The models can differ in the formulation of the objective function and constraints; in the specification of control variables (continuous, discrete, or mixed); in accounting for nonlinear effects, uncertainties, and stochasticity in the system; in selecting the optimization technique; and in many other respects.

Within the frame of the present effort it is not our intention to give an overview of various management models and optimization techniques. Instead we refer to well known books on water resources management, such as those of James and Lee (1971), Thomann (1972), Biswas (1981), Loucks *et al.* (1981), and Haith (1981). In this section two alternative management models are discussed whose objective is to work out the short-term "optimal" control strategy for Lake Balaton.

The long-term management action plan for the lake, elaborated in the mid-1970s (see Chapter 5) incorporated sewage diversion and pre-reservoirs, as well as other measures such as amelioration, dredging, etc. The progress of artificial eutrophication was then thought to be much slower than is observed today, so that a few large regional projects were planned with relatively slow schedules determined by economic constraints. Recent observations and scientific results, however, have revealed that the lake is now in a quite labile state. Uncertainty analyses (Figures 4.5 and 4.7) based on data from the mid- to late-1970s have shown that Chl-a concentrations could reach 150 mg/m^3 in the Keszthely basin (a hypertrophic value), which would be double those of 1977–81. In fact, during "unfavorable" meteorological conditions (high temperatures, light conditions conducive to algal growth, and heavy rainfall), extreme Chl-a concentrations of around 150 mg/m^3 were observed in 1982 (see Chapters 1 and 8).[6]

[6] At the same time the composition of phytoplankton has changed considerably, and a mass spreading of blue-green algae took place in August 1982. Admittedly, such changes cannot be predicted quantitatively at the present stage of ecological knowledge. This recognition should force the analyst to transmit gradually the available (and perhaps scientifically not entirely accurate) information to managers in order for them to make decisions on the basis of this knowledge rather than on that of "no" knowledge. The danger is that if the analyst waits for final "scientific" results, by that time a drastically different situation may have developed, and so his conclusions will be unusable in practice.

This acceleration of artificial eutrophication obviously calls for a set of protection measures that should be realized quickly. These could be provisional in nature, but should complement the long-term plan. As mentioned above, there are very few alternatives for a short-term control strategy for the coming years, apart from tertiary sewage treatment at existing plants, that do not require the development of the sewage network and pre-reservoirs on tributaries before they enter the lake. With these alternatives the aim of the management model is to provide an "optimal" distribution of investments for short-term measures.

The locations of potential major sewage control projects are given in Figure 4.9, together with the corresponding load data (multiannual average values, see Chapter 6). These cover about 85% of the total controllable nutrient load (the rest enters the lake via several small creeks and sewage discharges that could be involved in the analyses with no difficulty), which corresponds to about 60% of the lake's total BAP load.

The aggregated lake eutrophication model, LEMP (which describes short-term responses as required here), and the planning-type nutrient loading model, NLMP2, to be used in the management framework have already been outlined. We now consider control variables and the associated cost functions, the objective function and constraints, the solution technique adopted, and the evaluation of results.

Control variables

We have already discussed one class of control variable [see equation (4.8)] that is directly related to the nutrient load. These variables are basically continuous [although it is sometimes feasible to use (0,1) variables, e.g., for reservoirs]. Lower and upper bounds often differ from 0 and 1, respectively. For instance, the minimum reservoir size can be fixed, and lower bounds also frequently exist for tertiary treatment control variables: the effluent concentration standard is prescribed (2 g/m^3 in Hungary).

A second group of control variables consists of decision variables (called z_{MS}) associated with treatment plant management. Namely, not only is P precipitation missing at sewage plants in the Lake Balaton region, but also the efficiency of biological treatment is often unsatisfactory (Benedek and Szabó 1981). This automatically demonstrates the need for improved biological treatment, and this must be done before P precipitation can be effective. Using the variables z_{MS} related to the upgrading of biological treatment and logical constraints ($z_{DS} \geq z_{MS}$ and $z_{IS} \geq z_{MS}$ for all the plants) the necessity and sequencing of these actions (namely, first upgrading secondary treatment, followed by introducing tertiary treatment) can be prescribed in the model.

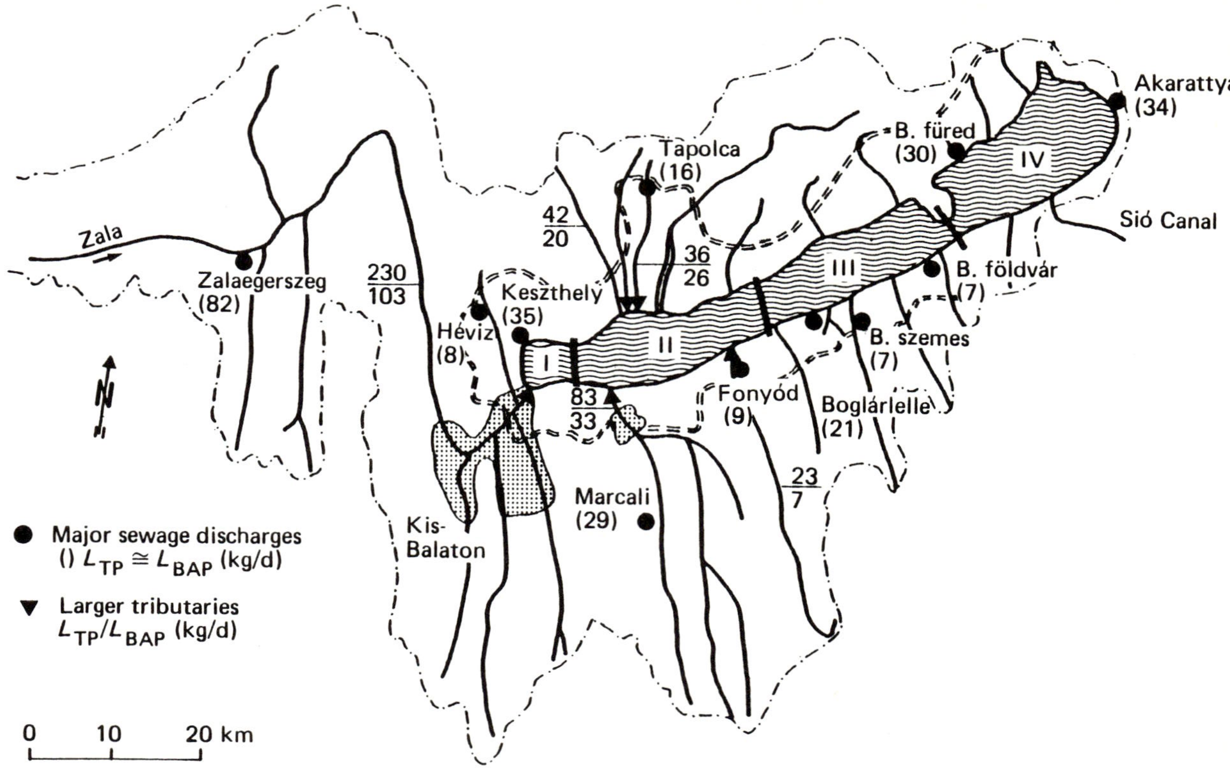

Figure 4.9. Major nutrient sources and control options.

Cost functions

Sewage Treatment

The costs of tertiary treatment grow exponentially with increasing removal rates and decreasing effluent concentrations (see, e.g., OECD 1982, Monteith *et al.* 1980, Schüssler 1981). Work on this particular problem was based on the report of Benedek and Szabó (1981), who surveyed the present status of sewage treatment plants in the Lake Balaton region, the investments required to upgrade biological treatment, and the costs of introducing tertiary treatment. From their data on plants of various sizes and technologies, cost functions were developed for the management model.

The procedure was based on an analysis of a variety of technological process combinations that lead to different removal efficiencies and/or effluent concentrations. For example, prior or simultaneous P precipitation can be used only if about 80% efficiency[7] (and/or an effluent concentration of about 1.5 g/m^3) is considered satisfactory. If, however, the requirements are more stringent, postprecipitation, followed occasionally by filtration, should be employed; this process calls for further equipment and additional (primarily investment) costs. When performing cost–benefit analysis, three groups of sewage treatment costs should be distinguished:

(1) Investment costs of upgrading biological treatment.
(2) Investment costs of tertiary treatment. These costs are considered independently of the control variable, z^i_{DS} or z^i_{IS}, up to a removal rate of 80% or to an effluent P concentration of not less than 1.5 g/m^3 (the more pessimistic case should be selected for the limit), while a rapid increase in investment costs should be expected for higher requirements.
(3) Running costs, which are assumed to be a linear function of the control variable, z^i_{DS} or z^i_{IS} (the cost of chemicals used in precipitation is particularly important).

Costs (1) and (2) depend on the capacity, inflow concentration, and other features of treatment plants, so that the cost functions elaborated are site-specific. An illustration of the costs for the largest treatment plant (Zalaegerszeg, capacity $Q = 15\,000$ m^3/d) is given in Figure 4.10.

If we were to go into details of sewage treatment plant design, integer variables should also be introduced for costs (1) and (2). For example, chemical injection equipment is either constructed at a certain size or not constructed at all; but since the problem is considered a global one, continuous variables can be satisfactorily used. The most straightforward method is then to approach the highly nonlinear cost functions by piecewise linearization[8]

[7] Values depend highly on how up-to-date treatment plants are and how satisfactorily they are operated (the literature contains more optimistic estimates than given here).

[8] The simultaneous use of continuous variables and piecewise linearized cost functions is realistic not only because of maintaining linearity, but because "corner points" appear most frequently in the optimum solution, thus some "discrete" features of the problem can be preserved.

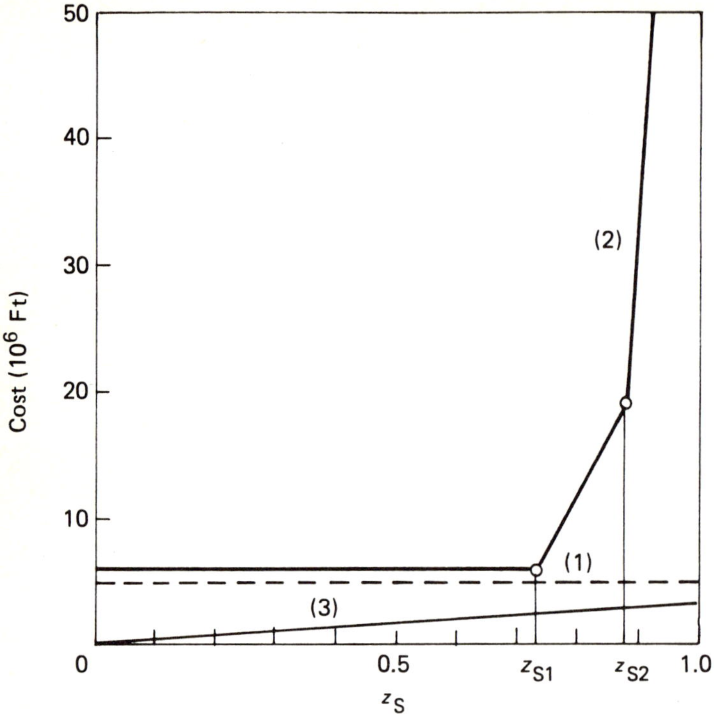

Figure 4.10. Costs of sewage treatment (Zalaegerszeg).

(see Figure 4.10 and e.g., Loucks *et al.* 1981), which requires the introduction of dummy variables (for both z^i_{DS} and z^i_{IS}) for segments (0, z_{S1}), (z_{S1}, z_{S2}), and (z_{S2}, 1) (see, e.g., Loucks *et al.* 1981). Costs (1) and a constant portion of (2) are related to the decision variables z_{MS}, while other costs are related to the control variables z_{DS} and z_{IS} that influence nutrient loads directly.

Reservoirs

The investment costs of reservoirs as a function of decision variables z_{r1} and z_{r2} (or P removal rates which are assumed to be proportional to size; see Section 4.5) are approximately linear, as has been established from information on the Kis-Balaton project at the mouth of the Zala River. Running costs are neglected, or are assumed to be compensated by the benefits of reservoirs (e.g., utilization of harvested reeds).

In relation to reservoirs (and to their cost-effectiveness) it is noted here that their application is restricted by several factors. First, the large size requirements imply that often they cannot be located at the mouth of rivers and thus pollution associated with downstream subwatersheds cannot be controlled. Second, reservoirs should be created in areas of low-quality land (which is not utilized intensively, as in the Kis-Balaton area), otherwise the dislocation or expropriation costs make them unrealistically expensive. Also, in many countries existing land protection laws control the conditions of land expropriation (and often *a priori* prohibit such actions).

Objective function

The water quality of the lake and its nutrient loads are given by stochastic variables [see equations (4.4) and (4.8)] with which the formulation of the management model can be performed in different ways (see, e.g., Loucks *et al.* 1981). Here we offer two possibilities, the corresponding models of which were implemented.

Model A. To start with, we define the water quality goal Y_g for all the basins ($1 \leq n \leq 4$), Y_{gn}, which we would like to achieve as closely as possible, taking account of stochastic changes and uncertainties [see Figure 4.11(*a*)]. This can be expressed by the following deterministic objective function (Somlyódy 1983):

$$\min \sum_{n=1}^{4} W_n E(\tilde{Y}_n - Y_{gn})^w \quad , \tag{4.10}$$

which states that if the exponent $w = 2$, then the weighted sum (see the weighting factors W_n, $W_n \geq 0$, $\sum_{n=1}^{4} W_n = 1$) of the expected square of the actual $(\text{Chl-a})_{\max}$ concentration minus the goal for all four basins is minimized. In a different formulation, simplified for a single basin, the sum of the square of the bias (B^2) and the variance ($V = \sigma^2$) is minimized [see Figure 4.11(*a*)], which has a direct statistical significance.

If $w > 2$, deviations from goals are penalized to a greater extent. Through W_n one basin can be preferred over others (more intensive tourism or need for urgent control, etc.). In general, the inclusion of parameters w, W_n, and partially also Y_g (see below) allows the analyst to handle the subjective judgment of users, the unquantifiable knowledge of decision makers, and various other aspects of the policymaking procedure.

In practice, water quality goals for eutrophication control often cannot be specified by fixed concentration values. While decision makers frequently decide that a lake should be shifted, say, from a eutrophic to an oligotrophic state, even the definition of trophic classes is not unambiguous – which is certainly no surprise (recently an effort was made to include uncertainty and the judgment of limnologists on how trophic terminology ought to be applied to reality; see OECD 1982).

From this explanation and the "soft" character of the management problem it follows that there is a large degree of flexibility (and arbitrariness) in the definition of the objective function in order to derive "feasible" solutions. In the present work, equation (4.10) was replaced by an analogous objective function:

$$\max \sum_{n=1}^{4} \left[W_n E(\Delta\tilde{Y}_n) - w\,\sigma(\Delta\tilde{Y}_n) \right] \quad , \tag{4.11}$$

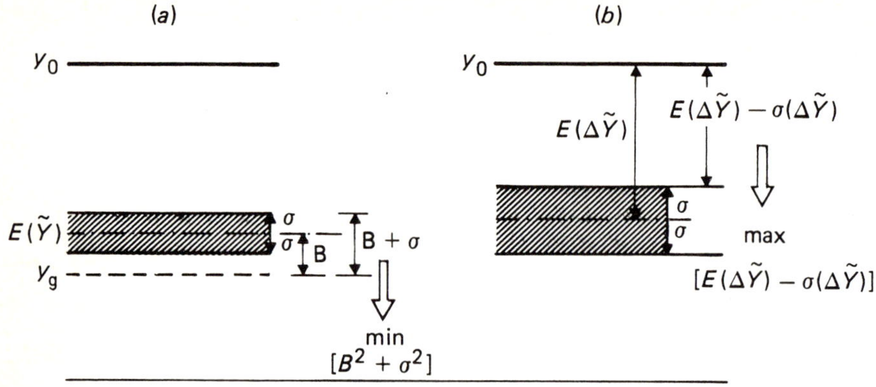

Figure 4.11. Definition of the objective function.

where $\Delta\tilde{Y}_n = Y_0 - \tilde{Y}$. According to this definition the distance measured from the original state "pessimistically" as the difference between the mean and the standard deviation [see Figure 4.11(*b*)] should be maximized (if $w = 1$) [for a more precise statistical explanation see Somlyódy and Wets (1985)].

Equation (4.11) can be set out as a function of control variables using equations (4.4) and (4.8) after expectations and standard deviations have been analytically derived using actual gamma, log-normal, normal, etc., distributions outlined above. Equation (4.11) still remains, however, nonlinear, since $\sigma(\Delta\tilde{Y}_n) \sim [\sum_i (z^i)^2(\sigma^i)^2]^{1/2}$, where σ^i is that part of the standard deviation which can be influenced by the decision variable z^i (e.g., the standard deviation of a tributary $\tilde{L}_{BAP}$ load). In order to arrive at a linear model, the objective function (4.11) is further modified and the term $\sigma(\Delta\tilde{Y}_n)$ is replaced by the sum $\sum_i z^i\sigma^i$ (Somlyódy 1984). The advantage of objective function (4.11) is that it retains the major features of function (4.10), but results in an LP model.

Model B. While model A is of an "engineering" type, model B is more precise in a statistical sense. Its objective function is similar to function (4.10) (see Somlyódy and Wets 1985):

$$\min \sum_{n=1}^{4} E\left[\Psi_n(\tilde{Y}_n - Y_{gn})\right] , \tag{4.12}$$

where the penalty function Ψ_n is zero if $\tilde{Y}_n \le Y_{gn}$, increases quadratically in the range $Y_{gn} \le \tilde{Y}_n \le Y_{un}$, and becomes linear subsequently (where Y_{un} is the upper limit of the quadratic domain). Thus, Ψ_n is a piecewise linear–quadratic–linear function of two parameters. Note, if $(Y_{un} - Y_{gn})$ is large, model (4.12) is equivalent to equation (4.10) if we consider positive deviations only in the latter objective function. The model formulated is thus nonlinear and results in a stochastic optimization problem (for details see Somlyódy and Wets 1985).

Note that, if the assumptions $a_{11}, \ldots, a_{nn} = 1, a_{21}, \ldots, a_{34} = 0$ had been made in equation (4.4) then this would have been equivalent to a formulation of the objective functions (4.11) and (4.12) in terms of the volumetric loads.

Constraints

In order to choose between management alternatives of different investment costs (C_{I}) and operational, maintenance, and repair costs (C_0), the total annual cost (C_{TA}) term is used

$$C_{\mathrm{TA}} = \sum_i C_0^i + \sum_i q^i C_{\mathrm{I}}^i \ , \tag{4.13}$$

in which the costs associated with all the decision variables ($1 \leq i \leq I$) are summed and q_i is the capital recovery factor that depends on the discount rate and lifetime of the project (see Loucks *et al.* 1981). This factor can be different for "small" and "large" projects (the introduction of chemical precipitation is small and building reservoirs is large in the case of Lake Balaton); for "large" investments governments often guarantee finance at low ("pure") interest rates. For this reason, as pointed out also by Thomann (1972), q_i should be considered as a model parameter of a certain range and its influence on model performance should be tested. In most cases the C_{TA} is limited by economic conditions. Also, it is possible to limit the total investment or operational costs. Other constraints (upper and lower bounds, as well as logical and physical conditions) were discussed earlier.

Solution techniques and evaluation of results

Model A, corresponding to objective function (4.11), was solved using a conventional LP routine. Model B, objective function (4.12), led to a problem which required a stochastic program with recourse (SP). As far as the solution procedure is concerned we refer to Rockafeller and Wets (1983) and Somlyódy and Wets (1985). The models produce, among other things, expectations and standard deviations of the basins' water quality according to the "optimal" policy. Water quality distributions are then evaluated by Monte Carlo simulations (1000 runs in each case). At this step of the analysis the influence of smaller tributaries (which were disregarded in the optimization) is also taken into account (see Section 4.5).

4.7. Results for Lake Balaton: Short-Term Control

In this section first we specify for convenience a "basic situation" that is close to the real case, in order to obtain an impression of the character of the problem and the behavior of the solution. Next we analyze the sensitivity of the model to major parameters. Finally a "realistic" solution is given using the actual information available for Lake Balaton. The discussion is based primarily on the use of the LP model; SP results – since they were in harmony with those of the LP model – are presented for "basic" and "realistic" situations, respectively. The basic situation is defined by the following features and assumptions:

(1) All control variables are continuous ($0 \leq z_i \leq 1$).
(2) No effluent standards are prescribed.
(3) No P retention takes place in rivers ($r_i = 0$).
(4) The capital recovery factor is the same for all projects, $q_i = q = 0.1$ (e.g., 15 years' economic life, a 6% interest rate).
(5) Equal weighting is adopted ($w = 1$, $W_n = 0.25$; $n = 1,\ldots,4$).

With these assumptions optimization was performed under different total annual cost (C_{TA}) conditions ($C_{TA} = 0.5–25 \times 10^7$ Forint, Ft; 1 US$ ≈ 50 Ft). Statistical parameters (expectation, standard deviation, and extremes) of the water quality indicator obtained from the Monte Carlo procedure are shown in Figure 4.12 for the Keszthely basin as a function of the total annual cost[9], for both LP and SP models.

Figure 4.12 can be analyzed in a multiobjective fashion since the two axes represent different objectives, namely economic and environmental. It is not our intention to look for a "best compromise solution", since the selection depends primarily on the resources available and the judgment of the decision maker. Instead, our intention is to specify typical total annual cost ranges with solutions different in character, the recognition of which is important for the decision maker. On the basis of the analysis of optimization results, four such domains can be distinguished (see also Figure 4.13, which illustrates the change of decision variables associated with the largest treatment plant, Zalaegerszeg, z^1_{IS}, and the reed-lake section of Kis-Balaton, z^1_{r2}, respectively):

(1) In the range $C_{TA} = 0.5–5 \times 10^7$ Ft/yr, it appears that sewage treatment can be intensified and tertiary treatment introduced. The expected $(\text{Chl-a})_{max}$ concentration will decrease considerably, but not the fluctuations. With very low costs ($C_I \sim 3 \times 10^7$ Ft) it turns out that only the sewage of Zalaegerszeg (the largest city in the Balaton region) should be treated, even if retention on the Zala River is greater than 50%. With an increased budget the investments are made from west to east, i.e., Zalaegerszeg, Keszthely, Héviz (basin I); then Tapolca and Marcali (basin II), followed by other treatment plants in basins III and IV.

[9]Running costs range between $0.5–1.0 \times 10^7$ Ft; thus with $q = 0.1$ the total investment cost is about ten times larger than the total annual cost.

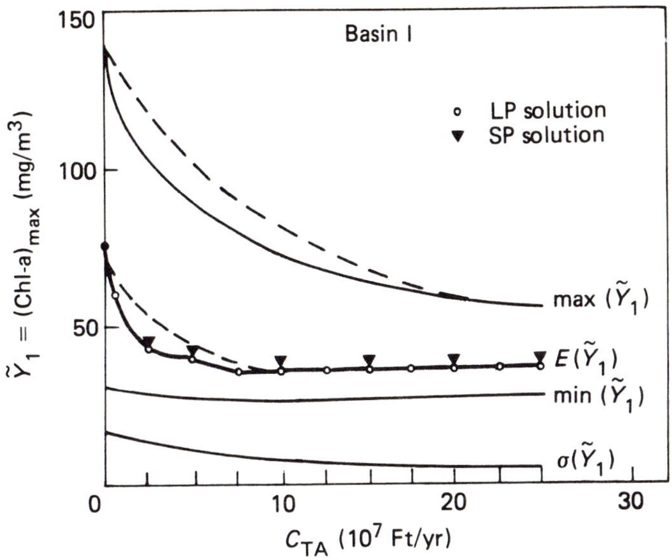

Figure 4.12. Water quality as a function of the total annual cost C_{TA} (solutions obtained from LP and SP models). —— basic case; --- with prescribed effluent standard (2 g/m^3).

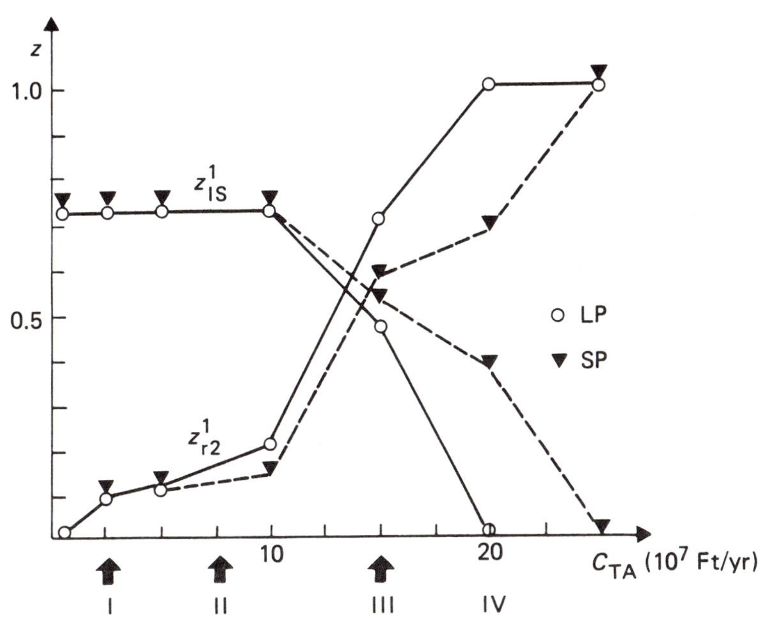

Figure 4.13. Change of major decision variables (LP and SP solutions).

(2) In the range $C_{TA} = 5 \times 10^7$–10×10^7 Ft/yr, the effectiveness of sewage treatment cannot be increased further but reservoir systems are still too expensive.

(3) At about $C_{TA} = 15 \times 10^7$ Ft/yr the solution is a combination of tertiary treatment and reservoirs. Fluctuations in water quality are reduced by the latter control alternatives.

(4) Finally, at around $C_{TA} = 20 \times 10^7$ Ft/yr, tertiary treatment is dropped in regions where reservoirs are an alternative. After creating all the reservoirs no further water quality improvement can be achieved.

As far as the comparison of LP and SP solutions is concerned (Figures 4.12 and 4.13) practically the same expectations, standard deviations, and extreme values were produced (for details, see Somlyódy and Wets 1985). The changes of decision variables, z^1_{IS} and z^1_{r2}, which are strongly coupled, are slightly different for the two cases: under the realistic parameters selected for objective function (4.12), the LP model gives more emphasis on the reservoir project influencing both expectation and variance of the load and water quality.

Next we turn our attention to the sensitivity of the solution. In the basic case we do not assume any prescribed effluent standard for sewage treatment, although in practice this is not the case, since standards are set by government agencies. In Figure 4.12 we also illustrate the solution for a situation noted earlier (the standard is 2 g/m^3 in Hungary). It is apparent from the figure that fixed water quality standards – which do not reflect the properties of the system in question (spatial nonuniformities) – do not result in an optional short-term strategy since the distribution of a portion of the budget is determined *a priori* by the given standards. From Figure 4.12 one could read the extent of water quality improvements in basin I if standards were relaxed under the condition that the money saved is spent somewhere else, according to the model formulation.

In Figure 4.14 we analyze the influence of the river retention coefficient (r) with and without prescribed effluent standards. As can be seen, with greater P retention, water quality improvements are less pronounced than in the basic case (Figure 4.12). The worst (albeit largely unrealistic) situation occurs if all the P is removed along the river (see below) and treatment is still required. In this case part of the budget should be allocated to investments that will have no influence on the lake's load. In summary, Figure 4.14 demonstrates the importance of river retention coefficients.

If we assume coefficients $a_{11},\ldots,a_{44} = 1$ and $a_{21},\ldots,a_{34} = 0$ in matrix $\mathbf{A}$ [see equation (4.2)] and a unit volume for the lake basins, the combined absolute load reduction for the four basins is maximized by the model. Since the loads for the four basins are of the same order (see Chapter 6), this policy differs drastically from the optimal one shown before. For lakes divided into segments of different water quality (spatial nonuniformity) the volumetric load – and not the absolute load, as is usually the case – should be minimized if the optimization problem cannot be formulated in terms of water quality (e.g., because the relationship between load and water quality is too complex).

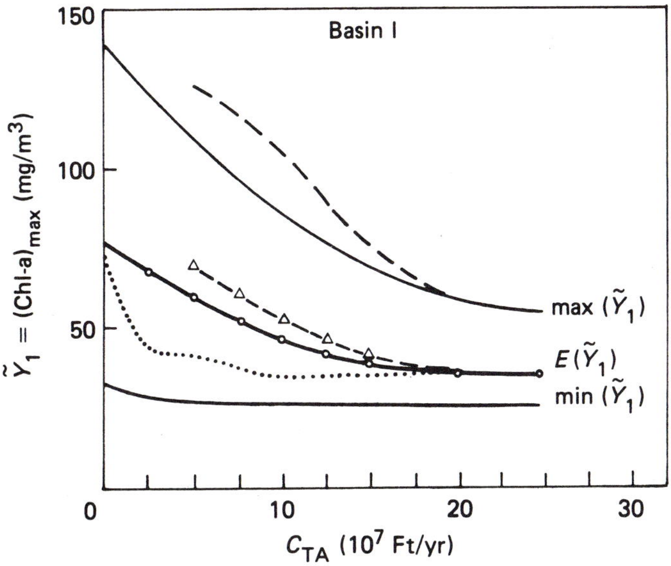

Figure 4.14. Sensitivity of the solution to P retention and prescribed effluent standards. ——, basic case; – – –, $r_i = 1$; ...., $r_i = 1$ with a prescribed effluent standard (2 g/m^3).

If only deterministic effects are considered ($w = 0$), reservoir projects enter the solution under much larger budget conditions (as compared with the case $w = 1$). In this case one of the twofold effects of reservoirs, namely the impact on water quality fluctuations, is deleted from the model.

As shown in Chapter 3, the slope of the load–response curve [$\alpha_{11}, \ldots, \alpha_{44}$ in equation (4.2)] is subject to some uncertainties (see also Chapter 13). In this case only, however, the management model solution has a negligible sensitivity to this factor. In extreme cases the slope of the load–response curve can be very small for some basins (internal load $\gg$ external load), calling incorrectly for no management action in the associated subregions, but this would be unrealistic (a large load reduction leads to slight water quality improvements, which is true only for the lake's short-term response). In such situations the matrix $\mathbf{A}$ should be replaced by $\mathbf{A}^*$ using the coefficients $k_1, \ldots, k_4$ [equation (4.3)] as formal parameters within a sensitivity analysis. Another possibility is simply to use the volumetric load for optimization. This could also be done with acceptable accuracy for the present problem because the influence of mass exchange between basins (interbasin coupling) is not very strong. In such cases the LEM can or should be completely dropped at the level of management.

The influences of some other factors are discussed later, and we give the "realistic" solution for the Lake Balaton management problem using actual retention coefficients (ranging between 0.3–0.5, see Chapters 6 and 14); an

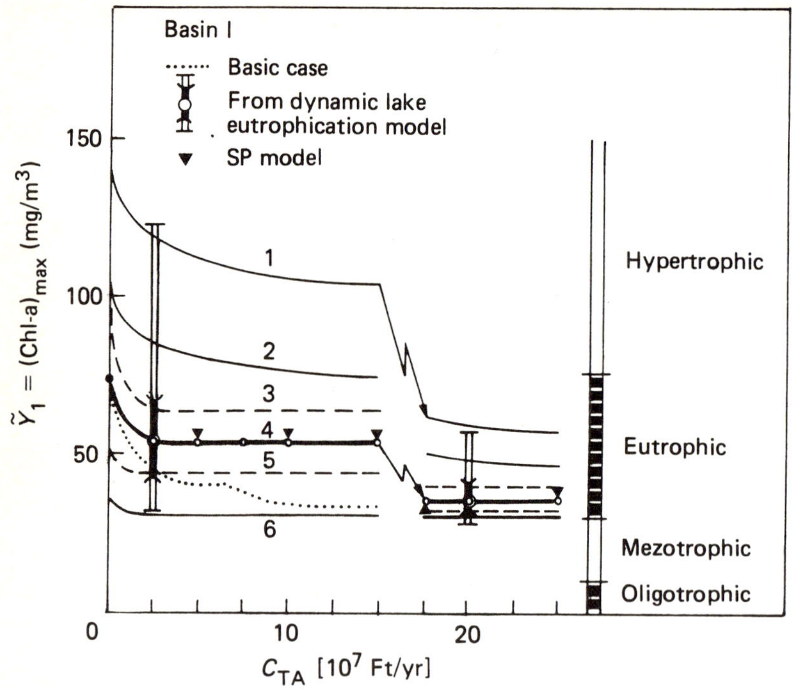

Figure 4.15. Solution of the management model for basin I. 4, expectation; 3 and 5, ± standard deviation; 2, 95% confidence level; 1 and 6, extremes (all from 1000 Monte Carlo simulations).

upper limit of 0.9 for the P removal rate of reservoirs (control variables, z_r); and "integer" variables for reservoirs (which take the value of the lower limit or the upper limit).

Figure 4.15, which again refers to the Keszthely basin, is remarkably different from Figure 4.12.[10] The drastic effect of reservoirs on expectation and their even stronger influence on water quality fluctuations are both stressed. Reservoirs enter the solution when the C_{TA} = 15–17.5 × 10^7 Ft/yr, resulting in a reduction in the mean (Chl-a)$_{max}$ concentration from about 55 to 35 mg/m^3, with extreme values of more than 100 to about 60 mg/m^3. As suggested by the figure, the SP model produced the same conclusions as the LP (similarly to the basic situation).

While Figure 4.12 offers several solutions for a decision maker according to the budget available, two feasible alternatives come to mind on the basis of Figure 4.15:

[10]In Figure 4.15 the ± standard deviation and the upper 95% confidence level are also shown. The distributions are bound towards small $\tilde{Y}_1$ concentrations and the lower 95% confidence level values are close to the minimum.

(1) If $C_{TA} \approx 2.5 \times 10^7$ Ft/yr, all sewage projects can and should be realized from west to east, with the same sequencing described above. Through this alternative the expectation of $\tilde{Y}_1 = (\text{Chl-a})_{max}$ is reduced to about 55 mg/m^3 (tertiary treatment affects the water quality to a slightly lesser extent than in the basic case due to P retention by tributaries), but extremes larger than 110 mg/m^3 can still occur (the hypertrophic domain according to the OECD classification). Any further budget increase ($C_{TA} \leq 17.5 \times 10^7$ Ft/yr) has no impact on water quality (under the alternatives included in the analysis).

(2) If the $C_{TA} \approx 20 \times 10^7$ Ft/yr, all the reservoirs can be created and tertiary treatment facilities can be installed for direct sewage sources. The mean $(\text{Chl-a})_{max}$ concentration is about 35 mg/m^3, with a maximum of about 60 mg/m^3 (eutrophic stage).

Figure 4.16 shows the results of the detailed simulation model, SIMBAL, for two optimal solutions ($C_{TA} = 2.5 \times 10^7$ and 20×10^7 Ft/yr); the satisfactory agreement indicates that the aggregated LEM is quite appropriate for our present purposes.

Figure 4.16 compares the typically skewed probability density functions of two strongly different solutions ($C_{TA} = 2.5 \times 10^7$ and 20×10^7 Ft/yr, respectively) for the four basins derived from Monte Carlo simulations (the uncontrolled state is also given). From this we can conclude that tertiary treatment is more effective than reservoirs (where they form alternatives) in reducing the mean concentration, but that fluctuations can only be controlled by reservoirs. In case (*a*) ($C_{TA} = 2.5 \times 10^7$ Ft/yr) basin I remains hypertrophic, basins II and III eutrophic, while basin IV is mesotrophic. In case (*b*) ($C_{TA} = 20 \times 10^7$ Ft/yr) the spatial differences and stochastic changes are much smaller: basins I–III are eutrophic and basin IV mesotrophic (it should be emphasized again that long-term improvements in water quality are more optimistic than the short-term improvement discussed here).

From the information provided by the management model, it appears that the optimal short-term eutrophication management strategy would be to introduce tertiary treatment of direct sewage discharges (from west to east) and, depending on the budget available, tertiary treatment of indirect sewage discharges or pre-reservoirs (again from west to east). In relation to the choice between the latter alternatives, the creation of reservoirs is much more expensive in terms of investment costs and time than is P precipitation, so that although the scheduling of decisions and investments was not considered in the model, it is obvious that tertiary treatment should be introduced quickly, irrespective of whether the sources are direct or indirect. This will lead to considerable improvements in mean water quality. Pre-reservoirs should also be built, resulting in a decrease primarily in the fluctuations in, but also in the expectation of, water quality.

After the reservoirs have started to operate, the P precipitation at the corresponding upstream treatment plants may be abandoned since running costs are relatively high. Such a decision obviously depends on the actual P

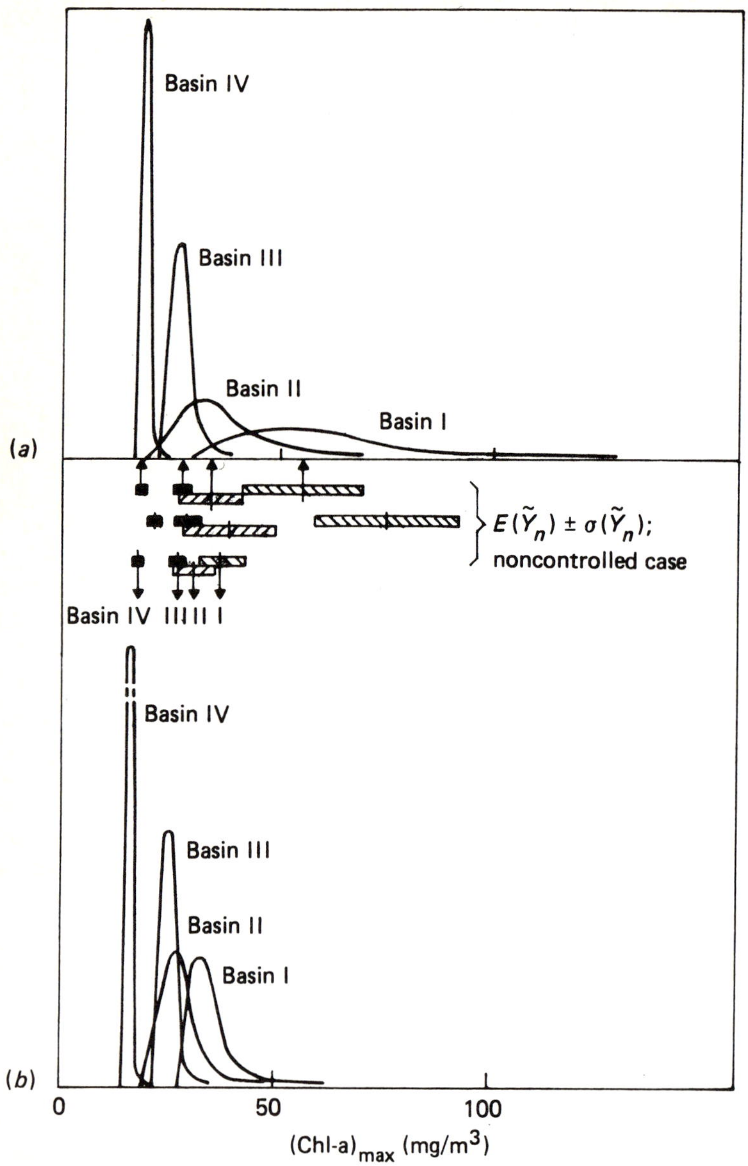

Figure 4.16. Probability density functions for two situations (*a*) $C_{TA} = 2.5 \times 10^7$ Ft/yr, and (*b*) $C_{TA} = 20 \times 10^7$ Ft/yr (from 1000 Monte Carlo simulations).

removal efficiencies of reservoirs which are estimated at the moment with considerable uncertainty and which should be monitored with the actual systems. Even if values close to those expected are found, the long-term maintenance of tertiary treatment facilities would be justified for at least three reasons:

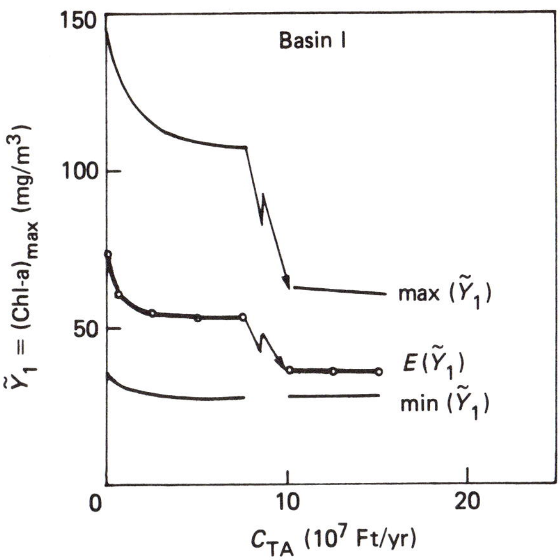

Figure 4.17. Model solution for capital recovery factor q = 0.05 for reservoirs and 0.1 for sewage treatment plants.

(1) To protect river water quality.
(2) To prevent P accumulation on river beds; sediment is often stirred up by floods, thus enabling considerable amounts of P to reach the reservoir or the lake.
(3) To "protect" the reservoir in order to avoid the rapid nutrient accumulation in the sediment that might necessitate frequent dredging.

In short, eutrophication management should focus on tertiary treatment from west to east, with a faster implementation of reservoir projects already planned.

At the end of this section we discuss the importance of the capital recovery factor and the efficiency of reservoirs. Figure 4.17 shows the solution if the capital recovery factor q = 0.05 for reservoir projects and = 0.1 for sewage treatment plants. This is a reasonable assumption since borrowing conditions are more advantageous for larger investments (such as the Kis-Balaton system; see Chapter 14) than for smaller ones. As can be seen, the only difference compared with Figure 4.15 is that reservoir projects start to be feasible "sooner", around C_{TA} = 7.5–10 × 10^7 Ft/yr.

As mentioned previously, the P removal mechanisms and the efficiency of reservoirs have not been adequately explored (see Uhlmann and Benndorf 1980, Burton *et al.* 1979, Fetter *et al.* 1978), and costs can be estimated only with some uncertainty. Both types of errors have the same effect on model performance as that of the capital recovery factor shown in Figure 4.17; they shift the particular total annual cost value where reservoirs appear in the solution. This reveals that cost–benefit analyses are as important as studying different technologies or various biological, chemical, etc., subprocesses in the lake and watershed.

4.8. Other Management Models

The major features of models developed for Lake Balaton are summarized in Table 4.1. The starting point of model 1 (see Chapter 15) is the cause–effect relation of eutrophication; a direct relationship is established on the basis of historical data between water quality and factors that influence nutrient loads (e.g., fertilizers, tourism, sewage discharges). The approach distinguishes four lake basins and seven connected subwatersheds.

As was shown in Chapter 1, the quality of the lake water is a result of long-term, artificial changes in the region and short-term fluctuations caused mainly by meteorological factors. Model 1 emphasizes the long-term development of eutrophication, and is thus a useful (simple) tool for long-term planning, although economic factors are not explicitly considered. The results of model 1 clearly indicate that protection measures taken in the Zala watershed would be about six times more effective than any in the eastern subwatersheds. A remarkable feature of the approach is that relatively few, mainly statistical, data are required, and these are already generally available at the beginning of such a study.

The economic objective of model 2 is to minimize the sum of costs and losses of the various control measures (such as the control of point source discharges, fertilizer use, erosion, land use, and runoff, as well as sediment trapping), and the environmental objective is to minimize the biologically available P load. Results obtained for a small agricultural subwatershed to the south of Lake Balaton are mainly of methodological interest, but they stress the importance of the stochastic nature of the P load. For this particular situation, combined control measures (storage, sediment trapping, and tertiary treatment) appear to be the best compromise; similar conclusions were drawn in Section 4.7.

Model 3 considers the development and planning of the sewer system in the Balaton recreational area. Results on the question of local versus regional sewerage systems are of interest. According to the model, two or three regional treatment plants are preferred on the southern shore, but about 10 are needed on the northern shore due to the high cost of network installation (mainly because of the topography), and the relatively high percentage of existing sewerage.

Model 4 has approximately the same, though simpler, structure as model 5, which is discussed in detail in this chapter. It incorporates the deterministic load response model given by equations (4.1) and (4.2), and uses (0,1)

Table 4.1. Lake eutrophication management models developed for Lake Balaton.

Model	Technique	Application	Comments
1. Watershed development approach (Dávid and Telegdi 1982; Chapter 15)	Regression and trend analysis	Entire lake–watershed system	No budgetary considerations
2. Multiobjective analysis (Bogárdi *et al.* 1983)	Stochastic nutrient load model and multiobjective programming	Tetves subwatershed (~ 300 km^2)	Lake's water quality is not included. Emphasis on stochastic features and methodology.
3. Sewer system planning (Kovács *et al.* 1983)	(0,1) integer programming	Recreational area	Engineering-type planning of sewer network with waste-water treatment plants. Constraints for effluent concentrations, but lake water quality is not considered.
4. Optimal spatial configuration of tertiary sewer and reservoir projects (Hughes 1982)	(0,1) integer programming	Entire lake–watershed system	Deterministic load–response model is incorporated
5. Short-term control strategy (Somlyódy 1983; Chapter 4)	Linear programming	Entire lake–watershed system	Stochastic load–response model is used

decision variables. The conclusions reached are similar to those discussed in detail in Section 4.7:

(1) Control measures should start at the western end of the lake.
(2) Sewage projects (tertiary treatment) are more cost-effective than reservoirs where they form alternative options (an exception is perhaps the largest reservoir system, Kis-Balaton).
(3) The objective of minimizing the absolute load is quite different from maximizing improvements in lake water quality (through nutrient load reductions).

4.9. Conclusions

(1) Based on sound scientific analysis it has been possible to establish (two alternative formulations of) a eutrophication management optimization model (EMOM) for decision making which preserves the influence of the most important in-lake and watershed subprocesses in an aggregated manner. The advantage of such an approach is that the level of lake water quality to be achieved through control measures can be directly involved in model formulation, the importance of which arises from interbasin differences in trophic state (spatial nonhomogeneities), incorporating the effects of sediment; and the presence of various stochastic factors that are important in shallow lakes.

(2) When describing the various biological, hydrophysical, etc., subprocesses several conceptual shortcomings appear. In the course of developing the water quality management model many details of these processes are ruled out; for instance, algal dynamics are less important, and the annual peak Chl-a concentration, $(\text{Chl-a})_{max}$, can be used as an indicator of eutrophication. Thus, the usefulness and effectiveness of a management model as described under (1) depends on the extent to which the scientific gaps are preserved.

(3) For Lake Balaton a stochastic, linear load response model was derived for $(\text{Chl-a})_{max}$ using a dynamic lake eutrophication model (LEM). The most important parameters of such an aggregated model represent sediment release (internal loads), interbasin mass exchange (control measures taken on subwatersheds within one basin can influence the water quality of others), and stochastic fluctuations in water quality due to hydrological and meteorological changes.

A lack of information on the behavior of sediment and interbasin transport can hamper the use of such an aggregated model and the formulation of management objectives in terms of lake water quality. The minimization of a combination (e.g., the sum) of volumetric loads – but not the loads themselves – of individual basins is the best that can be done under such circumstances. At the same time, if the interbasin transport is negligible it is sufficient to minimize the basin loads separately and there is no need to use a lake eutrophication model from the point of view of management.

(4) For Lake Balaton the objective was to establish an "optimal" short-term control strategy. For this purpose the model describing the "immediate" response of the lake to load reductions under various stochastic influences and uncertainties was used. Handling of interbasin transport did not cause difficulties since the basins are unidirectionally coupled and only the neighboring segments influence each other, but not very strongly. The sensitivity of the solution of the management model to the actual internal load was not significant.

About the "new" future equilibrium concentration of the lake we can only state that it will be lower than suggested by the short-term response. The time required to reach this state can be only roughly estimated. Much more information on the behavior of sediment is required before we can answer these questions. The stochastic properties of nutrient loads and meteorological factors and uncertainties (e.g., related to infrequent sampling on tributaries) are extremely important at the decision-making level. The situation should be the same for other shallow lakes.

(5) EMOM was developed in two alternative versions that differed from each other in the formulation of the objective functions and optimization techniques employed (linear programming and stochastic programming with recourse, respectively). The performances of the two models were practically the same. In agreement with EMOM, various other management models also clearly indicate that control measures should be undertaken from west to east. Short-term strategies should focus on the introduction of tertiary treatment at existing plants (within an unchanged sewerage system). Tertiary treatment is effective mainly in reducing the expectation of water quality, but to a lesser extent with regard to fluctuations. The reduction of random changes in water quality requires the creation of pre-reservoirs to equalize variations in loads derived mainly from nonpoint sources, although investment costs are considerably larger in this case.

(6) Apart from parameters representing in-lake processes several other, equally important factors are involved in the management model. Of these, river retention coefficients, effluent standards, cost functions and their uncertainties, and capital recovery factors, which can differ for sewerage and reservoir projects, should be mentioned. The prescription of overall water quality standards for lakes of spatially nonuniform water quality often results in a far from optimal, short-term control strategy. The subjective factors involved in policymaking procedures are also significant (e.g., how to rank reductions in the mean and variance of concentration; whether various areas of the lake are equally important from the viewpoint of water use).

(7) Structural changes in freshwater ecosystems cannot be satisfactorily predicted with the present state of knowledge, but the gradually increasing pollution of lakes causes more and more water quality problems, which should be solved within relatively short periods of time. Since in the solution "microscopic" (mostly "scientific") and "macroscopic" (practical, economic, etc.) issues appear to be approximately equally important, the most feasible approach is to start studying the problem at the same time from both angles. Results should then be gradually integrated and updated for application. Otherwise the danger can occur that by the time of completion of the study the system has already undergone structural changes, and so the conclusions are of no practical use.

Postscript

A question that comes to mind is whether the study has had any practical impact on the environmental problem of Lake Balaton. For many similar environmental studies reported in the literature the answer is "no", but this is not the case for the Lake Balaton problem. The study has had a definite, positive influence on policymaking because of several factors.

During the second half of the four year effort a growing interaction was established through meetings and discussions between scientists and decision makers (including their support staff). Owing to proper scientific coordination there was also a coincidence in the timing of completion of the research (mid-1982) and the period of policymaking (scheduled for 1982 by the government). An "unfortunate" coincidence, however, was that in August 1982 (at the time of the closing workshop of the study) previously unobserved extreme water quality conditions (but nevertheless prognosed by the study) occurred due to "unfavorable" meteorological conditions. Finally, the results of the study (which gave a more comprehensive picture of our knowledge of Lake Balaton than had existed before) served as a "scientific" basis for policymaking, and the scientists themselves were strongly involved in this procedure.

All of these factors then led to a government decision (January 1983) on controlling the eutrophication and regional development of Lake Balaton, and subsequently to the modification of the existing management plan. The procedure is discussed in detail in the next chapter.

References

Benedek, P. and Szabó, S. (1981) Proposal for protecting the water quality of Lake Balaton by improving the efficiency of sewage treatment. *VITUKI Report No.* 7783/3/22 (Budapest: VITUKI) p 43 (in Hungarian).

Biswas, A.K. (Ed.) (1981) *Models for Water Quality Management* (New York: McGraw-Hill) p 329.

Bogárdi, I., Dávid, L., and Duckstein, L. (1983) *Trade-offs Between Cost and Effectiveness of Control of Nutrient Loading into a Water Body.* Research Report RR-83-19 (Laxenburg, Austria: International Institute for Applied Systems Analysis) p 36.

Burton, T.M., King, D.L., and Ball, R.C. (1979) *Utilization of Natural Ecosystems for Wastewater Renovation.* EPA Report 90/3-89-003 (Washington, DC: EPA) p 154.

Dávid, L. and Telegdi, L. (1982) *The Influence of Watershed Development on the Eutrophication of Shallow Lakes: Modeling Approach Applied to Lake Balaton.* Collaborative Paper CP-82-42 (Laxenburg, Austria: International Institute for Applied Systems Analysis) p 47.

Fetter, C.W. Jr., Sloey, W.E., and Spangler, F.L. (1978) Use of a natural marsh for wastewater polishing. *JWPCF* February:290–307.

Haith, D.A. (1981) *Environmental Systems Optimization* (New York: Wiley) p 306.

Haith, D.A. (1982) *Models for Analyzing Agricultural Nonpoint-Source Pollution.* Research Report RR-82-17 (Laxenburg, Austria: International Institute for Applied Systems Analysis) p 34.

Hughes, T. (1982) *Lake Eutrophication Management Optimization Modeling – Approaches with Application to Lake Balaton.* Working Paper WP-82-10 (Laxenburg, Austria: International Institute for Applied Systems Analysis) 53pp.

James, L.D. and Lee, R.R. (1971) *Economics of Water Resources Planning* (New York: McGraw-Hill) p 624.

Kovács, L.B., Boros, E., and Inotay, F. (1983) A two-stage approach for large scale sewer systems design with application to the Lake Balaton resort area, in L. Somlyódy, S. Herodek, and J. Fischer (Eds.) *Eutrophication of Shallow Lakes: Modeling and Management. The Lake Balaton Case Study.* IIASA Collaborative Proceedings Series CP-83-S3 (Laxenburg, Austria: International Institute for Applied Systems Analysis) pp 315–33.

Lam, D.C.L. and Somlyódy, L. (1983) A comparison between the Lake Erie and Lake Balaton water quality studies, in L. Somlyódy, S. Herodek, and J. Fischer (Eds.), *Eutrophication of Shallow Lakes: Modeling and Management. The Lake Balaton Case Study.* Collaborative Proceedings Series CP-83-S3 (Laxenburg, Austria: International Institute for Applied Systems Analysis) pp 287–313.

Loucks, D.P., Stedinger, J.R., and Haith, D.A. (1981) *Water Resources Systems Planning and Analysis* (Englewood Cliffs, NJ: Prentice-Hall) 559pp.

Monteith, T.J., Sullivan, R.C., and Sonzogni, W.C. (1980) Phosphorus control strategies at municipal waste-water treatment plants in the U.S. Great Lakes basin. *GLEPS*, Contribution No. 14, 43pp.

OECD (1982) *Eutrophication of Waters: Monitoring, Assessment, and Control* (Paris: OECD) 154pp.

Rákóczi, L. and Varsa, E. (1976) Characteristics of streamflow and suspended solids load of the Zala River at Zalaapáti. *Proc. Balaton Ankét* 17pp (in Hungarian).

Rockafeller, R.T. and Wets, R. (1983) A dual solution procedure for quadratic stochastic programs with simple recourse, in V. Pereyra and A. Reinoza (Eds.) *Numerical Methods*, Lecture Notes in Mathematics 1005 (Berlin: Springer Verlag) pp 252–265.

Schüssler, H. (1981) *Kosten der Phosphatelimination in kommunalen Kläranlagen.* gwf-wasser/abwasser 122.H.6, pp 251–7.

Somlyódy, L. (1983) Lake eutrophication management models, in L. Somlyódy, S. Herodek, and J. Fischer (Eds.) *Eutrophication of Shallow Lakes: Modeling and Management. The Lake Balaton Case Study.* Collaborative Proceedings Series CP-83-S3 (Laxenburg, Austria: International Institute for Applied Systems Analysis) pp 207–50.

Somlyódy, L. (1984) *Water Quality Control of Shallow Lakes.* Doctoral Thesis, submitted to the Hungarian Academy of Sciences (in Hungarian).

Somlyódy, L. and Wets, R. (1985) *Stochastic Optimization Models for Lake Eutrophication Management.* Collaborative Proceedings CP-85-16 (Laxenburg, Austria: International Institute for Applied Systems Analysis).

Spofford, W.O., Russell, C.S., and Kelly, R.A. (1976) *Environmental Quality Management/an application to the Lower Delaware Valley.* Research Paper R-1 (Washington, DC: Resources for the Future) p 188.

van Straten, G. (1981) Aspects of nutrient loading in relation to lake eutrophication modeling. *Proc. MAB5 Project* (Amsterdam: UNESCO) pp 127–53.

Thomann, R.V. (1972) *Systems Analysis and Water Quality Management* (New York: McGraw-Hill) p 293.

Uhlmann, D. and Benndorf, J. (1980) The use of primary reservoirs to control eutrophication caused by nutrient inflows from non-point sources, in N. Duncan and J. Rzoska (Eds.) *Land Use Impacts on Lake and Reservoir Ecosystems.* Proc. MAB Project 5 Workshop 1978, Warsaw, Poland (Vienna: Facullas-Verlag) pp 152–89.

CHAPTER 5

Impact on Policymaking: Background to a Government Decision

I. Láng

5.1. Introduction

The artificial eutrophication of Lake Balaton accelerated rapidly in the late 1970s and early 1980s, thus forcing the government to consider once again the water quality problem of the lake and its region, and possible control alternatives. A new policymaking procedure was initiated and an expert committee set up, headed by the author of this chapter, to establish recommendations and proposals for restoring the lake. Several preparatory methods were elaborated and discussed at various institutional levels, and public debates were held. The final report and resolutions were discussed and approved by the Council of Ministers in January 1983.

The purpose of this chapter is to outline the procedure and recommendations made (see also Láng 1983), following the structure of the original report. Although this will result in some overlapping with other chapters of this book, the scientific elements and conclusions that played the most important roles in policymaking will be identified.

5.2. The Need for Action

The first signs of man-made eutrophication in Lake Balaton were recognized as early as the mid-1940s by Entz and Sebestyén (1942, 1946). The change in the ecosystem was indicated in 1944 by the invasion of a fibrous green algae, which recurred in 1956 (Tamás 1958). The unfavorable change in trophic state was signaled by the mass proliferation of *Ceratium hirundinella* (Sebestyén 1960) and was later clearly demonstrated by systematic observations of primary production, biomass, and chlorophyll-a (Chl-a) (see Chapters 1 and 8).

The Hungarian public only became aware of the danger threatening the lake much later, in the 1970s, following two major fish kills in 1965 and 1975. In spring 1975 dead fish were observed floating on the water surface, although the rate of kill was considerably lower than that in 1965 (according to conservative estimates 500 tons of fish were killed in 1965, but no more than 50 in 1975). Whereas in 1965 pesticide poisoning was identified as the most probable cause of the kill, investigations in 1975 attributed the event to a coincidence of several adverse factors, including a very low oxygen content of the water. The winter had been relatively mild, with long calm periods and low precipitation so that in spring the water level was below average. The biomass of diatoms increased and their microbiological decomposition consumed much of the lake's dissolved oxygen. The public was shocked by this phenomenon, which clearly showed that the stability of the existing ecosystem was endangered. The fish kill received wide media publicity, and public reaction was strong in calling for immediate action.

Within the Balaton region agricultural production has been intensified, food processing has increased, and many more tourists now use the lake for recreation than was the case 10–15 years ago. The domestic water supply has been expanded much more rapidly than the sewerage network and sewage treatment facilities, so that more nutrients (N and P) now find their way to the lake, thus accelerating the proliferation of water weeds and algae. The water has assumed a greenish hue, and in some places decaying debris forms repulsive blankets. Tourists are uneasy about the unsightly waste that accumulates along the shore. Although their impressions are gained from only a narrow section of the lake shore, so that opinions are often exaggerated and biased, nevertheless such generalizations are warranted. The average quality of the open water is also deteriorating, but this is not often visible to tourists. In short, the eutrophication problem threatening Lake Balaton has become widely recognized, and the restoration of the lake has become a national concern.

5.3. Historical Development of Research and Decision Making

Simultaneously with the more and more frequent warning signs of artificial eutrophication several efforts have been focused on comprehensive, more concentrated, and better coordinated scientific research and on accelerating the decision-making process. A special research program was compiled and included in the prominent national project "Protection of the Human Environment" in 1975. Disciplines not previously involved in environmental protection, such as law and economics, as well as mathematical modeling, were also incorporated into the program, and regular studies have continued on limnology, primary and secondary production, the nutrient cycle, and sediment chemistry; a quantitative survey of the lake's flora and fauna was also undertaken. The exploration of the dynamics of plant nutrients in the lake has also been continued. The interrelationships between the sanitary situation, tourism, and economics were studied under a separate program.

In 1976 the Coordinating Council of Environmental Research on Lake Balaton (CCER) was founded with the aim of harmonizing the activities of various research institutions. The council has also successfully served two other important functions: (*a*) to make scientifically sound recommendations for action at both national and regional levels based on research experience; and (*b*) to elaborate the foundations for an integrated environmental monitoring–observation system.

Expanded international relations have opened new avenues for research on Lake Balaton. In 1978 IIASA, supported by the scientific organizations of 17 countries, including the Hungarian Committee for Systems Analysis, started a research project on the eutrophication of shallow lakes focusing on Lake Balaton as a case study. IIASA signed a contract with CCER (comprising the Hungarian Academy of Sciences and the National Water Authority) to carry out a four-year research program to resolve both scientific and practical problems. As seen from Chapter 2, a systems approach was chosen to integrate various disciplines such as biology, chemistry, hydrology, economics, etc., and to harmonize research tools such as experimentation, data collection, and mathematical modeling.

By the time the cooperation with IIASA had begun in 1978, a considerable amount of data and expertise had already been accumulated in Hungary, primarily at the Balaton Limnological Research Institute and at the Research Center for Water Resources Development. The Hungarian scientists participating in the international effort were recruited mainly from these institutes and the Computer and Automation Institute. Cooperation with IIASA involved the adoption of international scientific findings, the utilization of data specific to Lake Balaton, and the development of new methodologies.

In addition to purely scientific findings, other important achievements should also be mentioned. Many communication and education problems were resolved which enabled meteorological, hydrological, and biological information to be analyzed in an integrated way in order to answer questions related to such a complex, multidisciplinary problem as eutrophication. In fact, this was the first time that the abundant information available on Lake Balaton had been collected, distilled, and selected according to its importance. The coordination of research thus improved considerably and many fruitful feedbacks were realized among various modes of research, as well as between research and management. Through independent contracts international cooperation was later expanded to include the Ralph M. Parsons Laboratory of the Massachusetts Institute of Technology, USA, and the Soviet Academy of Sciences.

The concern for Lake Balaton and the need for further research was again recognized in 1981, when the Regional Environmental Research Program on Lake Balaton was included as one of 17 projects in the national medium-range R&D program for the period 1981–85. The program included several subprograms, such as analysis of biological processes within the lake, land-use and regional development, water quality control, and socioeconomic interests and conflicts, etc.

As far as the historical line of decisions is concerned, a number of important measures have been taken at government level. The Central Development Program on Lake Balaton was approved by the government in 1969, and an action plan was adopted for the recreational area, comprising large-scale measures to control water pollution. In addition, these measures have been coordinated with the five-year national economic plans; the importance of water pollution control received special emphasis in the fifth five-year plan (1976–80), and a decree by the Council of Ministers contained provisions for updating and expanding the Water Management Development Program in accordance with the new regional development plan of the Balaton recreational district.

The updated version of the Water Management Development Program was completed in 1979, with a time horizon of 30 years (1980–2010), and a budget of about 36 500 million Ft (at 1978 price levels) was allocated. According to the schedule adopted, the objectives of the sixth and seventh five-year plans (1981–90) are to arrest any further deterioration of water quality, followed by projects to conserve or improve water quality. The implementation of the program will be evaluated at five-year intervals to ascertain whether the measures taken have produced the desired results, and these will be amended if experience, scientific advances, and general technological developments warrant solutions other than those originally envisaged.

Regardless of the expected impacts of these measures, other opinions have also been voiced, which anticipated a faster rate of improvement in water quality. This philosophy was reflected in the act approving the sixth five-year national economic plan, in which an improvement in the lake water quality was proposed, although experience so far has shown that previous opinions were overoptimistic. The effectiveness of past protection measures and actions contemplated for the period 1981–85 have been discussed in various forums and by different institutions, and have revealed several shortcomings in the implementation of earlier decisions. It was concluded that the overall state of the environment has not yet improved, and has even deteriorated in some areas.

As mentioned above, in January 1982 the Council of Ministers considered the situation and problems within the Lake Balaton recreational area. A decision was made to prepare water quality prognoses and control measures that would bring about the desired changes in water quality. Responsibility for the project was assigned to various ministries and institutions, with a deadline of the end of 1982.

The research project performed in cooperation with IIASA was completed by mid-1982. The coincidence of the termination of the case study due to accurate timing and the compilation of the Balaton action plan shortened the period over which information and data (comprising about 30 research papers) were made available to decision makers, which then enabled them to make well-founded decisions. Additionally, participating scientists of the cooperative research played an active role in the work of the expert committee and in preparing the committee's final report.

5.4. The Process of Eutrophication

Socioeconomic activities in the surroundings of the lake can be considered as the fundamental cause of eutrophication in a broad sense. As a direct consequence of such activities large quantities of nutrients, particularly N and P compounds, enter the lake through tributaries, effluent discharges, surface runoff and erosion, and atmospheric pollution. The deterioration in lake water quality is the resultant effect of these processes. The lake is fundamentally influenced by processes and activities within the catchment, and this has led to the realization that nutrient loads must first be controlled in the direct surroundings of the lake and subsequently in the entire catchment area (see Figure 1.1).

P has now been recognized as the most important factor in eutrophication, as well as its control. Most eutrophication models are based on P balance considerations, since P is usually the limiting factor in algal growth. N deficiency can be compensated biologically by specific nitrogen-fixing organisms, such as some bacteria and blue–green algae, which take up N from the atmosphere. Even if N does play a partially or completely limiting role, which is often the case in advanced stages of eutrophication (see Chapter 8), P is the particular nutrient that can be reduced most easily in practice, thus leading to P deficiency and limitation.

Research over the past few years has succeeded in giving a reliable picture of the P loads of the lake and of the contribution of various sources of pollution. According to Chapter 6 the total P entering the lake on average in 1975–81 was 865 kg/d (314 tons/yr), while the biologically available P load, L_{BAP}, is about 465 kg/d (169 tons/yr). Of the P discharged to the lake, 60–70% is associated with solid particles and only a small proportion enters directly through algae (the nutrient cycle), the major part being fixed in the sediment. The short-term behavior of the lake is therefore governed by the amount of biologically available P.

No more than 5–10% of the P leaves the lake, implying that 90–95% accumulates in the sediment. With advancing eutrophication, due to changes in chemical and biological processes, some of this may become remobilized to produce an internal P load. Over long periods of time, therefore, variations in water quality are governed by the total P load reaching the lake. This fact justifies the simultaneous use of total and biologically available P load data (see Chapter 6).

The percentage distribution of the total P load by sources is given in Table 5.1. The category "other sewage loads" includes effluents discharged to the lake from fish ponds, reed lakes, and marshes. The table does not give the contribution of sewage to the lake's total load, since the tributary load also includes indirect sewage discharges.

According to Chapter 6, about 30% of the P load of tributaries is of sewage origin, and thus sewage forms 28% of the lake's total P load (L_{TP}). Since the P content of sewage discharges can be considered completely available for algal growth (while the availability ratio of total P entering the lake through streams and leached directly from agricultural areas is no more than

Table 5.1. Distribution of total P loads by source.

Source	Total P load (%)
Tributaries	53[a]
Direct sewage discharges	11
Other sewage loads	3
Urban runoff	18
Direct runoff from agricultural areas	9
Atmospheric pollution	8
Total	100

[a]Including the effect of indirect sewage discharges; see text.

Table 5.2. Distribution of total P loads among basins.

Basin	Total load (%)	Volume-related total load (considered unit for basin IV)
I	30	13.4
II	26	2.9
III	21	1.3
IV	23	1.0

0.1–0.3%), the sewage contribution to L_{BAP} is considerably larger than to L_{TP}, at about 52%. The load of agricultural origin can be estimated as 47% and 33% for L_{TP} and L_{BAP}, respectively. These figures lead to the conclusion (as justified by the detailed analysis in Chapter 4) that short-term water quality control should focus on reducing the P load in sewage, whereas long-term strategies should be directed at minimizing the P load of agricultural origin.

Table 5.2 shows that the total P loads of the four lake basins are quite uniform, but due to differences in the dimensions of individual basins, the volumetric loads that determine water quality show a marked gradient: as illustrated also by Table 5.2 the load related to unit water volume is more than 13 times larger in basin I than in basin IV. This feature leads to the spread of eutrophication from west to east, as shown, for example, in Figures 1.2–1.4. Since a close correlation between volumetric P load and Chl-a concentration can be demonstrated, Table 5.2 suggests that P load reductions would be more effective at the western end of the lake (see Chapter 4).

The sediment is an important factor in the lake's nutrient cycle and thus also in the dynamics of eutrophication. Considerable amounts of nutrients (including P) accumulate in the bottom sediment, the upper layer (a few centimeters thick) of which has been found to contain most of the potentially mobilizable nutrients. The removal of this layer by mechanical methods, such as dredging, is impracticable in the short term, since this would incur a number of other problems, such as the safe disposal of the soil. Even in the longer term dredging should be confined to particular areas such as bathing beaches, bays, etc., unless a new technology is developed with which the top polluted layer alone can be removed.

One of the main results of the research was the demonstration of the "delayed response" of the system (Chapter 7), which means that, owing to the existing amounts of accumulated P, the lake responds to external load reductions with a lag of only a few years. The P accumulated over the years in the sediment, in fact, produces an "internal load", which limits short-term improvements in water quality following an external load reduction. Any major improvement can be anticipated to have a delay of a few years, depending on the "renewal" of the sediment.

Another important conclusion was that any postponement of major and concentrated remedial measures would probably result in an extension of the period of delayed response. In other words, environmental protection measures, particularly those that aim to minimize the nutrient load, must not be postponed, since primary production will proceed at successively faster rates, and will entail other adverse secondary changes as well, such as undesirable changes in the color and physical properties of the lake water.

The overall impression gained by visitors to Lake Balaton depends to a great extent on meteorological conditions during the summer, as shown in Chapter 4. An apparent improvement in water quality will be observed in years with a cool, rainy summer, whereas in years with hot summers, and especially during hot and rainy periods, biomass and primary production may increase substantially, as happened in 1982.

The results obtained from modeling work (see Chapter 4) have led scientists to the conclusion that earlier measures should be revised as regards both location and order of priority. Apart from accelerating construction work on the system of retention reservoirs already under way, P removal is the most urgent task in all areas, particularly at the sewage discharges in the Zala River catchment, and the subcatchment that drains to the Keszthely basin. Long-term improvements in water quality depend mainly on the renewal of the sediment in the lake. The random effects of weather conditions are superimposed thereon. It follows therefore that it will take a few years (say 3–5) of major fluctuations before the quality of lake water reaches a new "equilibrium" condition.

5.5. Water Pollution Control Strategies

In addition to the new policymaking procedure and the completion of the cooperative, international research on the eutrophication of Lake Balaton, the drastic change in water quality in 1982 made that year a remarkable one (see also Chapters 1, 4, and 8). The hot, rainy summer of 1982 stimulated primary production and the pronounced color change caused by the bloom of algae in late summer spread from the Keszthely basin to large areas of the lake. The annual average and peak values of eutrophication indicators exceeded all records. The color change was caused by the bloom of heterocystic blue–green algae, which indicated a disruption of the former autocontrol and stability of the aquatic ecosystem. This event triggered the public demand for scientifically sound control, as well as for urgent and effective intervention.

The elaboration of a scientific concept of water quality control raised a number of methodological problems, and the approach provoked complicated debates. However, an agreement was reached to outline first the target conditions, that is an outline of the water quality conditions required in Lake Balaton. The ultimate goal is to attain and preserve water quality levels that prevailed in the early 1960s, which are identified in terms of several parameters (such as Chl-a concentration), but it has also been realized that this would take up to several decades to attain. Thus the process of realization has been subdivided into several stages to identify separate targets.

The first target is to arrest the water quality deterioration observed during recent years, to a point at which the process of restoration assumes perceptible proportions, i.e., when a decrease in the degree of eutrophication becomes observable. At each stage the response time of the lake to sediment renewal must be taken into account.

In view of the fact that restoration may take several years, it would appear logical to implement remedial measures at the earliest possible date. But the current economic situation presents obstacles to and imposes limitations on the realization of this aim. The availability of financial means alone, moreover, is insufficient, since technical equipment, materials, power, labor, etc., are also necessary. It should also be appreciated that the highly complex institutional system has to be considered when scheduling the work toward the target condition, at the same time as coordinating these conditions with the defined objectives and the times of execution contemplated. A synthesis of this kind is impossible to accomplish unless assistance is received from professionals working in the various domains affected. The most effective approaches to the solution of such problems can be found:

(1) By repeatedly comparing objectives with possibilities.
(2) By continuous monitoring and checking of the effectiveness of measures already taken.
(3) By establishing continuous decision feedbacks.
(4) By the unconstrained exchange and debate of professional opinion.

This method was adopted in 1982 for elaborating the new policy. In order to work out the policy outlined above in detail the targets had to be specified first. The targets should express not only the goals of the water quality control as a stepwise procedure, but they should also be easily understood by outsiders. Eventually, three levels of water quality target conditions (A, B, and C) were defined, as follows:

(1) *Level A.* Conservation of the water quality representative of the late 1970s and early 1980s, i.e., prevention of further deterioration. This, however, would not exclude the possibility of exceptionally high algae production levels under adverse conditions, such as hot, rainy summers.

(2) *Level B.* A period of gradual improvement, in which the appearance of conspicuously high algae production levels can be excluded with high probability.
(3) *Level C.* Restoration of the water quality representative of the early 1960s.

In addition, the basic requirement of each level is evidently that the hygienic parameters of inshore water should meet the quality criteria of "clean, class I water". Although trophic states can be unambiguously assigned to levels A to C (e.g., for basin I hypereutrophic, eutrophic, and mesotrophic), the use of technical jargon was avoided in the final report and in public discussions. The definition of A, B, and C levels has proven successful and effective.

In connection with the measures contemplated for realizing the target levels, three fundamental requirements were also formulated. When considering alternative measures, priority must be accorded to:

(1) Those that are likely to result in rapid improvements in hygienic conditions for communities along the shore.
(2) Those that will remove the largest amounts of nutrients at the lowest cost–benefit ratio.
(3) Those that will provide relief to the most highly polluted western basin of the lake.

Different opinions have been voiced by professionals concerning the best engineering–technical alternatives. By continually comparing various ideas, concepts, and designs it is hoped that the most effective, economically efficient solutions, based on the most recent scientific advances, can be found. The schedule for the attainment of levels A, B, and C (1990, 1995–2000, and 2005–2100, respectively), was felt to be realistic when taking into account the delayed response of the sediment.

The possible control measures to achieve the various water quality levels can be classified into three major groups:

(1) Technical control alternatives, e.g., sanitary measures, sewage treatment and diversion, pre-reservoirs, fertilizer control, dredging, etc., with the aid of which the primarily defensive management strategies, such as those discussed in Chapter 4, can be implemented. The overall feature of these control options is that they require investment costs (money is directly involved), and cost-effectiveness can be relatively well evaluated.
(2) Control of tourism and limitation of regional development, associated with the management of the lake–watershed system, as discussed in Chapter 4. Such regulations certainly influence regional benefits, and additional investments are not needed.
(3) Indirect, nontechnical control tools such as legal regulation, inspection, and better training and information for those involved in management or associated with regional land use.

In relation to this classification and utilization of the available resources, it is stressed that reserves can and should also be mobilized in the application of scientific research results, while continued efforts must be made to develop practical techniques and methods that can be introduced as early as possible. Researchers also have the important task of checking measures for their effectiveness.

Returning to the target levels, the attainment of water quality level A means that a wide spectrum of measures must be implemented, mainly the "technical measures" outlined above. The most important steps are as follows:

(1) The level of hygiene of the lake should be further improved by disinfection of effluents, regular cleaning of the shore (water and land), increasing toilet facilities, etc.

(2) In compliance with the findings of Chapter 4, the biological treatment of wastewater should be upgraded, expanding the capacity of plants where necessary, and P precipitation should be introduced by 1985 at the major population centers, particularly Zalaegerszeg, Tapolca, Marcali, Keszthely, and Héviz (see Figure 4.9) at the western end of the lake.

(3) After biological treatment, sewage diversion system facilities should be expanded and/or implemented by 1985–87, primarily in the southwestern part of the recreational area and at the eastern end of the lake (both sides).

(4) Lake shore communities should be provided with sewage treatment facilities of adequate capacity, and the expansion of the sewerage network should be encouraged.

(5) The creation of reservoirs (and the introduction of other natural control methods to prevent sediment and nutrients reaching the lake via streams) should be rapidly accelerated in the catchments of basins I and II. The first and second stages of the Kis-Balaton pre-reservoir system (see Chapters 4 and 14) must be completed by 1987.

(6) The general changeover to farming methods that are not harmful to the environment should be further encouraged throughout the entire catchment area. The polluting effects of large, industrialized livestock farms must be stopped during the current five-year plan period. The necessary technical regulations must be formulated and enforced to control erosion caused by agrotechnical practices, especially in plantations along the lake. Further, soil conservation measures must be implemented, primarily in the northern part of the catchment area.

(7) The sediment should be dredged in highly polluted sections of the lake (mainly basins I and II) and along the bathing beaches over a total area of 8–10 km^2 by 1987. New silt dredging methods must be developed to allow the top sediment layer to be removed. The spoil must be disposed of in such a manner as to cause no further detriment to the environment. Sewage sludge disposal and utilization methods must also be devised to prevent the nutrients contained therein from finding access to the lake.

(8) Stores in the catchment area should be encouraged to sell only "soft" detergents of low P content to the public and these should be used exclusively by laundries.

These proposals form an "integrated program package". They will have to be realized by not later than the end of 1987; otherwise the progress of eutrophication cannot realistically be expected to be arrested, i.e., the state of water quality representative of the late 1970s and early 1980s will not be achieved by 1990. In contrast with earlier action plans, the P load reductions and the philosophy of implementing the principal measures as soon as possible have been emphasized more vigorously under these new proposals.

The exact methods for attaining water quality levels B and C have not been developed to the same level of detail as those for level A; this would have been futile. Nevertheless, it has already been accepted that in order to attain level B, the effective P removal at all sewage treatment plants (by chemical or other methods) is of paramount importance. At the same time, the implementation of regional soil conservation programs will have to proceed more vigorously in order to reach level C.

Apart from the technical measures [(1), bottom p 118], other means are also available belonging primarily to groups [(2) and (3), bottom p 118], which would require no additional finance and could be applied immediately. For example, statutory regulations could be more strictly enforced, and inspections could be carried out more frequently and extended to larger areas. The reed belt along the shore, which contributes significantly to the protection of the lake, could be conserved by appropriate measures, and fines levied on anyone destroying the reed stands. The reeds could also be harvested and farmed in such a manner as to promote improvements in water quality.

The types of measures described in this chapter have been formulated by experts, starting from the actual load conditions and assuming that the lake's nutrient load will be prevented from increasing further. For this reason any activity involving an increase in loads in special cases (sewage discharges, establishment of industrial, farming, service or tourist facilities, expansion of domestic water supply without simultaneous improvements in the sewerage system, increasing the rate of fertilzer applications, etc.) must not be tolerated. Any further expansion of tourist accommodation along the lake must be prohibited temporarily (this particular decree was issued in summer 1983); this last measure is expected not only to protect water quality, but also to improve – however slightly – the present overcrowded conditions.

As mentioned at the beginning of this chapter, the proposal containing measures and actions to improve water quality was submitted jointly by the appropriate ministries and organizations, and was approved by the Council of Ministers. Accordingly, the Water Management Development Program on Lake Balaton was expanded in 1983 jointly with the intercoordinated action program of the regional development plan.

5.7. A Prominent National Asset

Lake Balaton is an outstanding natural asset, and the entire country is concerned that the unique features of this lake be conserved. Lake Balaton is an organic part of the natural environment, so that any successful conservation can only be organized by recognizing correctly the biological laws that control it. Changes in water quality, regardless of whether they are beneficial or detrimental to man, are governed by particular features of the aquatic ecosystem. The processes involved are, however, accessible to control. Engineering, agricultural, and community development activities, guided by an understanding of biological processes, are capable of establishing a harmony between natural conditions and economic development.

Over the past 20 years the lake has suffered severe impacts and absorbed vastly increased nutrient loads with which conservation measures have failed to keep pace. As a consequence, the rate of man-made eutrophication has accelerated, and this could ultimately result in rendering the water unfit for recreational purposes. Fortunately this stage has not yet been reached; Lake Balaton is still suitable for water sport recreation. Restoration of the lake is still possible, but it will take at least another 20 years to reestablish the earlier conditions. Expensive projects, continuous care and maintenance, but above all active public support are essential if this goal is to be attained, and the funds required can be secured if the measures are scheduled for implementation in the correct sequence.

The scientific results obtained through international cooperation provide a sound basis for planning and implementing this ambitious program. The necessary decisions have been made at government level, and we are now faced with the task of realizing them in practice.

References

Entz, G. and Sebestyén, O. (1942) The life of Lake Balaton. ***Magyar Term. Tud. Társ. Budapest*** (in Hungarian).

Entz, G. and Sebestyén, O. (1946) The life of Lake Balaton. ***Magyar Biol. Kut. Int. Munk.*** 16:179–411 (in German).

Láng, I. (1983) Present and future of Lake Balaton. ***Természet Világa*** 5:194–9 (in Hungarian).

Sebestyén, O. (1960) On populations, with special regard to lake plankton (based on studies on Lake Balaton). ***Annal. Biol. Tihany*** 27:93–113 (in Hungarian).

Tamás, G. (1958) ***Mougeotia*** invasion in the plankton of Lake Balaton in autumn, 1956. ***Annal. Biol. Tihany*** 25:359–61 (in German).

PART TWO

Description of Processes and Subsystems

CHAPTER 6

Nutrient Loads

L. Somlyódy and G. Jolánkai

6.1. Introduction

There are several reasons for analyzing nutrient loads and the activities and processes that influence them. The two most obvious are:

(1) To understand the transport of nutrients (released from point and non-point sources) in the watershed.
(2) To manage the regions (of different sizes), taking into account trade-offs between agricultural production and environmental protection, as well as possible control mechanisms, economic aspects, etc.

The time scale of the analysis should be short for the first purpose (hours or days), but much longer (years) for the second. Certainly, item (1) should form a part of item (2), since no reliable management is possible without a proper understanding of the situation.

If we are considering only a particular environmental problem, in our case the eutrophication of lakes, our objectives can be specified differently:

(3) To study nutrient loads and processes, and subsequent activities at a level relevant to eutrophication.
(4) To describe loads of the water body in terms of understanding changes of in-lake water quality.
(5) To gather sufficient information on loads and possible control alternatives, which then allow the management of lake eutrophication.

In contrast to the general objectives (1) and (2), for (3) to (5) we are able to use more aggregated information. For instance, as discussed in Chapter 4, when describing in-lake chemical and biological processes, certain time-averages of loads (a month for Lake Balaton) can be satisfactorily used.

Additionally, aggregation in space is also possible since in most of the cases we are interested in temporal changes of average water quality of basins or segments (four basins or forty elements for Lake Balaton, Chapters 3 and 10).

The "accurate" derivation of the (aggregated) temporal and spatial distributions of nutrient loads, the major forcing function of the lake's ecosystem, is of primary importance: any error will influence the structure and calibration of the lake eutrophication model, LEM [see the various terms in equation (2.2)]. It should be noted again, that for objective (4) and, in particular, for calibration and (past) validation of the LEM the loads can be determined simply from *in situ* observations, which are not difficult if the monitoring network is well established around the lake. This implies, if only objective (4) is in mind, that "no" or only limited research effort would be needed.

It is primarily because of objective (5) that this is not true: we must have information on the contribution of various (point and nonpoint) sources to the load of the lake, together with their stochastic variabilities – at least on an annual basis (Chapter 4) – in order to plan a feasible management strategy, and this cannot be achieved without research.

In this chapter therefore, the problem of nutrient loads is considered from the viewpoint of eutrophication and its management. The discussion focuses on Lake Balaton, but the structure of the chapter reflects the strategy that we also suggest for other systems. In addition, we draw general conclusions as to the modeling of nutrient loads and sampling.

Section 6.2 deals with the estimate of the (multi-) annual loads for various sources, nutrient components, and lake basins. First, a classification of nutrient loads is given, followed by an estimation of the load of the lake. Here a summary is also presented on unit areal loads for the Balaton region and an attempt is made to determine the contribution of sewage loads and agricultural nonpoint sources in the load reaching the lake. Finally, in Section 6.2 it is shown how the information gained is utilized for the development of models LEM and EMOM (eutrophication management optimization model).

Section 6.3 considers the modeling of nutrient loads. Structural nonpoint source pollution models, time-series models, and regression models, and their application to subwatersheds of Lake Balaton, are discussed here. This section concludes with the description of a procedure for developing simple, planning-type nutrient load models for eutrophication management, bearing in mind practical needs. The final section of this chapter is devoted to sampling, a crucial topic in the field of nutrient load studies (sampling was done spontaneously in many cases; the resultant uncertainties should be recognized by the analyst).

6.2. Multiannual Average Nutrient Loads for Lake Balaton

Classification

As a starting point, the following classification can be made for estimating the load of a lake:

(1) Sewage load

- (i) direct,
- (ii) indirect, and
- (iii) others,

depending on whether the recipient is the lake (i), a tributary (ii), or a fish pond or reed lake (iii) (connected somehow to the lake).

(2) Tributary load, consisting of

- (i) indirect sewage load [identical with (1ii)], and
- (ii) the sum of indirect nonpoint source loads.

For both (2i) and (2ii) the recipient is the tributary.

(3) Direct nonpoint sources

- (i) urban runoff from towns and villages along the shoreline;
- (ii) rural runoff from the direct vicinity of the lake (areas which are not drained by permanent tributaries),
- (iii) atmospheric pollution, and
- (iv) groundwater infiltration.

Estimate of the multiannual average loads

On the basis of the available data (see Section 1.2 and later) our primary objective is to estimate the loads reaching the lake on a multiannual basis. The procedure consists of the following steps (Jolánkai and Somlyódy 1981; Somlyódy 1984):

(1) As a first step the load components listed above were calculated from the raw data. Estimates were made only when no or a very limited number of observations were available [classes (3i), (3ii), and (3iv)[1]]. For tributary and sewage loads (Figure 6.1) the observations of routine character (for the frequency see Section 1.2) taken between 1975–79 were used, as well as the measurements of Dobolyi (1980) reflecting the increased sewage discharge in the summer season.

The load associated with urban runoff was extrapolated from the measurements of Botond (1980) taken in three settlements of the Balaton region, while for the derivation of atmospheric load the monthly observations of Horváth *et al.* (1981), taken at four locations, were employed.

In this manner the first estimate of the load (average for 1975–79) was obtained for total P and dissolved reactive P as well as for total N and nitrate N.

[1]This load component was found to be negligible.

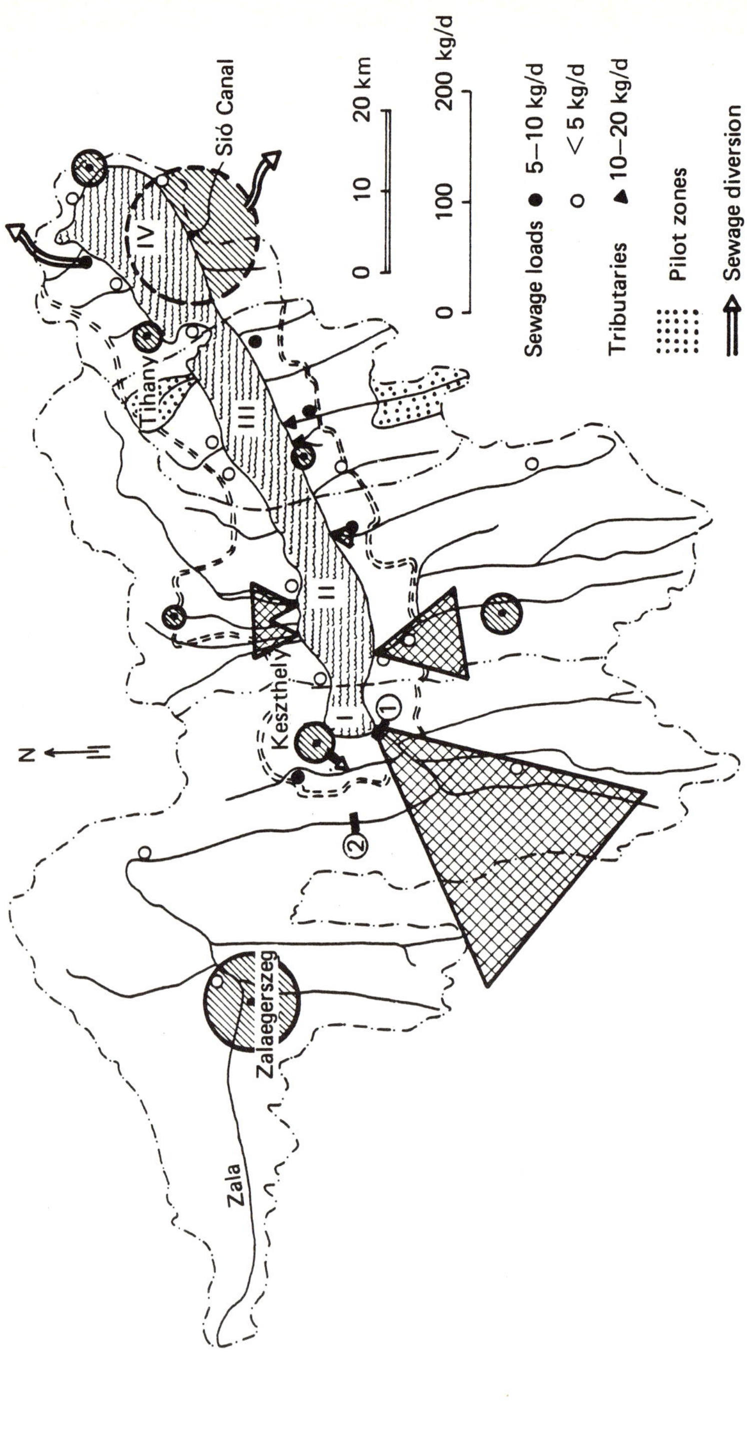

Figure 6.1. Major P sources in the Lake Balaton watershed (circles and triangles illustrate total P load values).

(2) Subsequently, the load estimate based on infrequent observations was corrected. For this purpose the following information was utilized:

(i) The event-based measurements for two pilot zones (Figure 6.1), the purpose of which was to study agricultural nonpoint source pollution (Jolánkai and Pintér 1982).

(ii) The conclusions of the uncertainty analysis performed on the daily Zala River data to evaluate the influence of infrequent sampling on the estimate of the annual mean load[2] (Somlyódy 1984; and see also later).

(iii) Data on population, water use, development of the sewer network, the seasonal changes of tourism, and the associated overloading of sewage treatment plants during the summer months, and finally the operation and nutrient removal capacity of fish ponds (Tóth 1979).

As a result of these corrections the nutrient loads of the lake and its basins were obtained (average for 1975–79). Later observations (1980 and 1981) did not show remarkable deviations from the above load estimate, and thus it can be considered as a multiannual mean for the period 1975–81.

(3) In the third step, the total load of the lake was derived for checking purposes by evaluating the various nutrient sources (fertilizer use, liquid manure, industry, sewage, etc.) and associated losses (van Straten *et al.* 1979; Jolánkai and Somlyódy 1981). In addition to the source data, literature data on unit areal loads (export rates) were used. These evaluations resulted in a wide range for the loads, within which the estimate of the second step lay, and contradictions between the two approaches were not found.

The lack of discrepancy can be demonstrated not only by comparing the loads of steps (2) and (3), but also by contrasting the unit areal loads of the Lake Balaton region with published data (see, e.g., Reckhow *et al.* 1980; Sonzogni *et al.* 1980; Heinonen *et al.* 1980; Rast 1981; Novotny and Chesters 1981; Jolánkai 1983; Whipple *et al.* 1983).

As can be seen from Table 6.1 all the unit areal total P loads obtained for Lake Balaton occur within the range offered in the literature: this illustrates, too, that it is difficult to arrive at a reasonable load estimate for a lake if there are no actual observations. In addition to the derivation of the annual export rates, an effort was made to relate them to the annual runoff and to the average slopes of the subwatersheds. The domain of about 100 (unit load; runoff) data pairs obtained for agricultural watersheds is shown in Figure 6.2. As can be seen, southern and northern subwatersheds separate clearly: the smaller load from the northern catchments is explained by their forest character. Data pairs of the Zala watershed are located in the domain

[2]Based on this information the multiannual mean load was increased by 20%: the routine observations exclude most of the flood events.

Table 6.1. Unit areal total P loads.

	Lake Balaton	Literature*	Comments
Municipal sewage (kg/capita yr)	0.47	0.2–1.1	
Agricultural nonpoint sources (kg/ha yr)	0.03–0.7 (see Figure 6.2)	0.06–2.9 0.007–0.88	Rural cropland Rural noncropland
Urban runoff (kg/ha yr)	0.19	0.1–7.6	
Atmospheric pollution (kg/ha yr)	0.30	0.12–0.97	Rural–agricultural regions

*Taken from Sonzogni and Loehr (1985).

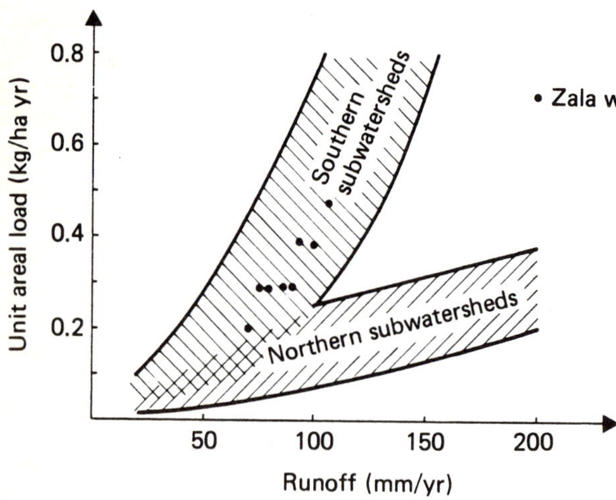

Figure 6.2. Unit areal total P load for agricultural subwatersheds in the Balaton region (1976–83).

of southern rivers (though the subtraction of the sewage load would lead to a decrease of about 0.07–0.1 kg/ha yr in the unit areal load).

For total N, larger unit loads were obtained for the northern subwatersheds than for southern regions (Pintér and Jolánkai 1982). Finally, the export rate was found to depend on the 0.4th power of the average slope for N, while no conclusion could be drawn for P.

Referring to the load estimate for Lake Balaton, the multiannual average is thus obtained from step (2). This is the most reliable load characteristic as the influence of random effects and sampling uncertainties is much weaker than for the monthly or annual averages. However, the estimate does not indicate trends, which can be derived from the analysis of sufficiently long data series of tributary loads (see, e.g., Jolánkai 1980). The multiannual mean values (1975–81) for the most important load components, total P and biologically available P (see later), for the entire lake and for its basins are summarized in Table 6.2. The total P loads from larger sewage sources and tributaries

Table 6.2. Multiannual average P loads [total P (TP) and biologically available P (BAP)] of Lake Balaton [in tons/yr (left) and kg/d (right)].

Load components		Total lake		Basin I		Basin II		Basin III		Basin IV	
Tributaries	TP	166	455	84	230	56	153	22	62	4	10
	BAP	92	255	47	130	35	95	9	26	1	4
Direct sewage* TP = BAP		34	95	1	3	2	6	3	8	28	78
"Other" sewage TP = BAP		9	25	–		–		9	25	–	
Urban runoff	TP	58	160	4	12	13	35	14	39	27	74
	BAP	17	45	1	3	4	10	4	10	8	22
Direct rural runoff	TP	29	80	4	11	8	21	12	33	5	15
	BAP	9	25	1	3	3	7	4	11	1	4
Atmospheric pollution	TP	18	50	1	3	4	12	6	16	7	19
	BAP	8	20	–	1	2	5	3	7	3	7
Total external load	TP	314	865	94	259	83	227	66	183	71	196
	BAP	169	465	50	140	46	123	32	87	41	115
Volumetric load (mg/m^3d)	TP	0.45		3.16		0.55		0.31		0.24	
	BAP	0.25		1.71		0.30		0.15		0.14	
Load per lake surface area (mg/m^2d)	TP	1.46		6.81		1.58		0.98		0.86	
	BAP	0.78		3.69		0.85		0.47		0.50	

*Without the load of sewage diversions (see Figure 6.1).

are illustrated in Figure 6.1. The total and biologically available P loads are used jointly for the following reasons. 60–70% of P reaching the lake is in particulate form, the majority of which is deposited in the sediment of the lake and only a small part enters the nutrient cycle through algae. Thus, the short term behavior of the lake is primarily determined by the biologically available P load. From experience, the total dissolved reactive P load and, depending on the composition of water, 10–30% of the particulate load of tributaries (see, e.g., Verhoff *et al.* 1978; Logan *et al.* 1979), as well as the sewage P load, can be considered directly available for algal growth. For the Balaton watershed no experiments are available, so in Table 6.2 the 20% ratio was assumed for the particulate P load.

As 90–95% of P reaching the lake accumulates in the bottom layer, the total P load plays a dominant role in determining long-term changes in water quality. As apparent from Table 6.2 the total P load is about 314 tons/yr (865 kg/d), of which 169 tons/yr (465 kg/d) is considered available. The load of different basins is more or less uniform, but the relative loads (last two lines in Table 6.2) show a drastic decrease from west to east, as referred to in Section 1.2. The ratio of volumetric total P load [mg/m^3d] is 13.2:2.3:1.3:1.0 for basins I to IV. Considering the spatial changes of load components, note that the tributary load decreases from west to east, while the direct sewage load has the opposite character (practically negligible for basin I, see also Figure 6.1).

Table 6.3. Multiannual average N total loads of Lake Balaton [in tons/yr (left) and kg/d (right)].

Load components	Total lake		Basin I		Basin II		Basin III		Basin IV	
Tributaries	1736	4757	916	2510	491	1345	239	655	90	247
Direct sewage	309	846	12	32	18	50	32	88	247	676
"Other" sewage	73	200	–	–	–	–	73	200	–	–
Urban runoff	117	324	9	26	26	71	28	78	54	149
Direct rural runoff	310	849	47	129	81	222	127	347	55	151
Atmospheric pollution	603	1654	36	99	145	397	187	513	235	645
Total external load	3148	8630	1020	2796	761	2085	686	1881	681	1868
Volumetric load (mg/m^3d)	4.50		34.10		5.05		3.10		2.30	
Load per lake surface area (mg/m^2d)	14.60		73.50		14.50		9.80		8.20	

The actual load of the lake strongly depends on the hydrologic conditions. On the basis of data available for the Zala watershed, representing half of the total catchment area, the total annual P load of the lake can reach 550 tons/yr.

Figure 6.1 includes the external P load. To this the internal load and, if individual basins are also considered, the load associated with hydrologic throughflow should be added. Internal load is the subject of the next chapter, while interbasin flow results in a 15–20 tons/yr load for total P. Table 6.1 incorporates details for total P and biologically available P loads. The corresponding summary for total N is given in Table 6.3. The total N load of the lake is about ten times larger than the total P load. Its distribution among basins is similar to that of total P.

Sewage loads versus agricultural nonpoint source pollution

Regarding eutrophication management, it is of major importance to separate the sewage loads from agricultural nonpoint source pollution within the lake's external load. The most straightforward and reliable way to study this problem is to distinguish indirect sewage load within the load of tributaries observed at the mouth.

Tributary load forms about one half of the lake's external load (Table 6.2). The indirect sewage load is 69 tons/yr. The portion of this load which reaches the lake can be estimated on the basis of longitudinal water quality profile observations. The measurements of the Research Center for Water

Resources Development, VITUKI, and the Western Transdanubian Water Authority were made on an 80 km stretch of the Zala River and at the Tapolca creek (Chapter 14 and Jolánkai 1984) show that total P retention in the river section examined is about 30%[3]. With this value, and assuming 50% retention of the sewage total P load from the town of Keszthely, a load which enters the lake through a marshland (Figures 6.1 and 4.9), the contribution of the indirect sewage load to the lake's pollution is 46 tons/yr. Together with the direct load component (Table 6.2), the load of sewage origin is $L_{TP} = L_{BAP} = 89$ tons/yr, whilst that of agricultural nonpoint source pollution is 150 tons/yr and 58 tons/yr for total P and biologically available P, respectively.

These figures mean that the contributions of sewage are 28% and 52% to the total P and biologically available P loads, respectively. The ratio is opposite in character for nonpoint sources, 47% and 33%, respectively. From these relations an important conclusion can be drawn, that from the viewpoint of short-term management the sewage load is the dominant factor, while for long-term planning it is agricultural pollution.

The use of the load estimate for the models LEM and EMOM

(1) Figure 4.1 indicates that for the development of LEM the nutrient loads are the most important inputs and they can be derived from observation, as mentioned previously. In this sense Table 6.2 is the basis of the nutrient load model NLM shown in Figure 4.1, except that temporal changes should be added (at least on a monthly basis) and the values should be actualized for the particular year considered. In both respects the daily observations made on the Zala River (see later and Chapter 14), from which the load of the most critical basin, basin I, is, for practical purposes, accurately known (even on a daily time scale), served as guidelines. It was assumed that the annual and monthly mean values of load components (3i), (3ii), and (2) (for basins II–IV) follow the actual load pattern at the Zala River. Loads (1ii) and (3iii) were considered constant in time, while for the dynamics of components (1i) and (1iii) the reader is referred to Jolánkai and Somlyódy (1981).

(2) The management model EMOM directly uses the load values illustrated in Figure 6.1 (see also Figure 4.2).

[3]Note that observations are available for low flow conditions only (even if measurements had been performed, the separation of the influence of nonpoint sources from that of river transport processes such as resuspension would be very difficult). For this reason the establishment of the retention coefficient in terms of mass balance, e.g., for a year, is not yet possible. Consequently, from the management view, the "no retention case" – a pessimistic situation – should also be studied (see Chapter 4). Note also that the retention coefficient varies along the river. For the Zala River the highest value – mainly due to intensive deposition – was found on the downstream stretch between cross sections (1) and (2), see Figure 6.1 and Chapter 14.

6.3. Modeling the Nutrient Loads

There is no doubt that very few regional water quality problems can be solved nowadays without considering nonpoint source pollution, and this is especially true for eutrophication management. Consequently, the modeling of nonpoint source pollution should form a crucial part of developing nutrient load models.

Agricultural nonpoint source pollution models: a brief overview

A large number of models have been developed in this field during the past ten years, reflecting the growing importance of nonpoint source pollution problems. These models differ significantly in their objectives and structures, and there are several ways to classify them. Here, the classification of Haith (1982) is accepted as a basis (see also, e.g., Novotny and Chesters 1981; Shvytov 1980, and their references).

Haith distinguishes:

(1) Chemical transport models.
(2) Planning and management models.

Chemical Transport Models

These models consider the transport and transformation of nutrients in a particular region via the major pathways of the hydrologic cycle, such as surface runoff, subsurface runoff, and percolation (the latter two processes influence the motion of dissolved materials, while surface runoff can carry both dissolved and particulate nutrients). Of the 37 models reviewed by Haith (1982), 25 are field models, in which a "field" is characterized by constant slope and uniform soil type, and a watershed is composed of fields. Watershed models can be developed and used – at least in principle – to describe nonhomogeneous drainage areas and the distribution of nutrient sources from different fields.

More than half of the models consider just one process, namely percolation. Haith (1982) makes a distinction between ***structural models*** and ***empirical models*** (though he uses the word "functional"). Structural models attempt to capture details of the various physical, chemical, and biological processes, while empirical models give a rough estimate of chemical losses without studying the underlying processes.

Haith (1982) found nine models that include complete hydrologic models. Among these sophisticated field models, CREAMS (Chemicals Runoff and Erosion from Agricultural Management Systems), the Cornell Nutrient Simulation (CNS) model, or the Agricultural Runoff Management (ARM) model, should be mentioned (see also Novotny and Chesters 1981). To illustrate the complexity and structure we note that, for instance, CNS uses the Stanford Watershed Model as its hydrologic component.

Many similarities in modeling philosophy exist between nutrient transport models and the first generation of water ecosystem models (see,e.g., Park 1978; Scavia and Robertson 1979). In ecological modeling efforts have been made to establish "general" models not requiring calibration, but this concept led to extremely large models. Recent developments have clearly shown, however, that "general" models should still be a subject of calibration and validation, but progress is hindered by model structures that are too complex, by the large number of parameters involved, and by the fact that there are far more state variables than measurement variables.

Quite the same symptoms can be observed for existing nutrient transport models. In practice all of them have to be calibrated, but to date no complete validation studies are known (including field models). Model structures are much more heterogeneous than in ecosystem models (which are based in most cases on ordinary differential equations derived from mass balance considerations); here it suffices to mention the Stanford Watershed Model as a possible submodel of a chemical transport model. This heterogeneity then makes it difficult to perform even a systematic numerical sensitivity analysis – one of the basic requirements of model development.

Two major problems of nonpoint source pollution modeling are as follows:

(1) Not only should the water pathways be described (which is not straightforward at all, e.g., for subsurface runoff), but also the nutrient pathways are only partly determined by water movement.
(2) It is not yet known how to make the transit from the field to the watershed level. Particular problems are how to link neighbor fields, how to incorporate river transport, and how to aggregate fields to larger areas, since otherwise models of unrealistic size and data requirements will result (as far as aggregation is concerned the reader is referred to Chapter 2).

These problems with structural models led to the application of empirical models. Most of them compute [see, e.g., the application of Bogárdi and Duckstein (1978) on a subwatershed of Lake Balaton] the volume of surface runoff and the sediment yield for a precipitation event [e.g., on the basis of the US Soil Conservation Service (SCS) formula and the Universal Soil Loss Equation (USLE), respectively]. The dissolved P loading of the event is obtained by multiplying the runoff volume by the dissolved P concentration of overland flow, C_1, and the particulate P loading is obtained from the sediment yield and the sorbed P concentration of the sediment, C_2. The annual total load is obtained by summing the loads of events for the particular year.

The empirical models can give a first estimate of nutrient loads with minimal computational effort. This feature has, however, several disadvantages. First, we note that USLE was originally developed for predicting long-term average soil losses based on the US experience (for the historical derivation of the method see Wischmeier and Smith 1978; and for further refinement see, e.g., Williams and Hann 1978). Furthermore, the application requires a good knowledge of the field and the continuous intuitive judgment of the expert when selecting parameter values from different tables and charts; however, this involvement is partially lost if the method is not used in the classical engineering way, but with the aid of computers.

Second, concentrations C_1 and C_2 are clearly lumped parameters subjected to calibration; they basically determine the total load from the field or watershed. Seemingly the model incorporates many detailed properties of the region (slope, erodibility, infiltration, etc.). If, however, a watershed is considered, it will be composed of various fields and thus the above parameters should also be determined using a calibration procedure. Thus, in reality, we just increase the number of parameters without assuring a reliable calibration, as in most cases observations are available only for the mouth section of the river receiving the entire watershed and for a few additional cross sections. Consequently, the problem of integration from field to watershed level is not solved with these simpler models, either. This step requires, among other items, flood, sediment, and nutrient routing from fields or small subwatersheds of a large watershed to obtain an estimate for the entrance intake of a lake or a river; a sophisticated procedure about which very limited experience is available at present (see, e.g., Williams and Hann 1978).

In general, it is felt that the primary advantage of these models is that they include parameters which express the influence of cropping management. The determination of these parameters is, however, quite subjective and therefore the advantage is only illusory. Thus the conclusion is that the application of empirical models first of all requires an overall practical knowledge of the watershed under study.

The state of the art of chemical transport models and the contradiction between structural and the rather practice-oriented empirical models are well reflected by the fact that Shvytov (1980) listed in his literature review watershed models for "predicting" P losses, but no models were found for "understanding" P transformation processes at the watershed level.

Planning and Management Models

Most planning and management models are optimization models (Haith 1980; Novotny and Chester 1981; Williams and Hann 1978). They have been developed for evaluating the economic consequences of alternative management actions for controlling nonpoint source pollution (including trade-offs between environmental and agricultural objectives). Because of the complexity of structural nutrient transport models and the limited understanding of watershed transport, management models incorporate only empirical transport models or simply the equation of USLE as an indirect tool.

Finally, note that an empirical P transport model (see above) was applied to the southern (agricultural) pilot zone ($\sim 70\ km^2$) of Balaton (see Figure 6.1); the rainfall input was generated in a Monte Carlo fashion (Bogárdi and Duckstein 1978; Bogárdi and Bolla 1980). As a next step, the above – partially calibrated, but not validated – model was incorporated in a multiobjective management model [considering now a little larger southern subwatershed of the lake ($\sim 310\ km^2$) which includes the pilot zone, see Bogárdi *et al.* (1983)].

This model gave conclusions of methodological importance, as discussed in Chapter 4 (Model 2, Table 4.1).

In the next two sections attention is turned to the application of a time-series model and a regression model to the Zala watershed. Compared with the methods discussed previously none of these models utilize structural knowledge of the watershed, but still they can provide useful information for our understanding and for the derivation of planning-type nutrient load models discussed in Chapter 4.

Time-series analysis

Based on the daily observations of Joó (1980 and Chapter 14) for Q, suspended solids, total P, and total N at two sections, Fenékpuszta and, 25 km upstream, at Zalaapáti on the Zala River [see Figure 6.1, points (1) and (2), respectively], as well as on precipitation data, a time-series analysis was performed using a class of discrete-time multiple-input single-output models, in particular the recursive instrumental variable (IV) estimates (Beck 1982).

The catchment area belonging to Zalaapáti is about 1000 km^2 less than the total (2622 km^2). The total P and dissolved reactive P loads of the river at the mouth are 230 and 105 kg/d respectively; the sewage contribution is estimated, using the method of Section 6.2 (p 128), to be about 80–90 kg/d. The dynamics of dissolved reactive P are much less pronounced than those of total P (the dissolved reactive P load ranged in 1978, for instance, between 50 and 300 kg/d, while total P varied between about 100 and 2000 kg/d). At low flow conditions sewage dominates in the total P load, while during floods nonpoint sources dominate: a feature clearly demonstrated by the decreasing ratio from 1 to about 0.1 of dissolved reactive P to total P with the increasing streamflow rate (Jolánkai and Somlyódy 1981).

The aim of the study was to estimate from analysis of the data the portions of loading that derived from point and nonpoint sources, the distribution of P fractions among dissolved, particulate, available and unavailable forms, and the dynamic relationships between these fractions and meteorological variations.

Models have been developed for both cross sections at the Zala River, for Q, suspended solids, total P, and total N. An example of the structure of models analyzed for Fenékpuszta is as follows (Beck 1982):

$$y(t_k) = a_1 y(t_{k-1}) + \underbrace{b_{11}u_1(t_{k-1}) + b_{12}u_1(t_{k-2}) + b_{13}u_1(t_{k-3})}_{(1)} \tag{6.1}$$

$$+ \underbrace{b_{20}u_2(t_k)}_{(2)} + \underbrace{b_{30}u_3(t_k)}_{(3)} + \underbrace{b_{40}}_{(4)} + \xi(t_k) \ ,$$

where, for example, y is the total P load (kg/d), u_1 is the observed precipitation input (mm/d), u_2 is the observed suspended solids load at Fenékpuszta (kg/d), u_3[4] is the load corresponding to y at the upstream station (kg/d), b_{40}

[4] Note that no complete statistical independence is assured among input variables.

is the base load, $\xi(t_k)$ is a sequence of zero-mean, random errors (or disturbances), and t_k refers to discrete time steps of the model. Note that not necessarily all the terms of equation (6.1) are involved in a model.

Assume for the moment that term (3) is excluded from the model. In this case the interpretation of terms in equation (6.1) is as follows:

(1) Term (1) indicates total P load associated with surface runoff that is essentially not dominated by particulate fractions (dissolved fraction).
(2) Term (2) refers to load dominated by particulate P.
(3) Term (4) reflects primarily the contribution of point source sewage discharges.

If term (3) is also incorporated, all the loads estimated from terms other than (3) in equation (6.1) are dominated by nonpoint source pollution (and tributaries) and processes in the river (such as deposition or resuspension) between Zalaapáti and Fenékpuszta. The interpretation of various terms in equation (6.1) clearly shows how the (descriptive) objective of the study can be achieved by a logically developed model sequence. For details, we refer to the original publication of Beck (1982).

The model was calibrated for the year 1978, a year felt to be typical. For validation purposes the observations of 1975–77 were utilized[5]. Calibration results are shown in Figure 6.3 [equation (6.1) was employed, without term (3), for total P]. As Figure 6.3 shows, the pattern of discharge and load variations during the year are characterized by four major precipitation–runoff events, all from April to July. The influence of these heavy rainfalls is dominant in the streamflow and in the loads of suspended solids, total P, and total N (for total P see Figure 6.3). The behavior of the derived models is quite realistic, although peaks are consistently underestimated – a general feature of such models. For total P the model without term (3) in equation (6.1) gave the "best" performance: the standard deviation of error is 102 kg/d, while the coefficient of determination is 0.88 (similar "fitting" was achieved also for Q and total N, but a much weaker one was obtained for suspended solids, except if the upstream station was involved in the analysis).

The validation step showed (Beck 1982) that the total P load of the Zala River for the years 1975–78 exhibits few features that are regular. This variability is due primarily to considerable variations in the quarterly precipitation patterns of the period. By far the wettest quarter occurred in 1978. In addition, years other than 1978 showed the influence of hydrologic events, such as snow melt, on a relatively large time scale with respect to precipitation-induced runoff. Thus, 1978 was not really typical, and the model calibrated clearly reflects this feature. It is an acceptable estimator as far as short-term precipitation events are concerned, but otherwise the model can be judged rather more "invalid" than "valid". For instance, it overestimates the base loads of 1976 and 1977, does not work properly in the winter

[5] Note that the precipitation over the selected period was 10–15% less than average.

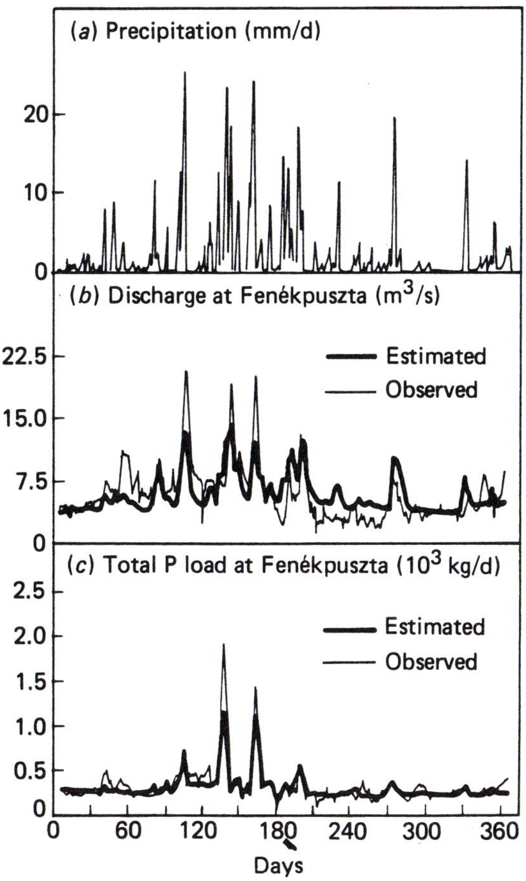

Figure 6.3. Calibration of the time-series model: (*a*) observed precipitation sequence for 1978; (*b*) observed and estimated stream discharge; (*c*) total P load at Fenékpuszta.

period of 1976–77, and underestimates the load in July and August, 1975, when a long-lasting flood of extremely high loads (~ 10000 kg/d for total P) took place (the monthly mean total P load in July was about 30 times larger than the long-term monthly mean).

From the modeling exercise and the associated statistical analyses the following conclusions can be drawn:

(1) It has been surprisingly difficult to identify consistently point and nonpoint sources for the various years. Referring to equation (6.1), the base load, composed mostly of dissolved P forms (sewage origin), contributes 40–60% of the annual average. The particulate P load associated with suspended matter is strongly dependent upon the major precipitation events, and is 20–40% of the yearly mean (smaller than expected).

The precipitation-induced dissolved load (presumably of agricultural origin) is about 15%. Its dynamics resemble a first-flush effect. Finally, the nonpoint source contribution between Zalaapáti and Fenékpuszta[6] is about 15–20% (the uncertainty of this estimate is considerable). It is suspected that this component is dominated by dissolved P fractions.

In summary, the important conclusion is that the dissolved and thus the available P load of the river seems to be larger than derived in Section 6.2 (see Table 6.1).

(2) In harmony with the findings of Joó (1980; see also Chapter 14), the areal unit load of nonpoint source origin is significantly smaller from the watershed downstream of Zalaapáti than from the upstream area. Erosion and transport of eroded soil to the river is dominant upstream from Zalaapáti, but the runoff from the lower catchment area contains a significantly smaller portion of particulate P [see item (1)].

(3) Comparison of records for Zalaapáti and Fenékpuszta shows that spatial changes in the particulate P load are primarily determined by deposition and scouring, but not by soil erosion from the watershed. The analysis suggests that below a threshold discharge of about 10–12 m^3/s deposition dominates (the net effect is deposition: the suspended solids load is larger at Zalaapáti than at the mouth section).

For details and results in respect of Q, suspended solids, and total N the reader is referred to Beck (1982).

The experiences gained clearly show that refinement of the model is necessary in two respects:

(1) Description of seasonal changes in hydrology.
(2) Incorporation of in-river processes.

Regression analysis

Depending on our objectives, different analyses can be performed on the same, detailed data set. The time-series analysis discussed before was designed for improving our understanding of the dynamic behavior of the system. If we are interested only in monthly changes of the loads, as is the case for eutrophication, these can easily be derived by aggregating the daily model output. If, however, we intend to produce a model which computes the monthly average loads as a function of stochastic inputs (e.g. precipitation, suspended solids load) derived by synthetic time-series generators, we would face the major difficulty of generating the inputs on a daily basis. This recognition leads to the suggestion of aggregating analytically the time-series model for deriving monthly average values. Since, on this time scale, autocorrelative properties are generally ruled out (see Figure 6.4 for an illustration of the

[6]Note that this forms a part of the two previous load components.

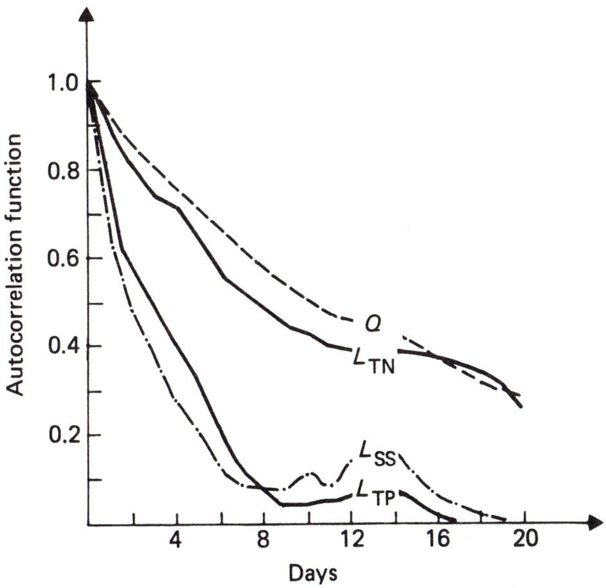

Figure 6.4. Autocorrelation function of different variables for the Zala River.

situation for Zala River, Fenékpuszta), the following relationship can be derived from equation (6.1) [without term (3)]:

$$\bar{y} \simeq \frac{1}{1 - a_1} [(b_{11} + b_{12} + b_{13})\bar{u}_1 + b_{20}\bar{u}_2 + b_{40}] \, , \tag{6.2}$$

in which the bar denotes monthly averages [notations are the same as in equation (6.1)]. In this case the monthly average precipitation and suspended solids load ($\bar{u}_1$ and $\bar{u}_2$, respectively) need to be generated; a more realistic task than that of the daily values.

Regression model versions

Equation (6.2) offers another possibility that we follow here: ***a priori*** aggregation of monthly averages from the data available and then a simple regression analysis. In addition to simplicity, an advantage of this procedure is that $\bar{u}_1$ can be replaced by the discharge which is expected to be more closely related to the load (or nutrient concentration in the river).

The regression analysis (in which data for the period 1976–79, Fenékpuszta, were involved) between the total P load (kg/d) and discharge $\bar{Q}$ (m^3/d) led to the equation

$$\bar{L}_{TP} = 38.5 + 0.303 \times 10^{-3}\bar{Q} \tag{6.3}$$

with the correlation coefficient $R = 0.83$, and $\sigma = 77$ kg/d for the standard deviation of errors [which proved to be of Gaussian distribution (Somlyódy 1984)]. According to equation (6.3), the base load (which is defined now as the load belonging to low flow conditions, $Q \leq 2\text{–}3\ \mathrm{m^3/s}$) is about 100–120 kg/d (it is assumed to be of sewage origin).

From the analysis of calibration results (Figure 6.5), two periods (May–July 1978 and May–August 1979) can be identified for which equation (6.3) gives an underestimate. According to the original data, the reason for this behavior is the occurrence of floods of a few days duration (see Figure 6.3). These floods influence strongly the suspended solids and total P loads (erosion, resuspension, etc.), but only a damped effect is expressed by the monthly mean discharge, and consequently by equation (6.3). This conclusion is justified in Figure 6.6 which shows the calibration results of a model corresponding to equation (6.2)

$$\bar{L}_{TP} = 72.8 + 0.14 \times 10^{-3}\bar{Q} + 2.36 \times 10^{-3}\bar{L}_{SS} \ , \tag{6.4}$$

where L_{SS} is expressed in kg/d. As shown – and in harmony with the findings of the time-series analysis – the model performance was significantly improved ($R = 0.92$ and $\sigma = 53$ kg/d). The first two terms of equation (6.4) can be considered as representing dissolved P loads, while the last term represents particulate P loads (their contributions to the loads of 1976–79 are 32%, 38%, and 30%, respectively). On average, the base load can be estimated to be 46% of the total P load, while the runoff-induced dissolved load and the particulate load are about 25% and 30%, respectively[7] (see also later).

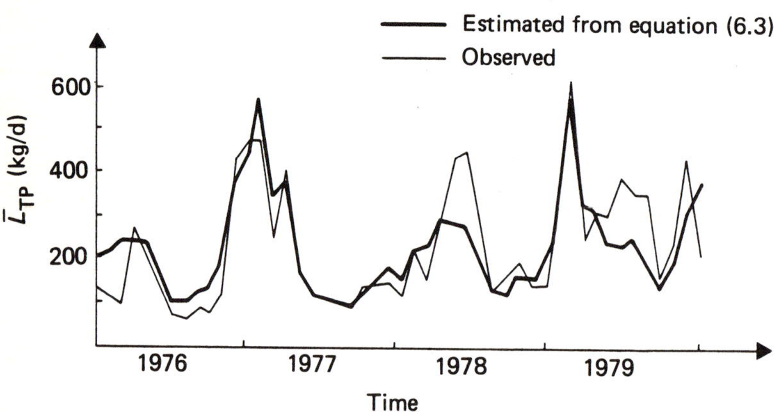

Figure 6.5. Calibration of the regression model for the Zala River.

Although equation (6.4) led to a considerable improvement with respect to equation (6.3), its application was hindered by the fact that for $\bar{L}_{SS}$ sufficiently long records were not available and thus its stochastic changes could not be reliably described.

[7] Note that above 6 $\mathrm{m^3/s}$ the slope dT_{SS}/dQ increases suddenly and therefore during large floods particulate P dominates total P load.

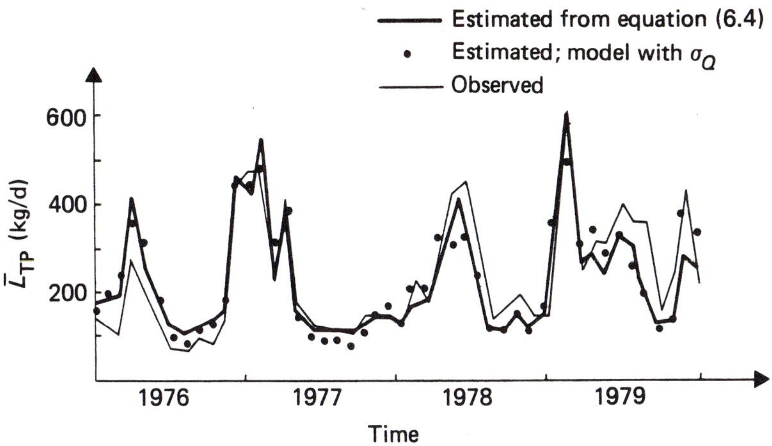

Figure 6.6. Calibration of the regression model with the inclusion of the suspended solids load for the Zala River.

It was found, however, that the effect of floods could be properly characterized on the monthly time scale by replacing in equation (6.4) the $\bar{L}_{SS}$ load with the monthly standard deviation of Q, σ_Q, computed from the daily data [the coefficients in equation (6.4) will certainly change]. Calibration results of this "model" are also shown in Figure 6.6: the fitting is nearly as close as in the previous case (R = 0.89 and σ = 61 kg/d). The advantage of this method is that the probability distribution of σ_Q – together with that of $\bar{Q}$ – can be generally derived from the available historical discharge records.

Validation

Models (6.3) and (6.4) were calibrated for 1976–79 and validated for 1980–83 [see Figures (6.7) and (6.8)]. As apparent from the figures the quality of the "prediction" is about the same as that for the calibration stage: base load estimates are acceptable, but larger peaks are underestimated, especially by equation (6.3). Nevertheless, this equation offers an unbiased estimate for the four-year average total P load, while the bias is about –8% for equation (6.4). Standard deviations for the two models are 97 kg/d and 42 kg/d, respectively (the latter value is smaller than for the calibration phase).

Comparison of the time-series model and the regression model

The time-series model was calibrated for 1978 (365 data) while parameters of the corresponding regression model [equation (6.4) including discharge instead of precipitation] were determined on the basis of the monthly average characteristics for the period 1976–79 (48 data). Because of these differences only an approximate comparison is possible, the starting point of which

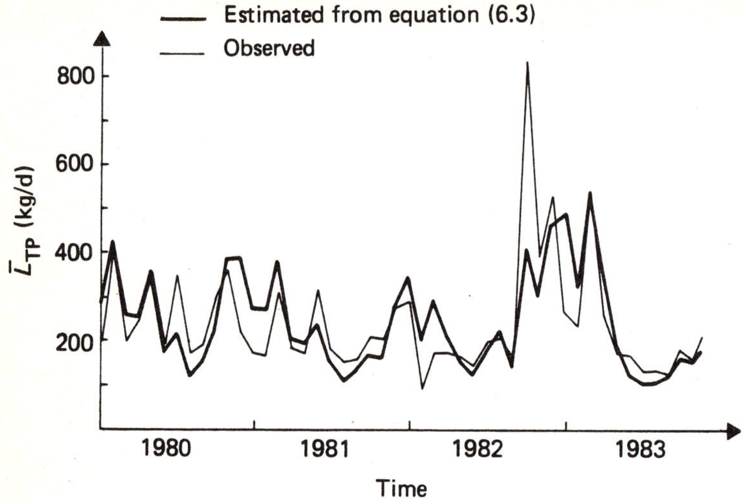

Figure 6.7. Validation of the regression model for the Zala River.

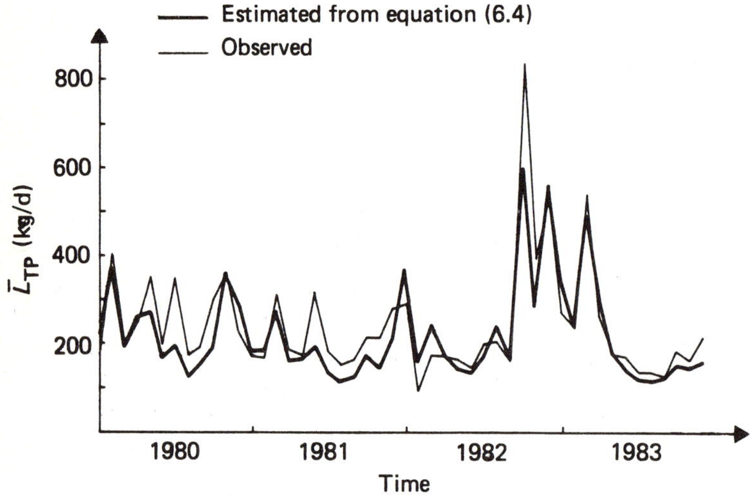

Figure 6.8. Validation of the regression model with the inclusion of the suspended solids load for the Zala River.

is equation (6.2). This can be written with the actual parameter values (see Beck 1982) as follows:

$$\bar{L}_{TP} = 144.7 + 18.5\bar{u}_1 + 2.14 \times 10^{-3}\bar{L}_{SS} \ . \tag{6.5}$$

The standard deviation of this estimate is about 20 kg/d (derived from σ = 102 kg/d of the original model, see p 138). From comparison of the corresponding coefficients in equations (6.4) and (6.5) one can see that the two models give practically the same particulate P load, but the constant load component is much larger for equation (6.5). From these it follows that the second term in equation (6.5) should be smaller than in equation (6.4). Indeed, on the basis of actual precipitation and discharge data for 1978, $0.06 \times 10^{-3} \bar{Q}$ corresponds on average to $18.5\bar{u}_1$, which is less than half of the same term in equation (6.4). The larger constant load component of the time-series model results – as shown by Figure 6.9 and mentioned above – in the predicted base load being consistently larger than the observed one.

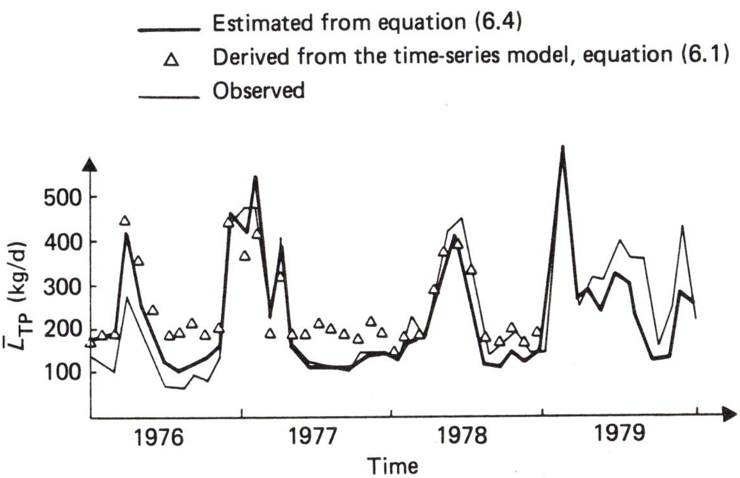

Figure 6.9. Comparison of the results of regression and time-series models.

As far as the calibration year of the time-series model, 1978 (with a high base load), is concerned, the agreement between monthly averages obtained from measurements and computations is fairly good (see Figure 6.9). A direct comparison between the two models is possible if we also calibrate the regression model for 1978 solely (which is certainly incorrect in a statistical sense because of the small number of data). Surprisingly, the coefficients of equation (6.4) are nearly unchanged (80, 0.15×10^{-3}, and 2.8×10^{-3} in the same order as in the equation). The multiple correlation coefficient is $R = 0.96$, and $\sigma = 30$ kg/d (larger than for the time-series model). If we perform the analysis by involving precipitation instead of discharge the parameters become 119, 24.6, and 2.6×10^{-3} (R and σ remain unchanged): again the base load is smaller than in the time-series model. This feature and the difference in the standard deviation of error are thus a consequence of differences in the two algorithms.

The final conclusion to be drawn from the comparison is that for practical purposes the regression model should clearly be preferred. No further refinement can be expected, however, from this approach; in this respect the time-series model is recommended.

Application of the results

To derive stochastic tributary loads in terms of models NLMP-1 and -2 (see Figure 4.1), regression models of type (6.3) were applied, as the lack of original discharge records hindered the establishment of statistical distributions for σ_Q. Thus, the monthly average stochastic load (total P and dissolved reactive P) of a river is expressed as follows

$$\tilde{L}_j = a_0 + a_1 \tilde{\Psi}_j E(\tilde{Q}_j) + \tilde{L}_\xi \ , \tag{6.6}$$

where j refers to the number of months, $\tilde{L}_\xi$ is the error term (of zero mean and Gaussian distribution), and E is the expectation operator (a tilde indicates a stochastic variable). The distributions $\tilde{\Psi}_j = \tilde{Q}_j / E(\tilde{Q}_j)$ were determined on the basis of past records (in our case they were gamma and log-normal distributions). In models NLMP-1 and -2, equation (6.6) is completed by a term that accounts for uncertainties that derive from infrequent sampling; this question is discussed in Section 6.4. For a result of the model NLMP1 see Figure 4.2 and p 77.

Strategy for developing planning-type nutrient load models

When preparing a strategy for controlling eutrophication, the objective is to find the "best" combination of possible control measures in the watershed. This requires – among other things – the identification and description of:

(1) Major nutrient sources distributed in the catchment area.
(2) Transmission processes.
(3) Control alternatives in the watershed, together with the associated costs and benefits.

All this information should be contained in the planning-type nutrient model, NLMP, which then forms a component of a management (optimization) model, as discussed in Chapter 4.

It is obvious from what has already been stated in this section that at present we are not in a position to offer an NLMP which can be widely applied. Clearly, nonpoint source pollution models are not yet well developed, the reason being the short history of study of such problems and the scarcity of information in contrast to the large amount of detailed (and expensive) observations required: decades of experience are probably necessary to achieve any significant progress in this field.

As complex watershed models are far from the level of direct application, the following simple procedure is proposed for developing a nutrient load model operating on an annual time scale. For the purpose of illustration the example of a tributary and its catchment is used (Figure 6.10).

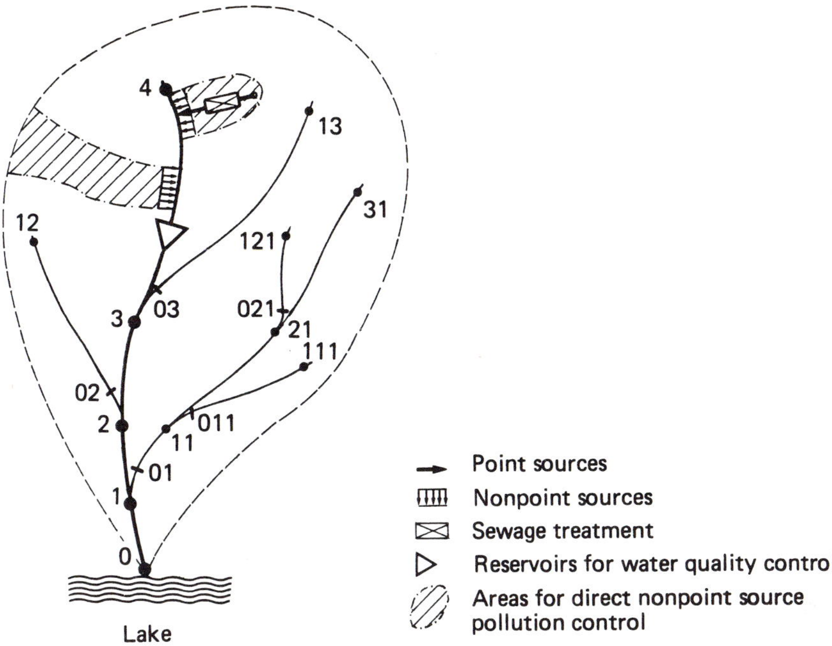

Figure 6.10. Development of nutrient load models for planning.

(1) As a first step, the "hydrologic tree" of the watershed is produced (Figure 6.10). In the figure the indices $0, 1, \ldots, 4$ refer to entry (or junction) points of tributaries and to terminal points, while $01, 02, \ldots, 011, 021, \ldots$ indicate mouth cross sections in the river system. If, for instance, the load in cross section $0j$ is L_{0j} and it is L_j on the stretch between j and $j+1$, then the load in the mouth section can be expressed as follows:

$$L_0 = L_1 + \sum_2^N L_j \prod_2^j t_{i-1} + \sum_1^{N-1} L_{0j} \prod_1^j t_i \ , \tag{6.7}$$

where t_j is the local transmission coefficient between j and $j+1$ ($r_j = 1 - t_j$ is the retention coefficient). Similar equations can be given for first – and higher – order tributaries (L_{0j} terms).

(2) Imposed on the hydrologic tree is another segmentation, showing areas according to different types and amounts of pollution. Both point and nonpoint sources are accounted for and thus, for example, the load of reach 3–4 is (see Figure 6.10)

$$L_n = \sum_m A_m l_m t_m + \sum_n L_{Sn} t_n \ , \tag{6.8}$$

where A_m are areas of "uniform" segments, l_m the corresponding unit areal loads, L_{Sn} the sewage loads, and t_m and t_n the transmission coefficients.

(3) By combining for the entire catchment all the equations of types (6.7) and (6.8) a single descriptive equation is obtained for the mouth section (see the model NLM in Figure 4.8, which distinguishes the origin and location of load components).

(4) In the next step major control alternatives (sewage treatment and diversion, reservoirs, urban runoff control, fertilizer use, erosion control, etc.) together with their locations are identified (Figure 6.10). Load components that are considered controllable by one of the possible measures are supplied with a decision variable, as shown in Chapter 4, while other terms of the descriptive model of step (3) are summed in a "noncontrollable" load term [$\tilde{L}_{NC}$ in equation (4.6)]. Thus a planning-type model is obtained (called NLMP2 in Chapter 4).

(5) Unit areal loads and transmission coefficients are estimated on the basis of literature data. Because of the large uncertainty in such estimates (see Section 6.2), the models NLM and NLMP2 derived thus can give only a first impression.

(6) In order to refine the "models", load observations in different river cross sections and longitudinal profile measurements are needed. These allow normalization of the unit areal loads in the corresponding upstream watershed segments. Measurements are especially important for the "controllable" components of the load.

(7) Stochasticity and uncertainty can be involved, as shown above.

(8) The model thus produced may be used with the desired accuracy for eutrophication management. Further improvement and development of structural watershed nutrient load models should be the subject of subsequent research.

6.4. Water Quality Sampling

Water quality monitoring plays a decisive role in both understanding and management; this statement is especially valid for problems related to nonpoint source pollutions. The development of monitoring networks requires the simultaneous determination of sampling locations and frequencies: a task in which costs and the required accuracy are dominant elements of the analysis and an "optimal" strategy, in some sense, is necessary.

In the final section of this chapter we consider only sampling frequency; for other aspects of monitoring see, e.g., Sanders (1979) and Schilperoort and Groot (1983). Reasons for observing water quality changes at a given station are threefold:

(1) To understand processes that influence water quality.
(2) To determine averages (monthly, annual, multiannual, etc.) of water quality parameters (concentrations or pollutant loads).
(3) To detect trends.

In all three cases accuracy or uncertainty of the information derived is a major issue. In the following the first two objectives are dealt with [for item (3) see, e.g., Schilperoort and Groot (1983)].

Reproduction of time-series from discrete observations

According to the classic theorem of Nyquist (1924), valid for a large variety of stochastic processes, a continuous time-series, with no spectral components above the critical frequency f_c, can be unambiguously reproduced from discrete observations if the sampling interval is smaller than $\Delta t_c = 1/(2f_c)$. The latter interval can be derived from the Fourier transform of the autocorrelation function (the spectral function).

Autocorrelation functions for the Zala River are shown in Figure 6.4, while spectral functions are displayed in Figure 6.11. Though determination of the component of largest frequency that could be detected is not quite straightforward from the figure [see Szöllösi-Nagy (1976) and Gauzer (1982) for various possibilities], the sampling interval can be estimated as approximately once a day for both total P and total N loads. For the discharge a slightly larger value is also acceptable, while for the suspended solids load less than once a day is suspected.

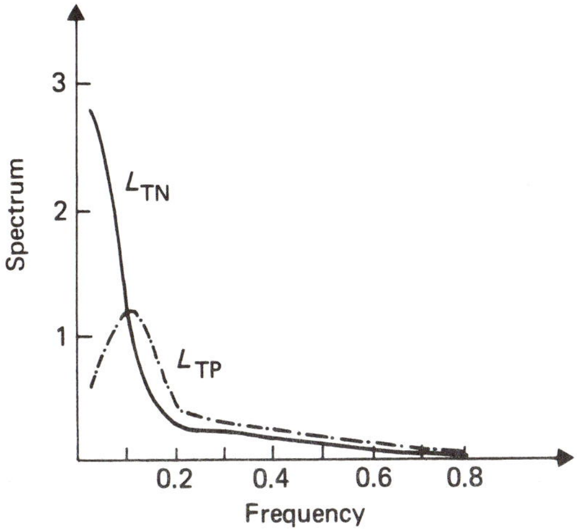

Figure 6.11. Spectral functions for the total P and total N loads, Zala River.

The conclusion of this analysis is that if we want to study processes influencing nutrient loads, daily sampling is needed for the relatively large Zala River (as has been done in practice). For smaller rivers of much faster dynamics a more frequent monitoring policy is desired.

Sampling for the determination of average values

Consider a finite population consisting of N uncorrelated elements $(y_1, y_2, \ldots, y_N)$. The exact sample mean is $\bar{Y}$, and the variance is $V_N = \sigma_N^2$. If we take a random sample of n elements $(1 \leq n \leq N)$, without replacement, the expectation of the sample mean, $E(\bar{y}_n)$, gives an unbiased estimate of $\bar{Y}$ (Cochran 1962). The variance of this estimate is

$$V_n(\bar{y}_n) = E(\bar{y}_n - \bar{Y})^2 = \frac{V_N}{n} \frac{N-n}{N} , \tag{6.9}$$

from which the standard deviation is

$$\sigma_n(y_n) = \left[\frac{V_N}{n} \frac{N-n}{N}\right]^{1/2} . \tag{6.10}$$

Specifically, if $n = N$, then $\sigma_n = 0$.

Accepting an error in the estimate of $\Delta y = \alpha \bar{Y}_N$, the required minimal sample consists of

$$n = \frac{N}{1 + N\left[\dfrac{\alpha}{y_t}\right]^2 \left[\dfrac{\bar{Y}_N}{\sigma_N}\right]^2} \tag{6.11}$$

elements. For example, assuming a 95% confidence level and a Gaussian distribution, $y_t \simeq 1.96$.

If elements of the population (observations) are correlated,

$$V_N^* = V_N \left[1 + 2 \sum_{k=1}^{N-1} \left[1 - \frac{k}{N}\right] R(k)\right] , \tag{6.12}$$

where $R(k)$ is the autocorrelation function [see, e.g., Sanders (1979) and note that statistical stationarity is assumed here].

On the basis of the simple expression (6.11), several useful conclusions can be drawn as to sampling. It is seen that n is primarily determined by the $\bar{Y}_N / \sigma_N$ ratio and, through N, by the length of the period over which averages are taken. For example, if $N = 365$, $y_t = 1.96$, $\bar{Y}_N = \sigma_N$ – a rather realistic assumption – and the desired relative accuracy of the annual load is 25%, then 52 samples per year are required (corresponding to weekly sampling). If, however, we wish to maintain the same accuracy also for monthly averages, 20 observations should be done per month.

Based on observations for 1976–79 on the Zala River, an average of 13 samples should be taken to estimate the monthly mean total P load with a 25%

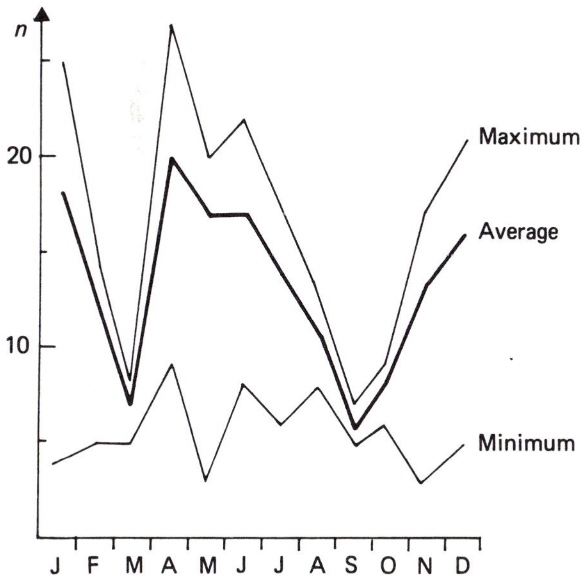

Figure 6.12. Number of samples for determining monthly average total P loads (Zala River, 1976–79).

accuracy [that is, fewer observations than for reproducing the time-series are satisfactory (see previous section)].

However, the ratio $\bar{Y}_N / \sigma_N$ is a function of the hydrologic regime and changes from month to month and year to year. This behavior is well reflected in Figure 6.12 for the Zala River [1976–79, see Somlyódy (1984)]. The maximum sample numbers consistently occur for 1979; the richest year in precipitation, and also the year when fluctuations in discharge and loads were the strongest.

As the actual values of $\bar{Y}_N$ and σ_N are unknown before sampling starts, the number of observations should be fixed on the basis of historical records. If for the Zala River a constant value of $n = 13$ is selected (a 25% average accuracy), the actual error of the monthly average load estimate varies between 9 and 80% (see Figure 6.12); because of this it is desirable to introduce stratified sampling (see Cochran 1962).

In harmony with the findings of the previous section, n is about the same for total N as for total P. The corresponding estimate of the suspended solids load requires more observations in comparison with total P, while n is smaller for Q. As in the general case for agricultural watersheds a close correlation exists between the ratios $(\bar{Y}_N / \sigma_N)_{\text{load}}$ and $(\bar{Y}_N / \sigma_N)_Q$ [for the Zala River $(\bar{Y}_N / \sigma_N)_{\text{load}} \simeq 1.65 \times (\bar{Y}_N / \sigma_N)_Q$ for both total P and total N]; hence discharge records available in most cases can be used to determine a first estimate of the number of samples. The $\bar{Y}_N / \sigma_N$ ratio generally decreases with decreasing size of the watershed, thus justifying the need for

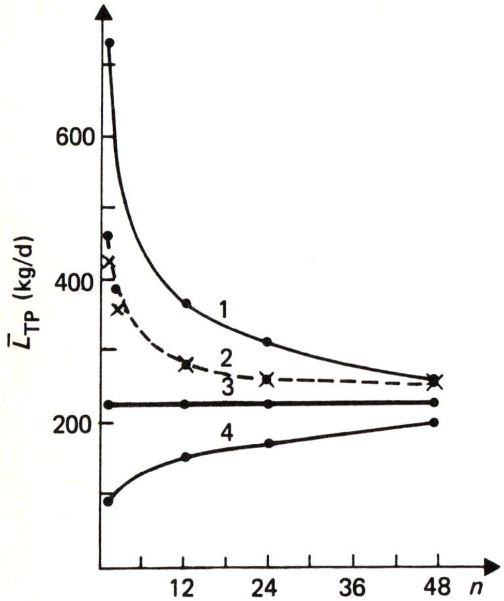

Figure 6.13. Influence of infrequent sampling on the estimate of the multiannual average total P load (Zala River, 1976–79): 1 and 4, max ($\bar{y}_n$) and min ($\bar{y}_n$); 2, $E(\bar{y}_n) + 1.96\,\sigma_n$; 3, $E(\bar{y}_n)$.

more frequent sampling for smaller rivers (if the accuracy requirement is unchanged).

The relative error of the estimate $\bar{Y}$ from equation (6.9) is

$$\alpha = y_t \frac{\sigma_N}{\bar{Y}_N} \left[\frac{N-n}{nN} \right]^{1/2}, \tag{6.13}$$

and it increases strongly with diminishing n ($n \neq 0$). For a more precise analysis of the error, Monte Carlo simulations were performed[8] using the Zala River data, corresponding to different sampling strategies (Somlyódy 1984). Results for the long-term annual and monthly averages (1976–79) are given in Figures 6.13 and 6.14.

Both figures show clearly that in the lower domain of n an increase in n pays off: the accuracy of the estimate significantly improves. As seen from Figure 6.13, equation (6.13) describes fairly well the error in the function n. The figures suggest typically skewed probability distributions for the error (especially for small n). The empirical distributions obtained from the Monte Carlo analysis were approximated by gamma distributions and then extrapolated to other rivers, depending on the actual number of samples. Finally, this uncertainty term was added to regression models of type (6.6).

[8]1000 simulations were done in each case.

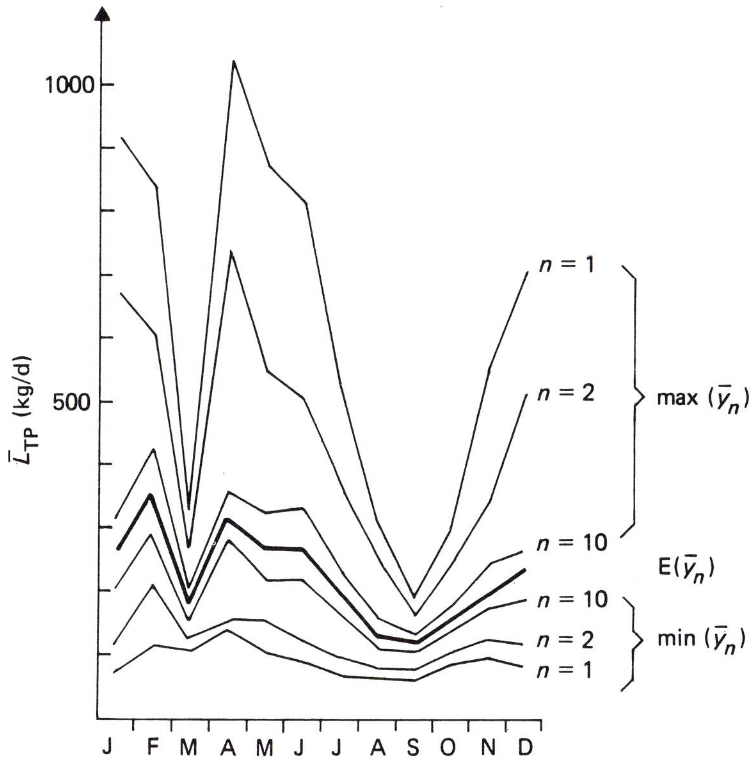

Figure 6.14. Influence of infrequent sampling on the estimate of long-term monthly average total P loads (Zala River, 1976–79).

To conclude this chapter, we note that the revision of existing water quality monitoring networks is now an important task in a number of countries, and this task can be achieved by using relatively simple, available statistical techniques, as shown here.

References

Beck, M.B. (1982) *Time-series Analysis of Zala River Nutrient Loadings*, Working Paper WP-82-116 (Laxenburg, Austria: International Institute for Applied Systems Analysis).

Bogárdi, I. and Bolla, M. (1980) Application of a stochastic phosphorus loading model for Lake Balaton, in G. van Straten, S. Herodek, J. Fischer, and J. Kovács (Eds.) *Proc. of the Second Joint MTA/IIASA Task Force Meeting on Lake Balaton Modeling* (Veszprém, Hungary: MTA VEAB).

Bogárdi, I. and Duckstein, L. (1978) Input for a stochastic control model of P loading. *Ecological Modelling* 4:173–95.

Bogárdi, I., Dávid, L., and Duckstein, L. (1983) *Trade-off between Cost and Effectiveness of Control of Nutrient Loading into a Water Body*, Research

Report RR-83-19 (Laxenburg, Austria: International Institute for Applied Systems Analysis).

Botond, Gy. (1980) Study on nutrient loads related to urban runoff in the Balaton region. *VITUKI Report* No. 7632/3/1 (Budapest: VITUKI) (in Hungarian).

Cochran, W.G. (1962) *Sampling Techniques* (New York: John Wiley).

Dobolyi, E. (1980) Nutrient loads of sewage origin in the Balaton region. *VITUKI Report* No. 7632/3/2 (Budapest: VITUKI) (in Hungarian).

Gauzer, B. (1982) Thoughts on determining sampling frequency. *Vizügyi Közl.* 3:496–501 (in Hungarian).

Haith, D.A. (1982) *Models for Analyzing Agricultural Non-point-source Pollution.* Research Report RR-82-17 (Laxenburg, Austria: International Institute for Applied Systems Analysis).

Heinonen, P., Eloranta, J., Kinnunen, K., Kohonen, T., Noukka, K., and Wallin, M. (1980) Water quality and human activities: monitoring and modelling effort in Finland. *Water Supply and Management* 4:339–69.

Horváth, L., Mészáros, A., Mészáros, E., and Várhelyi, G. (1981) On the atmospheric deposition of nitrogen and phosphorus into Lake Balaton. *OMSZ, Időjárás* 85(4):128–39.

Jolánkai, G. (1980) Evaluation of nutrient load measurements in the Balaton region. *VITUKI Report* No. 7631/3/82/J (Budapest: VITUKI) (in Hungarian).

Jolánkai, G. (1983) Modelling of non-point source pollution, in S.E. Jorgensen (Ed.) *Application of Ecological Modelling in Environmental Management, Part A* (Amsterdam: Elsevier Scientific Publishing Company) pp 283–379.

Jolánkai, G. (1984) A study on the longitudinal water quality profile of the River Zala. *Vizügyi Közl.* 1:91–103 (in Hungarian).

Jolánkai, G. and Pintér, Gy. (1982) *Non-point Source Pollution and Surface Runoff.* VMGT 131 (Budapest: VIZDOK) (in Hungarian).

Jolánkai, G. and Somlyódy, L. (1981) *Nutrient Loading Estimate for Lake Balaton.* Collaborative Paper CP-81-21 (Laxenburg, Austria: International Institute for Applied Systems Analysis).

Joó, O. (1980) Data for the eutrophication of Lake Balaton and considerations related to control activities, in G. van Straten, S. Herodek, J. Fischer, and J. Kovács (Eds.) *Proc. of the Second Joint MTA/IIASA Task Force Meeting on Lake Balaton Modeling* (Veszprém, Hungary: MTA VEAB) pp 139–67.

Logan, T.J., Verhoff, F.H., and De Pinto, J.V. (1979) Biological availability of total phosphorus. *Lake Erie Wastewater Management Study* (Buffalo: US Army Engineer District).

Novotny, V. and Chesters, G. (1981) *Handbook of Non-point Source Pollution: Sources and Management* (New York: Van Nostrand Reinhold Co.).

Nyquist, H. (1924) Certain factors affecting telegraph speed. *Bell System Techn. Journ.* 3(2):224–46.

Park, R.A. (1978) *A Model for Simulating Lake Ecosystems.* Report No. 3 (Troy, New York: Center for Ecological Modeling, Rensselaer Polytechnic Institute).

Pintér, Gy. and Jolánkai, G. (1982) *Case Study on Non-point Source Plant Nutrient Load Calculations.* IAHS Publ. No. 139, pp 207–14.

Rast, W. (1981) Quantification of the input of pollutants from land use activities in the Great Lakes basin. *Proc. UNESCO–MAB. 5. Workshop on the Impact of Non-point Sources on the Water Quality of Watersheds and Lakes* (UNESCO: Amsterdam) pp 98–139.

Reckhow, K.H., Beanlac, M.N., and Simpson, J.T. (1980) *Modelling Phosphorus Loading and Lake Response under Uncertainty: A Manual and Compilation of Export Coefficients* (Washington: Clean Lakes Section, US Environmental Protection Agency).

Sanders, T.G. (Ed.) (1979) *Design of Water Quality Monitoring Networks* (Colorado: Colorado State University).

Scavia, D. and Robertson, A. (Eds.) (1979) *Perspectives on Lake Ecosystem Modeling* (Michigan: Ann Arbor Science).

Schilperoort, T. and Groot, S. (1983) *Design and Optimization of Water Quality Monitoring Networks*. Delft Hydraulics Laboratory, Publ. No. 268 (The Netherlands: Delft).

Shvytov, I.A. (1980) Cropland phosphorus transformation and loss models, in G. van Straten, S. Herodek, J. Fischer, and J. Kovács (Eds.) *Proc. of the Second Joint MTA/IIASA Task Force Meeting on Lake Balaton Modeling* (Veszprém, Hungary: MTA VEAB) pp 186–204.

Somlyódy, L. (1984) *Water Quality Control of Shallow Lakes*. Thesis submitted to the Hungarian Academy of Sciences, Budapest, for the degree of Doctor of Sciences.

Sonzogni, W.C. and Loehr, R.C. (1985) *Estimating Nutrient Inputs*. Manuscript of a chapter for the Manual on Eutrophication Control, UNESCO (under preparation).

Sonzogni, W.C., Jeffs, D.N., Konrad, J.C., Robinson, J.B., Chesters, G., Coote, D.R., and Ostry, R.C. (1980) Pollution from land runoff. *Environ. Sci. Technol.* 14:148–53.

Szöllösi-Nagy, A. (1976) Influence of sampling on the information content of hydrologic time-series. *Annual Report of VITUKI, 1973* (Budapest: VIZDOK) pp 210–23.

Tóth, L. (1979) Phosphorus removal from fishponds. *VITUKI Report* No. 7631/3/81 (Budapest: VITUKI) (in Hungarian).

van Straten, G., Jolánkai, G., and Herodek, S. (1979) *Review and Evaluation of Research on the Eutrophication of Lake Balaton – a Background Report for Modeling*. Collaborative Report CP-79-13 (Laxenburg, Austria: International Institute for Applied Systems Analysis).

Verhoff, F.H., Heffner, M., and Sack, W.A. (1978) Measurement of availability rate for total phosphorus from river waters. *Lake Erie Wastewater Management Study* (Buffalo: US Army Engineer District).

Whipple, W., Grigg, N.S., Grizzard, T., Randall, C.W., Shubinski, R.P., and Tucker, C.S. (1983) *Stormwater Management in Urbanizing Areas* (Englewood Cliffs, NJ: Prentice-Hall, Inc.).

Williams, J.R. and Hann, R.W. (1978) *Optimal Operation of Large Agricultural Watersheds with Water Quality Constraints*. Technical Report No. 96 (Texas A/M University: Texas Water Resources Institute).

Wischmeier, W.H. and Smith, D.R. (1978) *Predicting Rainfall Erosion Losses – A Guide to Conservation Planning*. Agriculture Handbook No. 537 (Washington: Department of Agriculture).

CHAPTER 7

Sediment and Its Interaction with Water

L. Lijklema, P. Gelencsér, F. Szilágyi, and L. Somlyódy

7.1. Introduction

Lakes act as the dustbins of their watersheds. Major fractions of materials washed out of the soil or discharged to the rivers feeding a lake accumulate in the sediments, including undesirable substances, such as insecticides used in agriculture, lead from gasoline and washed off the streets, and nutrients discharged in agricultural, domestic, and industrial wastewater. Together with natural substances, such as organic decay products, silt, clay, sand, and salts, these substances form a complex mixture that settles on the lake bottom. During their transport in the watershed and residence in the lake and its sediment many physical, chemical, and biological transformations may change the amount and physicochemical nature of these compounds over varying time scales. Hence the lake sediments reflect the history of the watershed and the natural and cultural events over long periods of time.

Quantitatively, the amount of a substance stored in the upper few centimeters of sediment may be two or more orders of magnitude greater than that present in the overlying water. For phosphate this is always the case, so that sediments have a considerable potential for sustaining biological processes in the water column that depend on the supply of phosphates. Because P is a key element in eutrophication in many freshwater systems, including Lake Balaton (see Chapters 1, 3, and 8), its retention in the lake and its exchange between sediment and water are of paramount importance. The proportion of the external phosphate load that is retained in lake sediments (the retention coefficient) depends on the characteristics of the watershed (soil composition, hydrology, wastewater discharges, etc.), the hydraulic retention time, in-lake processes such as primary production and biogenic precipitation of carbonates, evaporation, water depth, etc. As a rule, more

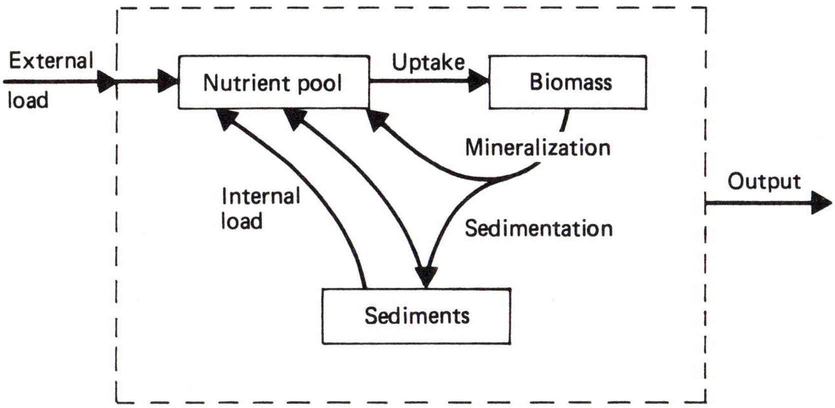

Figure 7.1. Main pathways of nutrients in a lake.

than 50% of the external load accumulates in the lake and much higher retention rates are not unusual; in Lake Balaton the retention is about 95%.

For eutrophication management knowledge of the relation between external and internal loads is a prerequisite. The internal load refers to the flux of nutrients from the sediments to the overlying water, which sustains the P cycle through its various stages (see Figure 7.1). The internal load is the difference between gross and net accumulation. In principle, the sum of external and internal loads is available for biological processes. Figures used for both external and internal loads are generally based on average values for prolonged periods (a month, a year), but fluxes may vary with time and also spatially. Also the dynamics of the internal load is related to dynamic in-lake processes, and vice versa.

This chapter concentrates on the factors that determine the internal load, with emphasis on phosphate. After a general discussion of sediment characteristics and the major processes that affect the internal load, specific conditions in Lake Balaton will be considered, including the experimental information available and the significance of the effects of various management measures on the future behavior of sediments.

7.2. Sediment Characteristics

Physical characteristics

The most important physical characteristics of sediments that affect the internal load of a lake are: grain size; grain size distribution, compaction, and porosity; density of the constituent particles; roughness of the sediment–water interface; vertical and horizontal (in)homogeneities; sediment mixing depth; and bottom slope. Sometimes sediments are extremely soft, in which case a clear, well defined interface between water and sediment may be absent, and a gradual increase in concentration of particulate material with depth is observed. However, there is usually a discontinuous transition, especially in sediments in shallow areas, on steep slopes, or on the windward side of the lake where sand and pebbles predominate. Clay, silt, and light organic material tend to accumulate in deeper zones, at the bottom of slopes, or on the lee side of the lake, a distribution due to internal sediment transport (Häkanson 1977). Besides the horizontal variations associated with such characteristics as water depth, bottom slope, and predominant wind direction, accidental variations in sediment composition on a much smaller scale can also be observed. When sampling sediments this should be kept in mind; a full characterization of a lake bottom may require a dense network of sampling sites. The vertical variation is a function of the vertical mixing rate, the consolidation rate, and may also reflect gradual or sudden changes in the accumulating material. For instance, the effect of a historic landslide has been traced using a band with different physical and chemical characteristics, which was at a depth that helped to identify the accumulation rate since that event (Bloesch and Evans 1982).

Chemical composition

In eutrophication management the main interest is in the quantity, composition, and stability of P- and N-containing compounds. N in sediments is mainly associated with organic substances transported to or produced within the lake. P can be present in several forms: absorbed onto clays; bound to metal oxides or hydroxides of Fe, Mn, and Al; in salts or mixed salts in which Ca often plays a dominant role; and in various organic compounds. It is important to note that the behavior of these substances may differ considerably from that of the pure chemical compounds studied in the laboratory, because of nonstoichiometric composition, amorphous or poorly crystalized forms, and inhomogeneities such as coatings. Also, organic phosphates may cover a wide range of compounds that are often associated with inorganic substances in different complexes. This generally means that determination of the chemical composition depends essentially on the analytical methods used. Several P extraction procedures are being used. Initially these methods were developed for soil fertility studies (Chang and Jackson 1957), but were later improved and applied to sediments (Williams *et al.* 1967). Gradually, the idea that pure chemical compounds could be identified was dispelled, and instead a rough classification of the forms of P present (related to solubility) was drawn up (Williams *et al.* 1976, Hieltjes and Lijklema 1980).

Biological characteristics

Organisms living in or on the surface of the sediments may have a great influence on the internal load, either directly or indirectly. Bioturbation by organisms, such as tubificids, may contribute to the mixing of sediment particles, enhancement of the porosity, and pumping of interstitial water (Petr 1976). Algae living at the sediment–water interface affect P and N concentrations, as well as pH and redox conditions. Bacteria that feed on organic and inorganic compounds in the sediment control environmental conditions and the regeneration of nutrients. Their concentration, activity, and ability to use a variety of substrates and electron acceptors are essential features in determining the composition of the sediment and the pore water.

7.3. Processes Related to the Internal Load

Accumulation and dilution of sediment

The well mixed top layer of sediment is generally about 10 cm deep, corresponding to the zone in which tubificids, chironomids, etc., are active. The net annual deposition of sediment may vary widely, with higher values in deep parts of a lake and almost nothing in the littoral zones, but the order of magnitude of the accumulation rate is usually a few mm/yr. Hence a change in the external nutrient load of a lake will cause a slow change in the concentration of this nutrient in the top layer, because the new equilibrium concentration will be attained only after several decades. A simple mass balance equation for the top layer of a sediment is (see Figure 7.2):

$$\frac{dc}{dt} = \frac{S}{h} - \frac{\Delta h}{h}c - kc \quad , \tag{7.1}$$

where

c	=	nutrient concentration in the sediment (g/m^3)
S	=	nutrient sedimentation rate (g/m^{2}yr)
h	=	sediment mixing depth (m)
Δh	=	deposition rate (m/yr)
k	=	first-order nutrient decay rate (yr^{-1}).

The assumptions underlying equation (7.1) are a homogeneously mixed top layer, a constant mixing depth h, first-order decay, and a constant accumulation rate of both the nutrient, S, and all other sedimenting material in the absence of other transport processes. The solution to equation (7.1) for an initial condition of $c = c_0$ at $t = 0$ is

$$c = \left[c_0 - \frac{S}{\Delta h + kh}\right] \exp\left[-\frac{\Delta h + kh}{h}t\right] + \frac{S}{\Delta h + kh} \quad . \tag{7.2}$$

The new equilibrium concentration, $c_{eq} = S/(\Delta h + kh)$, and the rate of attainment are controlled by the term $\tau = h/(\Delta h + kh)$. Note that c_{eq} is proportional to S and thus, in certain situations, to the external load. Since an increasing sediment nutrient concentration usually leads to a higher internal load, equations (7.1) and (7.2) also indicate the coupling between external and internal loads. It is apparent from equation (7.2) that for nonconservative materials ($k \neq 0$), the equilibrium concentration will be lower than for

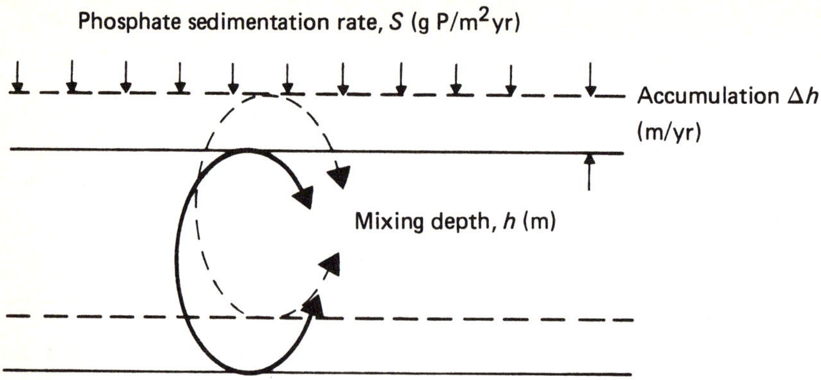

Figure 7.2. Dilution and accumulation in sediment.

conservative materials ($k = 0$), and that the rate of attainment of the new equilibrium is (much) higher for nonconservative substances. For our purposes this difference is important for N and P. Whereas N can be lost from the sediment by denitrification, resulting in gaseous N that can be considered as inert (nonreactive), P cannot be lost by any process other than transportation. Because Δh is generally low ($\sim 10^{-3}$ m/yr), a low decay rate of the order of 0.1 yr^{-1} combined with a mixing depth of 0.1 m brings about a substantial increase in ($\Delta h + kh$). Thus the typical time constants for the sediment equilibration for P and N may be several decades and several years, respectively. This means that, especially for phosphates, long-term effects of the internal load on lake eutrophication must be anticipated. However, because the sediment phosphate concentration as such is not the sole factor controlling the internal load, this aspect is discussed in detail below. The difference in behavior of conservative and nonconservative materials is also manifest in the N:P molar ratio; whereas in sedimenting material this ratio is of the order of 5 to 10, the ratio in sediments is generally much lower, normally around 2 or 3.

Physical transport processes

These processes can be divided into the transport of dissolved material by advection and diffusion, and the transport of particulate material by erosion, resuspension, and settling. Transport across the sediment–water interface is of special interest.

Advection

Horizontal variations in hydrostatic pressure of the groundwater may cause vertical percolation of pore water through sediments. The flow direction of this seepage may be upward or downward, depending on local conditions. The flow rate is proportional to the pressure gradient in the direction of flow and the perviousness of the soil (Darcy's law). Because the nutrient concentrations in pore water are usually higher than those in the overlying lake water, an upward flow of groundwater tends to enhance the internal load, but the actual flux across the interface is controlled by the intricate interaction between advection, diffusion, and reactions in the boundary layer (see below).

Diffusion

Molecular diffusion is a very slow process and the stagnant water film adhering to the sediment particles, combined with the low porosity of the sediment, further reduces the effective diffusion coefficient to about 10^{-9} m^2/s. However, this low value allows steep concentration gradients to build up near the interface due to reactions such as mineralization. Thus, according to Fick's law of diffusion, the ultimate flux of material may still be considerable.

It is worthwhile noting that in principle in shallow lakes the pressure variations near the bottom caused by waves may propagate into the boundary layer of the sediments and may cause an oscillatory flow pattern with a periodicity of about 4–5 s. It can be shown that for incompressible sediments the amplitude of the resultant vertical pore water velocity in the boundary layer is of the order of the grain size of the sediment particles and that the contribution to the (turbulent) diffusion coefficient is limited (Rutgers van der Loeff 1981). In compressible, soft sediments conditions are further complicated by oscillations of the (diffuse) sediment–water interface itself. However, analyses indicate that whenever wave action tends to become important, its effect is usually through fluidization and subsequent resuspension rather than through enhanced diffusion. An exception may be the situation around ripples.

Another physical process that should be mentioned is the slow and gradual compaction of sediments by the weight of overlying particles and the concurrent squeezing out of interstitial water. On an annual basis the volume involved is small and the flux of nutrients released by this process can be neglected. The importance of the process lies in its effect on sediment concentration, volumetric nutrient regeneration rates, concentration gradients, etc., and through these, the indirect effect on the internal load.

Advection, diffusion, and reaction

Mathematically, the transport and accumulation of dissolved nutrients in sediment can be described by the partial differential equation:

$$\frac{\partial c}{\partial t} = -u\frac{\partial c}{\partial z} + D_e\frac{\partial^2 c}{\partial z^2} \pm R(z) \tag{7.3}$$

for vertically homogeneous sediment. The concentration in the overlying water is generally used as the boundary condition at the sediment–water interface. More complicated descriptions include phenomena such as compaction, etc. (see Tzur 1971).

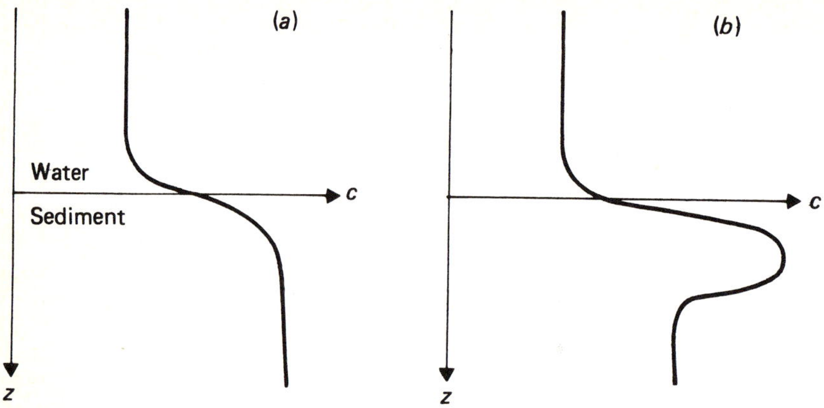

Figure 7.3. Concentration gradients in the sediments near the surface.

In equation (7.3), D_e and u are the effective diffusivity and water velocity in the vertical (z) direction, and R is the production of nutrients by mineralization, desorption or dissolution, or the removal by adsorption onto solid particles, or precipitation. The relative importance of advection and diffusion is indicated by the magnitude of the dimensionless Peclet number uh_0/D_e, in which h_0 is a characteristic length (e.g., the thickness of the topmost layer, $\simeq 10^{-2}$ m). With values of D_e of the order of 10^{-9} m^2/s this means that seepage rates of 1 mm or more per day will affect the concentration profile. The problems in applying equation (7.3) to real systems are the evaluation and prediction of the reaction rates taken together in the term $R(z)$. Especially near the interface ($z = 0$) a strong variation of R with depth is probable (see below). Another difficulty is the nature of the gradients $\partial c / \partial z$ and $\partial^2 c / \partial z^2$, which are related to the magnitude of $R(z)$. In the literature this is usually a profile as shown in Figure 7.3(*a*), but the profile in Figure 7.3(*b*) is more realistic for actively recycling sediment (Tessenov 1979, Lijklema 1983). Owing to the limited knowledge of rate constants and other parameters relevant to the conceptual descriptions of transport according to equation (7.3), the description and prediction of the internal load in practice often relies at least partly on empirical information.

Transport of particulate material

The transport of particulate material within a lake by erosion and resettling contributes to horizontal variations in sediment characteristics such as grain size, organic fraction, accumulation rate, porosity, nutrient content, etc. Of particular interest is the resuspension of sediment particles with a high potential for adsorption or desorption of nutrients. This applies especially to P because a substantial fraction of sediment phosphate is often in an adsorbed form and is readily exchangeable, whereas N is mainly present in organic debris from which the dissolved nutrient is released by slow and irreversible decay processes (slow in relation to the duration of a resuspension event). Resuspension and erosion or winnowing of sediments is restricted to shallow lakes or the shores of deeper lakes. Factors that control the rate and extent of resuspension are shear forces near the bottom induced by flow and/or waves; the density, shape, and size of sediment particles; and the cohesive properties of the sediment. The latter is a very complex factor related to the grain size distribution, the water and organic content, bioturbation, and to the charge and charge distribution of the particles causing electrostatic interactions. The pH and ionic strength of the interstitial water affect these charges. The intricacy of the factors controlling the cohesion and the virtual impossibility of prediction, together with the lack of an appropriate equation for the boundary condition at the interface for deposition and resuspension, has prompted sedimentologists to tackle the problem of predicting resuspension along more empirical lines. One approach has been to measure the rate of resuspension as a function of shear force and to use this empirical relation in combination with hydrodynamic models for waves and flow that generate data on bottom shear forces (Sheng and Lick 1979). A laboratory annular flume has been used to assess the rate of resuspension under steady-state conditions. Experiments indicate that particle size variation is a significant factor; that the amount of material available for entrainment at a particular bottom stress is finite; and that only particles below a critical size and not overlain by other, coarser particles can be entrained (Lick 1981, Fukuda and Lick 1980).

Another approach that has been applied to Lake Balaton is based on the overall concept that the actual observed suspended solids concentration is the outcome of two opposing dynamic processes: resuspension, which is controlled by wind; and settling, which is proportional to the suspended solids concentration. The analysis of field (time series) data on wind and suspended solids by parameter estimation techniques, combined with a simplified but conceptual model for these two processes, gives the relevant parameters for this model and thus a predictive tool for the rate of resuspension for the lake involved (Somlyódy 1980).

Although analytical expressions for the settling rate at the sediment–water interface have been presented (e.g., Lick 1981), the process of sedimentation has traditionally been studied in the field with sediment traps. However, the inappropriate design of traps has reduced the significance of many studies (Bloesch and Burns 1980, Hargrave and Burns 1979). Proper use of traps can give good estimates of local sedimentation rates, and comparisons of the material collected in traps with the composition of suspended material and that of the top sediment layer allows, in principle, the assessment of the contribution of resuspended material to the downward flux. However, interpretation becomes difficult in systems with variable sediment composition and internal sediment transport due to horizontal gradients in the suspended solids concentration.

Further information on the net accumulation rate of sediments can be obtained through dating techniques, but because this information relies on fairly long time scales (e.g., decades) the data obtained is reliable only if no major changes in the sediment accumulation rate have occurred, because of changes in the watershed, increased eutrophication, dredging, etc. Apart from the effect on the internal load, it should be noted that resuspension and settling affect the extinction coefficient of water and hence primary production.

Chemical processes

Although the composition and form of substances in the sediment containing nutrients may be nonstoichiometric and nonideal, sediment transformations can be described in the traditional terms used for chemical processes: adsorption and desorption, crystallization or precipitation and dissolution, complexation, etc. For phosphate, the formation of one of the most stable forms, apatites such as $Ca_{10}(PO_4)_6(OH)_2$, seems to be inhibited by kinetic barriers in crystallization (Martens and Harris 1970) to such an extent that interstitial (and surface) waters are frequently highly supersaturated with respect to such substances. In practice, their formation is only important when considering long time scales, but in recent deposits considerable quantities of phosphate are associated with Ca. In most cases this material is a mixture of Ca (and Mg) carbonates on which coprecipitation of phosphate has occurred during their formation in the overlying water. This process is sometimes due to a loss of CO_2 to the atmosphere, but the removal of CO_2 by algal growth is often responsible for the shift in the carbonate equilibrium and for exceeding the solubility product. The concentration effect of evaporation can also play a role, such as in the Dead Sea. The pH, which reflects the distribution of free CO_2, bicarbonate, and carbonate ions, therefore controls solubility. In the sediment the precipitated material may be dissolved by CO_2 produced during mineralization (see below). The pH in sediments is generally much lower than that of the overlying water. The dissolution of carbonates naturally also leads to dissolution of coprecipitated phosphates. An important feature of coprecipitation is the pH effect: at high pH freshly precipitating carbonates adsorb phosphates very effectively and can reduce the concentration in the surrounding water to very low levels.

The pH also controls the adsorption of phosphates by oxides and hydroxides of Fe, Al, and Mn. In particular, Fe (hydr)oxides, which often occur as coatings on sand, are responsible for the building of an important fraction of the phosphate. Hence the competition between hydroxy- and phosphate ions for adsorption sites causes a reduced adsorption of phosphate at high pH, when the concentration of hydroxyions is high. This behavior contrasts with that of the Ca-associated phosphates discussed earlier. The adsorption/desorption of Fe-bound phosphates due to variations in pH is not a fully reversible process. The Fe hydroxides are, in fact, amorphous polymers with oxo- and hydroxo-bridges, and these materials exhibit aging phenomena. Freshly formed Fe[III] hydroxide behaves differently from the aged material: it has a higher adsorption capacity and a higher (initial) rate of adsorption, but both materials are slow to arrive at equilibrium (Lijklema 1980). This aspect is relevant because reduction and oxidation in sediments show annual cycles, with reduction progressing from greater sediment depth towards the sediment–water interface during summer, when the temperature rises and readily degradable, freshly precipitated organic material accumulates in the top layer. The reduction of Fe enhances the solubility of both Fe and phosphate; their solubility is probably often regulated by the solubility product of vivianite [$Fe_3(PO_4)_2$] (Nriagu and Dell 1974, Hieltjes 1980). By diffusion the dissolved phosphate and Fe[II] migrate toward the aerobic top layer, where the Fe[II] is oxidized and the freshly precipitated Fe[III] hydroxide adsorbs the phosphate. This removal of phosphate is effective at neutral or slightly acidic pH, the normal condition within the sediment. However, at the surface (or very close to the interface) the pH may be higher due to photosynthesis, and under such circumstances the reduction–oxidation sequence will result in enhanced phosphate concentrations and a concomitant high flux (Lijklema 1980). In many lakes such a high internal load can be observed during the summer season.

Apart from Ca, Fe, Al, and Mn phosphates, some phosphates in the sediments are bound to clays, in complexes of humic substances and cations or in organic form. Generally their quantity and/or reactivity are less than those of the substances discussed above. An important class of organic compounds are the inositol phosphates, of which the hexaphosphate or phytic acid are extremely rich in phosphate. These metabolic end products have been identified in lake sediments (Weimer and Armstrong 1979), but their origin (autochthonous and/or allochthonous) is not yet clear. These compounds appear to be fairly stable with respect to hydrolysis and biodegradation (Potman and Lijklema 1983).

Biological processes

The biota in the sediment is active both in physical transport (mixing, pumping) and in transformations [mineralization, (de)nitrification]. At the sediment surface sessile algae find suitable habitats if the light conditions are sufficient to sustain growth. In shallow lakes thick algal mats have frequently been observed, even under ice cover; after melting and degradation in spring these may be a nutrient source for other algae. Also, macrophytes growing on the lake shore can be considered as a part of the sediment and, obviously, they contribute to the transport of material (including nutrients) across the interface (Barko and Smart 1980).

Biotransport within the sediment, however, is mainly due to the bottom fauna, whose activity can influence both the interstitial water and the sediment itself, but is restricted mainly to the top 10 cm. Depending on species, burrowing, and feeding activities, the zoobenthos can transport sediment particles in several ways, but usually toward the surface (Petr 1976). A normal mechanism is ingestion of sediment at some depth and deposition of fecal pellets at the surface. The activity of organisms is strongly temperature dependent; a higher concentration of utilizable detritus in the sediment causes a lower ingestion rate (per animal) and hence a lower turnover rate. The construction of burrows by chironomids, tubificids, and other organisms enhances irrigation of deeper layers and thus the uptake of oxygen. The redox potential and pH profiles are affected and subsequently also the microbiological transformations controlled by these environmental conditions. Both phosphate release (Holdren and Armstrong 1980) and the rates of nitrification and denitrification (Chatarpaul *et al.* 1980) have been observed to increase significantly with bioturbation. Denitrification is apparently partly due to reduction within the guts of tubificids. Water pumping also contributes to local advective transport of water and the dissolved material therein.

The most important microbiological transformations in the sediment are oxidation of ammonia (nitrification), which is restricted to the well oxygenated top sediment layer, and the degradation of organic matter (mineralization). Mineralization includes a multitude of organic compounds originating from algae, macrophytes, leaves, and other materials imported from the watershed, including intermediates resulting from turnover by grazing zooplankton, bottom feeders, and bacteria. The bacteria comprise different taxa with specific functions (enzyme systems) related both to the substrate and to the electron acceptor. A certain stratification in sediments is due to the preference for the use of O_2, nitrates, Fe[III], sulfates, and CO_2 as electron acceptors, in this order, and accordingly a vertical decrease in redox potential can be observed. The reduction of electron acceptors can induce secondary effects: the production of N_2, CH_4, and CO_2 (from the substrate) can lead to the formation of bubbles containing one or more of these gases, which disturb the sediment as they rise to the surface. This affects the transport of sediment particles and the exchange of interstitial and overlying water. Also the production of sulfides may mobilize phosphate indirectly because Fe[II] sulfide is less soluble (more stable) than Fe[II] phosphate. Nitrification, denitrification, and other redox processes also affect pH, with generally a lower value for the reduced sediment as compared with the oxidized layer. Further intermediate acid products may be formed, but their effect on pH is generally small. More important is the formation of complexing materials, including humic substances, but their role in nutrient mobility is probably limited unless their concentrations are very high (such as in peat areas).

7.4. Application to Lake Balaton

This discussion of experimental observations on Lake Balaton sediments in relation to the overlying water follows roughly the same order as the general reviews presented in previous sections. However, it is limited mainly to those aspects relevant to phosphate exchange. Also it should be kept in mind that the available information is incomplete or provisional in many instances, since the research focusing on the internal load is still in progress.

Physical characteristics of Lake Balaton sediments

Lake Balaton sediments are predominantly of mineral character, fine grained, and with a distinct interface with the lake water. However, along the northern shore sediments tend to be softer and richer in organic material due to the prevailing wind from the north–northwest. The water content of the sediments varies from 65–80% in the top 5 cm, and decreases with sediment depth to 50–60% at depths of 25–30 cm. In the Tihany Strait, with its exceptionally high water velocities, the winnowing and compaction of sediments have resulted in a much lower water content of about 30% (Dobolyi 1980). The water content of the sediments along the southern shoreline is similar. The grain size distribution was analyzed by, among others, Győrke (1978), and later by Máté (1981), who initiated a comprehensive study in 1978 to establish a detailed sediment map for the entire lake. The results showed that most of the sediment particles were smaller than 100 μm, with mean particle sizes of 10–40 μm.

Chemical composition of Lake Balaton sediments

The most striking feature of the chemical composition of the sediments is the high carbonate content, reflected in the Ca content of 170–235 mg/(g dry sediment), and the Mg content of 20–38 mg/(g dry weight). These values indicate that 45–65% of the sediment is carbonates. Generally, higher values are observed in basin I (Keszthely) where the Zala River enters the lake and eutrophication is most advanced. There is also a tendency for the carbonate concentration to decrease with sediment depth. At 25–30 cm the range for Ca is 130–170 mg/(g dry sediment), with a few higher values (Dobolyi 1980). Again, the Tihany Strait is an exception with a much lower carbonate content at all depths. The Fe content of the sediments varies from 1.2–1.7% by weight, a part of which is associated with the clay mineral chlorite (Gelencsér *et al.* 1982). About 30–40% is clay and silica.

The organic carbon content ranges from 5–26 mg/(g dry matter), with higher values in the upper layer and in the western part of the lake. Total N ranges from 2–4 mg/(g dry sediment). The total P concentration is fairly constant, between 0.4–0.7 mg/(g dry sediment), of which 0.3–0.6 mg/g can be extracted by acids. This absence of spatial variations in P content is a striking feature. When considering the load distribution in the lake and the recent progression of eutrophication, one would expect a pronounced decrease in concentration from west to east and from the top layer to deeper strata. However, the only spatial variation is lateral, with higher concentrations near the northern shore. An explanation is suggested in Section 7.5.

The form in which phosphate is present has also been the subject of research. Qualitatively, the presence of hydroxyapatite could be established (Dobolyi and Bidló 1980) after gravity separation and subsequent X-ray diffraction and electron microscopic examination. Its quantitative importance is difficult to assess; certainly both pore and surface water are supersaturated

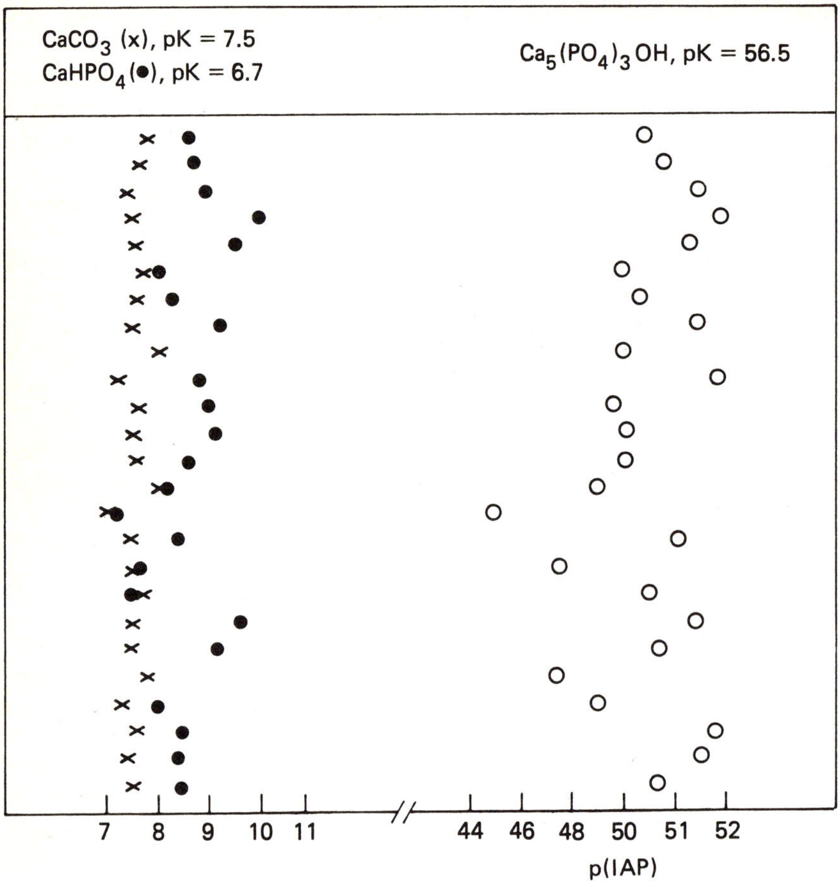

Figure 7.4. Ion activity products (IAP) of three minerals for lake water and pore water samples from Lake Balaton. Values expressed as a negative logarithm, p(IAP).

with respect to apatite; the presence of Mg^{2+} ions hampers the crystallization of apatites.

For a number of samples of both lake and pore water the ion activity product of calcite, brushite, and hydroxyapatite was assessed on the basis of chemical analyses (see Figure 7.4). A comparison with pK values from the literature indicates that all samples are supersaturated with respect to apatite, undersaturated with respect to brushite, and approximately in equilibrium with respect to calcite. Consequently, these phosphate-containing minerals do not appear to control the phosphate concentration according to thermodynamic equilibria. However, the carbonate equilibrium is controlled by the calcite formation and/or dissolution.

At least some phosphate is apparently associated with carbonates because in experiments in which carbonates were dissolved by flushing a sediment suspension with CO_2, variable amounts of phosphate also went into solution. During such experiments a redistribution of phosphate (e.g., adsorption onto Fe hydroxides) can also take place. The discrimination between Ca- and Fe-bound phosphates using different combinations of extractors indeed suggests a substantial contribution of Fe to phosphate sorption (Dobolyi 1980 found about 50% in one sample) but the extraction schemes are open to question (Hieltjes and Lijklema 1980). The organic P content is apparently rather low.

In summary, the sediment consists mainly of carbonates (up to 65%), sand and clay (30–40%) with some Fe hydroxide (coatings?), and a fairly low organic matter content. Phosphates are associated with carbonates and Fe hydroxides, and are also present as apatite and in organic form.

Samples of pore water from Lake Balaton sediments have also been analyzed. The various methods of sampling, separation of sediment and water, transport, and conservation of samples may cause changes in concentrations during the procedure (Brinkman *et al.* 1982), so that results obtained thus far are subject to criticism, but general trends can be obtained from the data.

The most important observations on the sediment are a lower pH with respect to the surface water (range 7.5–8.5; this measurement is sensitive to the squeezing operation and the loss of CO_2), and higher concentrations of phosphate (range 50–250 mg P/m^3), Ca, Mg, bicarbonate, and ammonia (range 3–8 g N/m^3) (Gelencsér *et al.* 1982). Under anaerobic conditions nutrient concentrations are higher, and Fe[II] can be identified. In the Keszthely basin, near an old sewage outlet, very high pore water nutrient concentrations have been observed (2 g P/m^3 and 16 g N/m^3), but otherwise the spatial variation is not great. The temporal variation is probably related to seasonal variations in the redox profile and to the occurrence of storms that disturb the top layer.

7.5. Processes Related to the Internal Load in Lake Balaton

Accumulation and dilution of sediments

About 95% of the external P load accumulates in the sediment, which amounts to a net sediment P load of about 0.7 g/m^{2}yr. This is a lake-wide average, but at Keszthely values are much higher, and there is a pronounced temporal variation related to annual variations in precipitation and tributary discharges. This strong decrease in the calculated P load from west to east is not reflected in a concomitant decrease in the P concentration in the sediment because the input of suspended solids is also high in the western basin. The Zala River carries about 100 tons of P and 10^4 tons of suspended solids/yr (Chapter 14). If the sedimentation were in the same proportion, the sediment would contain about 7 mg P/(g dry sediment). However, not all the phosphate is precipitated and in addition to the suspended solids entering via the Zala River, there is also a twofold quantity of carbonate that precipitates in the lake if the present sediment composition at Keszthely (with about 65% carbonates) reflects also the composition of settling material. This would result in about 2.3 mg P/(g dry matter). However, because sediment dilution is a slow process, both the carbonate and phosphate content may still be increasing. A fraction of 80% carbonates in the settling material would further decrease the phosphate equilibrium value to 1.4 mg P/(g dry matter); the actual concentration is presently about half of this. This very rough estimate does not take into account internal sediment transport and phosphate redistribution by the internal load and (repeated) precipitation. On the basis of lake-wide mass balances with an estimated annual retention of 3.4×10^4 tons of Ca^{2+} and 420 tons of phosphate P, the P:$CaCO_3$ ratio would be 1:200 on average, or 5 mg P/g $CaCO_3$. A value of about 3 mg/(g dry sediment) then seems reasonable, taking into account the proportion of $CaCO_3$. A somewhat lower value can be envisaged because the precipitation of Mg must also be considered. The available data clearly indicate that the settling material has a higher P concentration than the sediment itself. The acid-soluble P:Ca ratios in suspended solids are about 6.4×10^{-3} and 2.3×10^{-3} at Keszthely and Szemes, respectively (see Figure 7.5); whereas for the sediment these ratios are about 1.7×10^{-3} and 1.0×10^{-3} at the same sampling points.

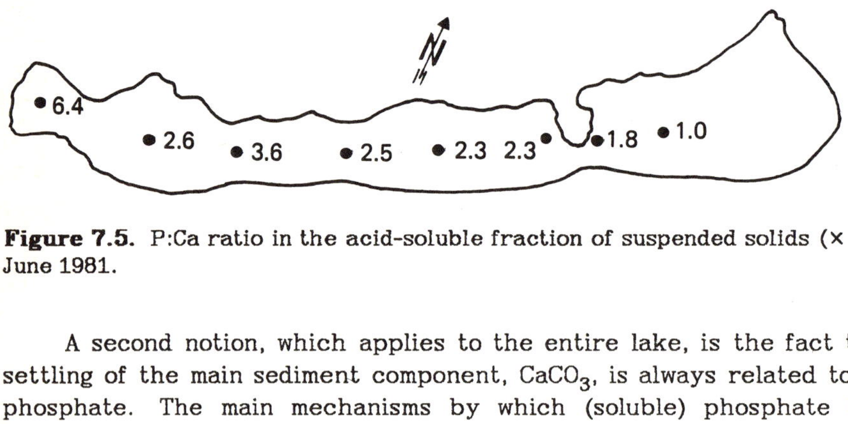

Figure 7.5. P:Ca ratio in the acid-soluble fraction of suspended solids ($\times 10^{-3}$). 6 June 1981.

A second notion, which applies to the entire lake, is the fact that the settling of the main sediment component, $CaCO_3$, is always related to that of phosphate. The main mechanisms by which (soluble) phosphate becomes incorporated in the sediment are uptake by algae during primary production and adsorption onto the carbonate formed as a by-product of primary production and evaporation. For each mole of CO_2 incorporated in the biomass a mole of $CaCO_3$ is also formed according to the equation

$$Ca^{2+} + 2HCO_3^- \rightleftarrows CaCO_3 + CO_2 + H_2O \ ,$$

when the water is saturated with carbonate and no pH shift occurs. Because these conditions are roughly met in Lake Balaton the availability of phosphate for primary production inevitably results in the formation and (co)precipitation of carbonates, carbonate-associated phosphate, and algae, etc. This will cause a certain proportionality between phosphate and $CaCO_3$, but at higher local phosphate concentrations more phosphate will also be incorporated. This explains the higher P:Ca ratio in suspended solids at Keszthely (6.4×10^{-3}) than at Siófok (1.0×10^{-3}) (see Figure 7.5). In this respect it is interesting to note that under most circumstances a modest primary production is sufficient to remove the phosphate added by the external load. A low production of 0.05 g C/m^2d or 4×10^{-3} mol/m^2d corresponds to the same amount of $CaCO_3$ (in moles), or 400 mg/m^2d. With an average load of less than 2 mg P/m^2d, this would require 0.5% by weight of phosphate P in the carbonate. This is not an exceptionally high value (Hieltjes 1980) and is of the same order as the phosphate released from sediments by dissolution with CO_2, which is the reverse process.

Another way of looking at Ca–Mg–P relations is to compare their mass balances. From the inflow of water (957 mm/yr) about 290 mm/yr is lost through more evaporation than precipitation; hence about 670 mm/yr leaves the lake (see Chapter 1). With a water depth of 3.1 m this means that the dissolved material in the water entering the lake will become substantially more concentrated during its retention over several years due to loss by evaporation. Indeed, for Mg a concentration increase from west to east can be observed, despite mixing and precipitation. For Ca, however, the concentration decreases from the Zala River through the Keszthely basin toward the "end" of the lake; the same is also true for most of the dissolved and particulate P fractions. A comparison of these concentration profiles, including a conservative tracer, in combination with the water balance should yield information on the composition and quantity of the sedimenting material. Sediment traps could add additional information but during periods of strong winds resuspension will blur the effects of formation of insoluble material.

The net accumulation rate and the mixing depth of the sediment are not well known; hence the application of equations (7.1) and (7.2) is not yet possible. Further research is needed in this respect.

Physical transport processes in Lake Balaton sediment

Advection and diffusion

No information is available on the process and rate of groundwater infiltration into Lake Balaton, but its role is probably insignificant.

Although some attempts have been made to assess experimentally the diffusion of phosphate out of a sediment core examined in the laboratory, no conclusive results have been obtained. Since the amount and reliability of available data on interstitial water concentrations are also limited it is difficult at this time to make estimates of the internal load due to diffusion, never mind the theoretical difficulties outlined in Section 7.3. A very rough approach is to assume a diffusion coefficient of 10^{-9} m^2/s, a concentration in the pore water of 50–250 mg P/m^3, and a depth of 2×10^{-2}–2×10^{-3} m at which this concentration can be maintained by desorption and mineralization. The latter estimate is very uncertain, but it can be envisaged that a higher activity of sediments and a higher phosphate concentration also leads to a steeper gradient. This would yield release rates in the range 0.2–10 mg P/m^2d or 0.07–3.5 g P/m^2yr. For comparison, the lake-wide biologically available external P load is about 0.8 g/m^{2}yr, but this also varies along the length of the lake. A provisional conclusion could be that the internal load (due to diffusion) is of the same order as or (somewhat) smaller than the external load. The algal C:P ratio is generally about 60, so that such an internal load could sustain a primary production rate of 0.01–0.6 g C/m^{2}d; actual rates are often much higher.

The (potential) internal load in many lakes around the world has been assessed, generally on the basis of mass balances over the lake or of column experiments. Both continuous and batch tests with respect to the overlying water have been reported in column studies. Low values in oligotrophic lakes are usually in the range 0.2–0.5 g P/m^{2}yr, whereas more eutrophic lakes may have values up to 10 g P/m^{2}yr or even higher.

One of the most striking and unique features of Lake Balaton water quality is the combination of (very) low dissolved reactive P concentrations and yet comparatively high productivity and high specific growth rates. This necessarily means a high phosphate turnover rate; either in the water itself, or in the interaction with sediment, or both. Without a rapid desorption–adsorption equilibrium between the particulate material in suspension or in the topmost sediment layer, such a high turnover rate would be improbable, although dissolved organic phosphate certainly also contributes to the phosphate cycle. In particular, phosphate associated with carbonates (by adsorption or rather chemisorption) can be considered to establish a rapid equilibrium, so that part of the research has concentrated on the sorption characteristics of the sediment (see below).

Diffusion or, rather, mixing of interstitial water with lake water during storms should also be considered. Somlyódy (1980) showed that the flux of stirring of sediment is approximately $0.034W$ kg/m^{2}d, where W is the wind velocity in m/s. Assuming a specific sediment density of 2.5 g/cm^3 and 75% water content, this would result in mixing of $0.04W \times 10^{-3}$ m^3 interstitial water/m^{2}d, or, with a dissolved reactive P content of 150 mg P/m^3, a release rate of about 30 μg P/m^{2}d at an average wind speed of 5 m/s; in other words, an internal load of 0.01 g P/m^{2}yr. This is a very low contribution, but a much higher flux can be envisaged because a much thicker layer of the sediment is fluidized by wind action. Szilágyi (1982) showed that in the upper 2 cm of sediment the chlorophyll-a (Chl-a) content is higher by almost an order of magnitude than in deeper regions, and if this is the result of wind-induced mixing the flux calculated above may be much higher. Nevertheless the role of pore water in this wind mixing is probably unimportant compared with that of the resuspended sediment because the volumetric phosphate concentration in the particulate material is about four orders of magnitude higher than in the pore water. Hence it is important to take into account the resuspension flux.

Resuspension

Upon realizing the pronounced dynamics of wind-induced interaction between water and sediment in Lake Balaton, the research team made daily suspended solids observations in the Szemes basin (open water) for six months, and hourly wind data were also collected (Somlyódy 1980, 1982). For analyzing the data a simplified, unsteady transport equation was set for the suspended solids concentration and subsequently integrated along the depth. The ordinary differential equation obtained incorporated the unknown deposition and resuspension fluxes, respectively. The deposition flux was assumed to be proportional to the depth-integrated suspended solids concentration, c_{SS}, while the resuspension flux was derived from simple energy balance considerations. This procedure led to equation (7.4) for the average suspended solids concentration (Somlyódy 1980)

$$\mathrm{d}c_{SS}/\mathrm{d}t = -k_1 c_{SS} + k_2 W^n \quad , \tag{7.4}$$

where W is the absolute value of the wind speed, and the coefficients k_1, k_2, and n are evaluated from observations. First a deterministic parameter estimation technique was employed, followed by application of the extended Kalman filter method, which also serves for model structure identification and validation (see Somlyódy 1980, 1982). The analysis resulted in the flux equations

$$\varphi_{\text{sed}} = 5.6\, c_{SS} \tag{7.5a}$$

$$\varphi_{\text{res}} = 0.034\, W \tag{7.5b}$$

(both in kg/m^{2}d), with the exponent $n = 1$. Equation (7.5a) implies a settling rate of 5.6 m/d, which is realistic.

Desorption studies indicated that at the prevailing lake water concentrations a desorption capacity of 5–10 μg P/(g dry matter) is not unusual, and hence the resuspension flux at average wind speeds (5 m/s) could maintain an internal load of 0.85–1.7 mg P/m^{2}d (0.3–0.6 g P/m^{2}yr). This value is of the same order as the external load and is also within the range calculated for the diffusion flux. It should be noted, however, that these mechanisms are not independent and that their effects cannot simply be added; sediment stirring will decrease the concentration gradient in the pore water and hence a high resuspension flux is associated with a low diffusion flux, and vice versa.

Finally, it should be stressed that the resuspension–desorption mechanism when coupled with primary production will be characterized by compensation mechanisms: a high rate of photosynthesis will tend to decrease the phosphate concentration, but this will stimulate the rate and extent of desorption from the suspended solids, whereas input of phosphate from external sources or decomposing algae will reduce the desorption (or, rather, will result in adsorption). This explains the remarkably small variations, both spatial and temporal, in phosphate concentration (dissolved reactive P).

Chemical processes: sorption of phosphates in Lake Balaton

Apart from the contribution of Fe the Balaton sediments have favorable sorption properties due to fine-grained carbonates with a high specific surface area and an Mg content that enhances the adsorption capacity (Jacobsen

Table 7.1. Desorption and adsorption of phosphate on sediments of the Szemes basin, 21 May 1981.

No.	Equilibration time (h)	Suspended solids (10^3 g/m^3)	Initial phosphate P (mg/m^3)[a]	Final phosphate P (mg/m^3)	(De)sorbed [μg P/(g dry sediment)][b]
1	3	5.9	132	53	+13.2
2	3	5.9	1	43	− 7.2
3	3	0.59	132	93	+65.0
4	3	0.59	1	25	−41.7
5	1	5.9	132	82	+ 8.3
6	1	5.9	1	40	− 6.7
7	1	0.59	132	130	+ 3.3
8	1	0.59	1	14	−23.3

[a] Average of three measurements.
[b] Negative values mean desorption.

1978). A first impression of the sorption characteristics of the sediment can be obtained from Table 7.1, which summarizes experimental results with sediments taken from the top 3 cm of the Szemes basin in the center of the lake. By systematic analyses the influence of the major parameters could be investigated: initial dissolved reactive P, suspended solids concentration, and equilibration time. The original sediment pore water concentration was 179 mg P/m^3, the pH was 8.2, and the conductivity 600 μSiemens/cm. The same parameters for lake water in which the sediment was resuspended were: <1 mg P/m^3, pH 8.7, and 520 μSiemens/cm.

Apparently the original pore water concentration is not a representative equilibrium concentration for the prevailing lake water conditions because adsorption was found in all experiments with an initial phosphate P concentration of 132 mg P/m^3. On the other hand, desorption occurred in all experiments with very low lake water P concentrations.

Similar experiments were performed on sediments from the Keszthely basin and the Zala River, which has a dissolved reactive P concentration of 190–310 mg P/m^3 (results are expressed in % phosphate removal; see Figure 7.6). Although not all the relevant parameters were measured, it is clear that the adsorption (initially) is fast, and adsorption capacities of >100 μg P/(g sediment) have been observed. Further experiments in a more realistic concentration range (50–2000 g suspended solids/m^3) with different sediment samples and filtered lake water enriched with phosphate to a level of 23 mg P/m^3 are presented in Figure 7.7. Here adsorption occurred under all conditions. Owing to the low concentration range and the small changes therein the uncertainty in the results is appreciable, as indicated by the error bars for Keszthely sediment. Desorption experiments were also performed in triplicate with the same sediment samples, with equilibration in lake water without the addition of phosphate and with 30 min equilibration time (see Table 7.2). It can be seen that desorption indeed occurs, but again the accuracy is not very high for the reasons mentioned.

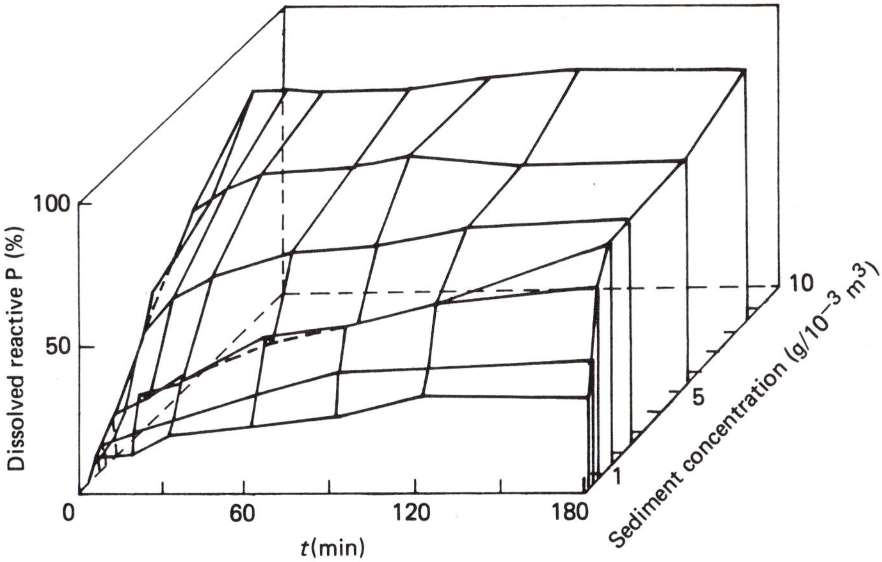

Figure 7.6. Phosphate adsorption as a function of time and suspended solids (Keszthely sediment).

Table 7.2. Desorption of phosphate from sediments in filtered lake water. Equilibration time, 30 min.

Keszthely		Szemes	
c_{SS} (g/m^3)	c_{eq} (μg/l)	c_{SS} (g/m^3)	c_{eq} (μg/l)
71.9	3.6 ± 1.0	122.1	3.3 ± 0.6
83.1	3.0 ± 0.7	391.2	4.9 ± 0.3
230.8	4.1 ± 0.3	751.6	7.1 ± 0.6
230.7	4.7 ± 0.1	1035.9	7.6 ± 1.6
553.9	6.6 ± 0.3	901.5	9.1 ± 0.6
1305.5	10.2 ± 0.3	2060.5	16.1 ± 7.5
Lake water	2.4 ± 0.3	Lake water	3.4 ± 0.9
Szigliget		**Siófok**	
c_{SS} (g/m^3)	c_{eq} (μg/l)	c_{SS} (g/m^3)	c_{eq} (μg/l)
90.3	2.4 ± 0.2	156.4	4.3 ± 0.1
184.8	4.7 ± 0.4	122.5	6.4 ± 2.7
225.4	10.1 ± 0.6	283.8	6.3 ± 1.8
777.5	6.3 ± 0.4	606.3	8.0 ± 1.0
837.8	6.1 ± 0.8	1031.0	7.0 ± 0.5
1598.5	7.4 ± 0.4	1642.8	9.1 ± 0.1
Lake water	1.5 ± 1.3	Lake water	4.1 ± 0.5

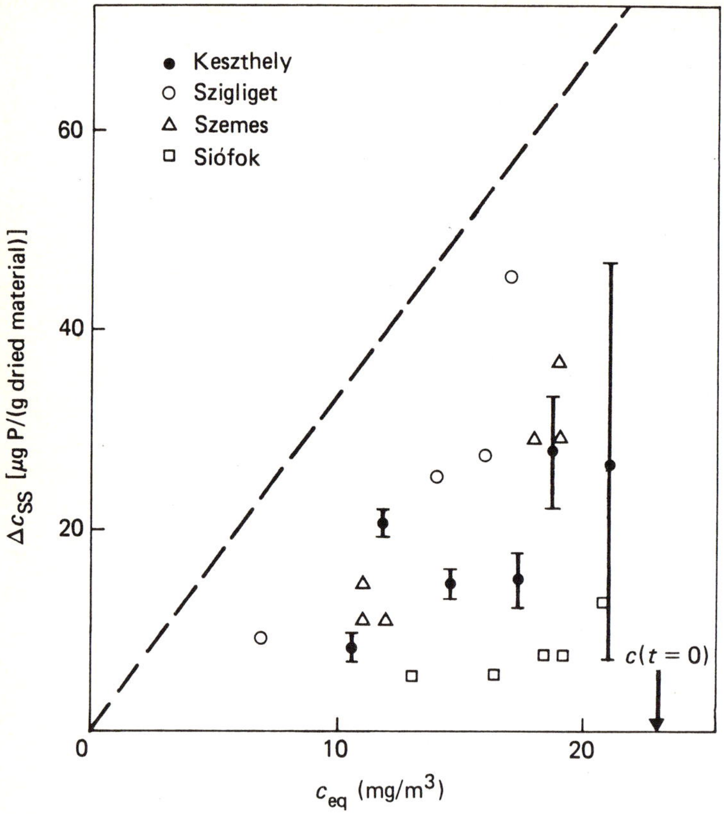

Figure 7.7. Phosphate adsorption of different sediments. Suspended solids = 50–2000 g/m^3; equilibration time, 24 h. Samples collected 28 October 1981.

In order to apply such results in calculations and models it is useful to represent adsorption and desorption processes in terms of adsorption isotherms as illustrated in Figure 7.8. Resuspension [Figure 7.8(*a*)] brings a sediment particle from an environment of high phosphate concentration (1) into a lower range (2), and hence desorption occurs until a new equilibrium is reached (3). This equilibrium is a function of the total amount of sediment resuspended: a high suspended solids concentration causes a higher equilibrium concentration (3″) than does a low suspended solids concentration (3′). An increase in external load [Figure 7.8(*b*)] causes a higher dissolved reactive P concentration [the system moves from (1) to (2)], and adsorption occurs until a new equilibrium is reached (3) with a higher equilibrium phosphate concentration. The full potential of this modeling approach requires further investigation, with emphasis on the effects of time, pH, and ionic composition, and more accurate analytical methods.

Finally, the mechanism of phosphate release from Fe-bound complexes under anaerobic conditions is important, at least for certain areas of the

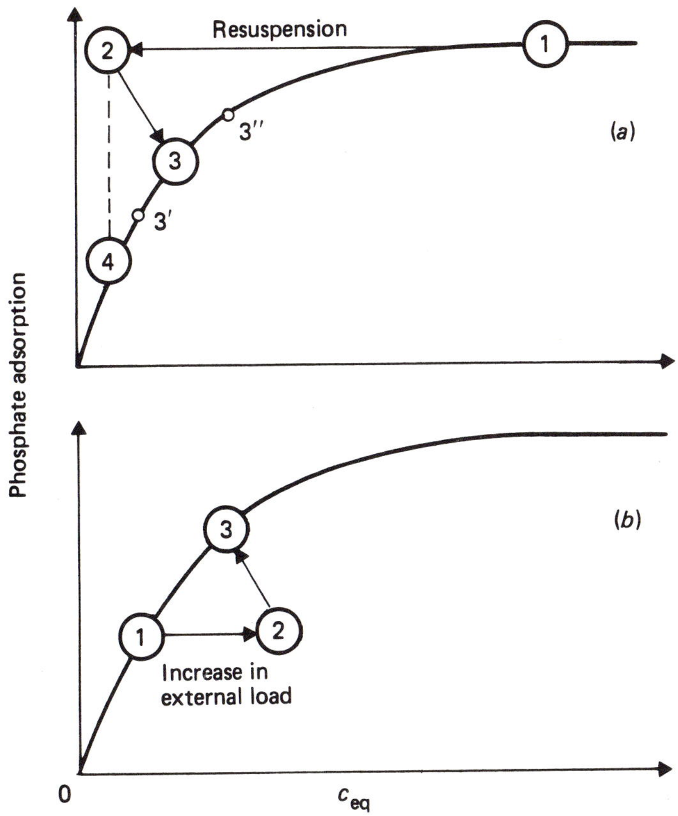

Figure 7.8. (*a*) Desorption and (*b*) adsorption related to sorption isotherms.

Keszthely basin and for shore regions with comparatively high levels of organic carbon. Indeed, anaerobic conditions have occasionally been observed near the interface but a systematic data set is not available. An extremely high primary production rate of 14 g C/m^{2}d, as was observed in the Keszthely basin in 1973, may have been sustained by a phosphate flux due to this mechanism.

Stripping experiments with Zala River and lake water (Keszthely basin) supported the conclusions on coprecipitation and dissolution of phosphate with carbonates upon removal of CO_2 by primary production or CO_2 production by mineralization (in the sediment). Water samples were stripped with air for 24 h. Zala River water was supersaturated with CO_2 and consequent stripping resulted in an enhanced pH, a shift towards CO_3^{2-} in the carbonate equilibria, coprecipitation of phosphate with Ca(Mg)CO_3, and a reduction in conductivity (Table 7.3). In lake water undersaturated with CO_2 due to intense primary production, the reverse trends could be observed (Table 7.4). The suspended material was the source of (small amounts of) phosphate going into solution.

Table 7.3. Stripping experiments: Zala River water.

	28 June 1982		13 July 1982		14 July 1982		5 August 1982	
	Initially	After 24 h	Initially	After 24 h	Initially	After 24 h	Initially	After 24 h
pH	7.53	8.30	7.50	8.68	7.70	8.61	7.75	8.58
Conductivity (μSiemens/cm)	688	512	746	620	524	509	558	516
CO_3^{2-} (mol/m^3)	0.01	0.60	0.01	0.98	0.01	0.44	0.01	0.65
HCO_3^- (mol/m^3)	6.96	3.21	6.58	4.46	5.02	4.24	5.81	5.62
Phosphate P (mol/m^3)	10^{-2}	1.3×10^{-4}	1.5×10^{-2}	1.4×10^{-2}	1.2×10^{-2}	1.2×10^{-2}	8.1×10^{-3}	7.8×10^{-3}

Table 7.4. Stripping experiments: lake water (Keszthely).

	13 July 1982		14 July 1982		5 August 1982	
	Initially	After 24 h	Initially	After 24 h	Initially	After 24 h
pH	9.11	8.58	9.20	8.53	9.20	8.60
Conductivity (μSiemens/cm)	532	476	434	463	460	469
CO_3^{2-} (mol/m^3)	0.87	0.22	0.83	0.22	0.60	0.44
HCO_3^- (mol/m^3)	2.12	3.48	2.16	3.26	2.34	3.80
Phosphate P (mol/m^3)	9.7×10^{-5}	2.9×10^{-4}	1.6×10^{-4}	4.2×10^{-4}	2.9×10^{-4}	3.2×10^{-4}

7.6. Prediction of the Effects of Reduced External Loads

Qualitatively, the effects of management measures can be gathered from historical data. In the past, before and during the first period of artificial eutrophication, most of the phosphate that reached the lake was presumably removed by adsorption on Fe and by biogenic lime coprecipitation. With increasing levels of phosphate the high-energy adsorption sites of amorphous Fe[III] hydroxides gradually became saturated with subsequent higher pore water concentrations and internal loads. The higher primary productivity and increased organic load of the sediment also stimulated phosphate cycling through its effects on redox conditions and resolubilization of precipitated carbonates. These slow transitions progressed gradually with the main input from west to east. It can be anticipated that this trend will be reversed when the external load is reduced. The time scale of this regeneration process is difficult to predict because several important characteristics, particularly the sediment dilution rate, are not well known for Lake Balaton. This quantity also affects the rate at which the actual working point shifts along the adsorption isotherm towards a lower equilibrium phosphate concentration (see Figure 7.8) (Lijklema 1983). It can also be expected that diagenetic processes in the sediment, notably transformations of carbonate-bound phosphate (which is thought to be a precursor of apatite) will tend naturally to decrease the solubility of the phosphate. From a thermodynamic point of view this is also a logical expectation, because the system tends to a lower free energy or higher stability. Another aspect to be considered is that, on the time scale of several years, the sediment adsorption capacity of the entire lake is involved, not just that of the most polluted area. Although the phosphate cycling rate is apparently high due to rapid equilibration of carbonate–phosphate, the anticipated reduction in equilibrium concentration will reduce the fluxes by a proportional reduction of the driving force.

As shown by equation (7.2) the equilibrium concentration (lake-wide average) is proportional to the external load: $c_{eq} = S/(\Delta h + kh)$. The sediment response time can be characterized by $\tau = h/(\Delta h + kh)$, where Δh is the (poorly known) sediment dilution rate. Although phosphate as such is conservative, the term kh can be interpreted as the conversion of reactive phosphate into more refractory material by the diagenetic processes discussed above. Hence it can be seen that such conversions tend to lower both the equilibrium concentration of "reactive" sediment phosphate and the sediment response time.

Taking all together it is felt that external load reductions will be effective in restoration, probably within a few years, when considering the factors discussed above and also experiences elsewhere, often with sediments with less favorable adsorption characteristics. A well known example of a fast response of a lake to management measures to reduce external loads is Lake Washington (Lorenzen *et al.* 1976).

7.7. Conclusions on Modeling and Management

(1) On the basis of data available for the sediment of Lake Balaton the internal and external loads can be estimated to be of the same order of magnitude.

(2) Sorption and diffusion are the major mechanisms that lead to P release from the sediment, although observations describing spatial and temporal changes are scarce. Consequently, it is not possible or realistic with our present, limited knowledge to develop a detailed sediment submodel in the frame of lake eutrophication modeling.

(3) Though not directly, the external and internal P loads are clearly coupled. An increase in the external load induces, with some time lag, an internal load. On the contrary, a reduction in the external load results in a decrease in the internal load (due to dilution of the top sediment layer and other processes).

(4) The relatively slow response of sediment to external load reductions (several years) has two important consequences from the viewpoint of eutrophication management:

   (i) Remedies should not be postponed, since the sediment will be further enriched in nutrients, which will then increase the length of the recovery process.

   (ii) After controlling the external load a spectacular improvement in lake water quality cannot be expected from one year to the next, since the new equilibrium of the lake and the time to reach it depend primarily on the refreshment of the sediment, a field where our knowledge is still limited.

References

Barko, J.W. and Smart, R.M. (1980) Mobilisation of sediment phosphorus by submersed freshwater macrophytes. *Freshwater Biology* 10:229–38.

Bloesch, J. and Burns, N.M. (1980) A critical review of sedimentation trap techniques. *Schweiz. Z. Hydrol.* 42(1):15–55.

Bloesch, J. and Evans, R.D. (1982) Lead-210 dating of sediments compared with accumulation rates estimated by natural markers and measured with sediment traps. *Hydrobiologia* 92:579–86.

Brinkman, A.G., van Raaphorst, W., and Lijklema, L. (1982) *In situ* sampling of interstitial water from lake sediments. *Hydrobiologia* 92:659–63.

Chang, S.C. and Jackson, M.L. (1957) Fractionation of soil phosphorus. *Soil Sci.* 84:133–44.

Chatarpaul, L., Robinson, J.B., and Kaushik, N.K. (1980) Effects of tubificid worms on denitrification and nitrification in streams. *Can. J. Fish. Aquatic Sci.* 37:656–63.

Dobolyi, E. (1980) Data on the bottom sediment in Lake Balaton, in G. van Straten, S. Herodek, J. Fischer, and I. Kovács (Eds.) *Proc. 2nd Joint MTA/IIASA Task Force Meeting on Lake Balaton Modeling*, Vol. II (Veszprém: MTA VEAB), pp 66–81.

Dobolyi, E. and Bidló, G. (1980) Some conclusions from the analysis of the sediment of Lake Balaton. *Hidrol. Közl.* 60:72–6 (in Hungarian).

Fukuda, M. and Lick, W. (1980) The environment of cohesive sediments in freshwater. *J. Geophys. Research* 85(C6):2813–24.

Gelencsér, P., Szilágyi, F., Somlyódy, L., and Lijklema, L. (1982) *A Study of the Influence of Sediment in the Phosphorus Cycle for Lake Balaton.* Collaborative Paper CP-82-44 (Laxenburg, Austria: International Institute for Applied Systems Analysis).

Györke, O. (1978) Changes in suspended solids concentration in shallow lakes and reservoirs. *VITUKI Report No.* 7411/2-323 (Budapest: VITUKI) (in Hungarian).

Häkanson, L. (1977) The influence of wind, fetch and water depth on the distribution of sediments in Lake Vanern, Sweden. *Can. J. Earth Sci.* 14:397–412.

Hargrave, B.T. and Burns, N.M. (1979) Assessment of sediment trap collection efficiency. *Limnol. Oceanogr.* 24(6):1124–36.

Hieltjes, A.H.M. (1980) *Properties and Behaviour of Phosphate in Sediments.* PhD Thesis, Twente University of Technology, The Netherlands (in Dutch).

Hieltjes, A.H.M. and Lijklema, L. (1980) Fractionation of inorganic phosphates in calcareous sediments. *J. Environmental Quality* 9(3):405–7.

Holdren, G.C. and Armstrong, D.E. (1980) Factors affecting phosphorus release from intact lake sediment cores. *Environmental Sci. Technol.* 14:79–87.

Jacobsen, O.S. (1978) Sorption, adsorption and chemisorption of phosphate by Danish lake sediments. *Vatten* 4:230–43.

Lick, W. (1981) Entrainment, deposition and transport of fine grained sediments in lakes. *Proc. 2nd Int. Symp. on Interactions between Sediments and Freshwater*, Kingston, Ontario 1981 (The Hague: Junk).

Lijklema, L. (1980) Interaction of orthophosphate with iron[III] and aluminium hydroxides. *Environmental Sci. Technol.* 14:537–41.

Lijklema, L. (1983) Internal loading. *Water Supply* 1:35–42.

Lorenzen, M.W., Smith, D.J., and Kimmel, L.V. (1976) A long-term phosphorus model for lakes: Application to Lake Washington, in R.P. Canale (Ed.) *Modelling Biochemical Processes in Aquatic Ecosystems* (Ann Arbor, MI: Ann Arbor Science Publishers) pp 75–92.

Martens, C.S. and Harris, R.C. (1970) Inhibition of apatite precipitation in the marine environment by magnesium ions. *Geochim. Cosmochim. Acta* 36:454–70.

Máté, F. (1981) *The Phosphorus Retention on the Sediment of Lake Balaton.* Lecture presented at the 3rd Task Force Meeting on Lake Balaton Modeling, Veszprém, Hungary, 24–27 August 1981.

Nriagu, J.O. and Dell, C.I. (1974) Diagenetic formation of iron phosphates in recent lake sediments. *Am. Mineral.* 59:934–46.

Petr, T. (1976) Bioturbation and exchange of chemicals in the mud–water interface. *Proc. Int. Symp. on Interactions between Sediments and Freshwater*, Amsterdam, September 1976 (The Hague: Junk).

Potman, W. and Lijklema, L. (1983) Inositol phosphates: Hydrolysis, stability and analysis. *Water Research* 17(4):411–4.

Rutgers van der Loeff, M.M. (1981) Wave effects on sediment water exchange in a submerged sand bed. *Neth. J. Sea Research* 15(1):100–12.

Sheng, Y.P. and Lick, W. (1979) The transport and resuspension of sediments in a shallow lake. *J. Geophys. Research* 84(C4):1809–26.

Somlyódy, L. (1980) Preliminary study on wind-induced interaction between water and sediment for Lake Balaton, in G. van Straten, S. Herodek, J. Fischer, and I. Kovács (Eds.) *Proc. 2nd Joint MTA/IIASA Task Force Meeting on Lake Balaton Modelling*, Vol. II (Veszprém: MTA VEAB), pp 26–49.

Somlyódy, L. (1982) Water quality modelling: A comparison of transport-oriented and ecology-oriented approaches. *Ecological Modelling* 17:183–207.

Szilágyi, F. (1982) A comparison of the methods of chlorophyll measurement. *Vízügyi Közl.* 64:85–98 (in Hungarian).

Tessenov, U. (1979) Die Wechselwirkungen zwischen Sediment und Wasser in ihrer Bedeutung für den Nährstoffhaushalt von Seen. *Z. Wasser- und Abwasser Forschung* 12(2):29–36.

Tzur, Y. (1971) Interstitial diffusion and advection of solute in accumulating sediments. *J. Geophys. Research* 76(18):4208–11.

Weimer, W.C. and Armstrong, D.E. (1979) Naturally occurring organic phosphorus compounds in aquatic plants. *Environmental Sci. Technol.* 13:826–9.

Williams, J.D.H., Syers, J.K., and Walker, T.W. (1967) Fractionation of soil inorganic phosphate by a modification of Chang and Jackson's procedure. *Soil Sci. Soc. Am. Proc.* 31:736–9.

Williams, J.D.H., Jaquet, J.M., and Thomas, R.L. (1976) Forms of phosphorus in the surficial sediments of Lake Erie. *J. Fish. Res. Board Canada* 33:413–29.

CHAPTER 8

Phytoplankton Changes During Eutrophication and P and N Metabolism

S. Herodek

8.1. Introduction

The immediate ecological consequence of nutrient enrichment in lakes is the increased growth of aquatic plants. The mass proliferation of algae may interfere directly with water uses so that most eutrophication models usually contain phytoplankton as the most important state variable. In eutrophication studies either the states of water bodies under different loads are compared, or the processes are studied in selected lakes.

This chapter deals first with changes in the composition, biomass, cell number, chlorophyll-a (Chl-a) concentration, and primary production of phytoplankton in Lake Balaton, and then reviews the processes of P and N metabolism in the open water. In shallow lakes, the open water is only part of the system. For a more complete picture of the nutrient cycles, the sediment–water interaction (Chapter 7) should be taken into account as well. Finally, the characteristics of phytoplankton and of nutrient metabolism are summarized from a modeling point of view.

8.2. Composition and Quantity of Phytoplankton

Quantitative phytoplankton studies were initiated in Lake Balaton by Sebestyén in the 1930s. Between 1936 and 1951 she counted cell numbers and determined the biomass of *Dinoflagellatae* by the cell volume method, and found a definite increase in the amount of *Ceratium hirundinella*, the dominant algal species (Sebestyén 1953, 1954). In 1945, 1947, 1949, and 1951 the cell numbers and the biomass of the other algal groups were also determined from basin IV samples collected at monthly intervals by the Limnological

Institute (Tamás 1955), from which data the total phytoplankton biomass was reconstructed. Its annual mean was about 0.3 g fresh weight/m^3 in the 1940s and 1.2 g fresh weight/m^3 in 1951. In these early studies a Kolkwitz chamber was used, so that smaller algal species were not detected. Thus, plankton cell numbers reported for this period are strong underestimates, but the biomass error is probably not more than 20%.

In 1965, 1966, 1967, 1974, and 1976 samples were collected every two or four weeks from all four lake basins, and were counted under a Utermöhl microscope (Tamás 1974, 1975, Vörös 1980). Similar cell number and biomass studies paralleled the primary production measurements in the four basins in the 1970s (Herodek and Tamás 1976, 1978, Herodek *et al.* 1982). In 1977 samples were collected weekly, and in 1978 biweekly for such algological studies of basins I and IV (Vörös 1982).

Cluster analysis showed five distinct groups of phytoplankton community composition, which are named according to the season in which they occur: "winter", "early spring", "late spring", "summer", and "autumn" phytoplankton (Vörös and Kiss 1985). Under winter ice there are small, slowly sedimenting algae, together with some motile species, while a strong diatom outburst usually starts a few weeks after the ice begins to melt. In May nitrates are depleted from the water, the phytoplankton biomass diminishes, and transient communities with rapidly changing structures can be observed. Summer phytoplankton start to develop in mid-June, and attain maxima in July or August. The autumn community appears around mid-September when temperatures begin to fall; its biomass remains low.

As the eutrophication of the lake progressed the structure of the phytoplankton community underwent definite changes. Early studies indicated that in the spring community the disc-shaped species (centrales), mainly *Cyclotella bodanica* and *C. ocellata*, prevailed, but since the mid-1970s thin, needle-like (pennales) *Synedra acus* and *Nitzschia acicularis* have been the most abundant. The changes in the summer community have been even more pronounced. Originally the whole lake was dominated by *Ceratium hirundinella*, but in the most polluted parts of the lake blue–green algae have appeared from time to time in large amounts since 1965, and since 1973 blooms of filamentous, heterocystic blue–green species (*Aphanizomenon flos-aquae, Anabaena spiroides*) have become regular phenomena. In basins III and IV the level of blue–green algae remained low until summer 1982, when the entire lake was invaded by a filamentous blue–green species, *Anabaenopsis raciborskii*.

The phytoplankton biomass of basin I (see Figure 8.1) was always below 10 g/m^3 in the 1960s, but in 1973, when very high primary production levels were attained it reached 13 g/m^3. Similar values were obtained in 1974 and 1976. In 1977 a second rapid increase occurred with a phytoplankton peak higher than 40 g/m^3, although in 1978 the summer maximum was somewhat lower, possibly due to low temperatures. Biomass maxima were 50 g/m^3 in 1979 and 60 g/m^3 in 1982.

Increases in phytoplankton biomass have occurred at all sampling stations (Figure 8.2). At Tihany (basin IV) the highest value was only 1 g/m^3 in

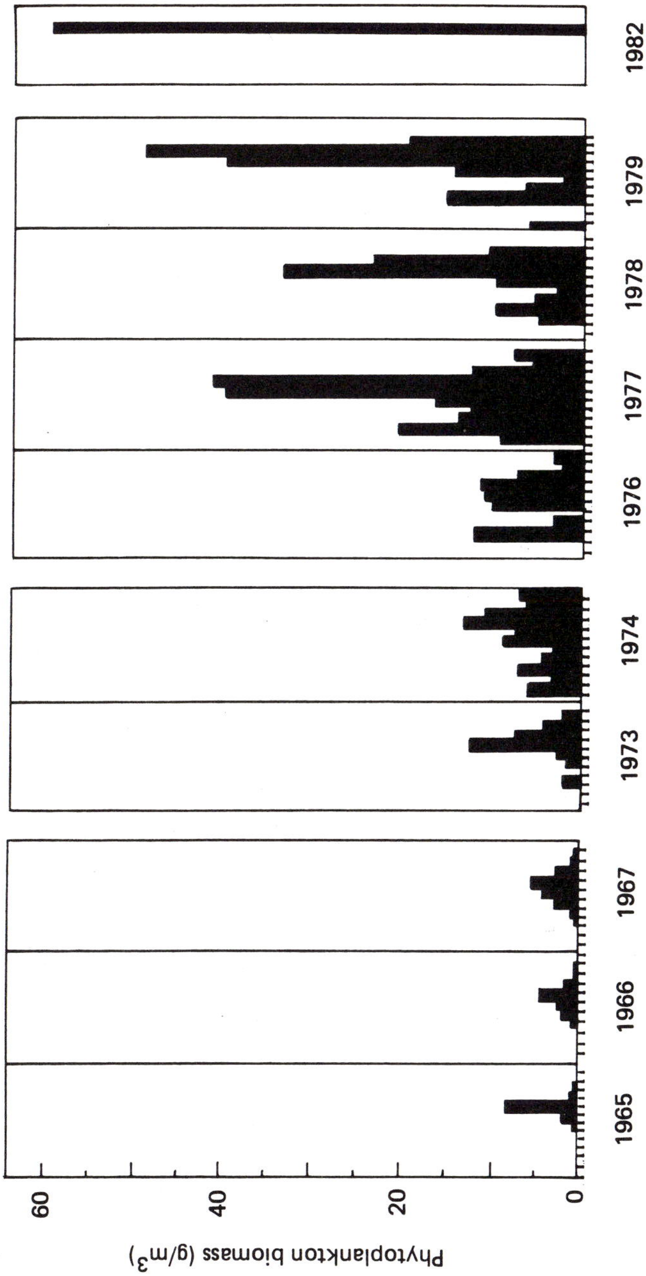

Figure 8.1. Phytoplankton biomass (in fresh weight) of basin I, 1965–82 (Vörös 1982).

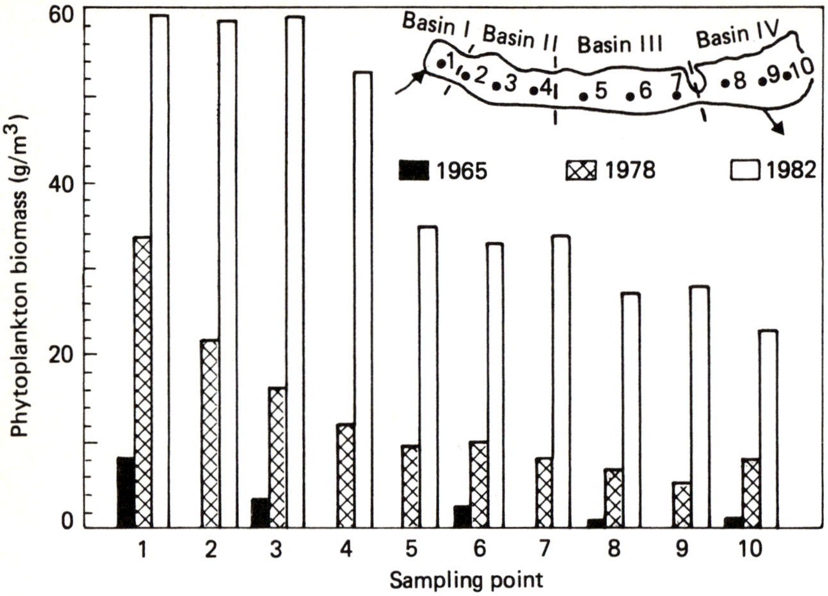

Figure 8.2. Phytoplankton biomass maxima (in fresh weight), August–September 1965, 1978, and 1982 (Vörös, unpublished).

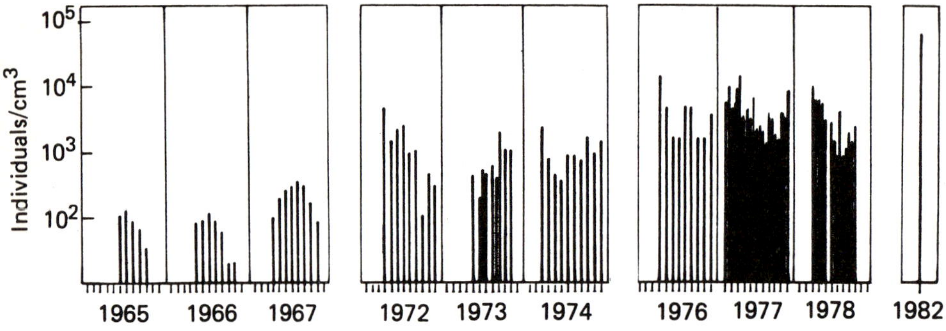

Figure 8.3. Total number of phytoplankton individuals in basin IV, 1965–82 (Vörös 1984).

1965, the summer peak of 1978 was 8 g/m^3, while during the 1982 *Anabaenopsis raciborskii* bloom the algal concentration was 27 g/m^3.

As eutrophication has progressed, species with smaller cell volumes have become dominant, so that the number of individuals has increased much more than the biomass. At Tihany the number of algae was only a few hundred individuals/cm^3 in the 1960s, several thousand in the 1970s, and more than ten thousand in 1982 (Figure 8.3; Vörös 1982).

Cell counting and the determination of biomass by the cell volume method are rather time-consuming; Chl-a concentration measurements are much

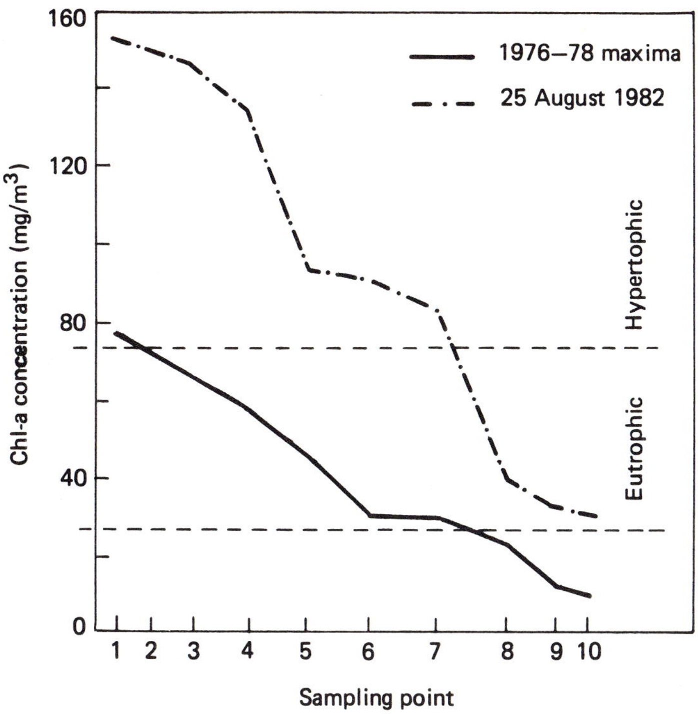

Figure 8.4. Chl-a concentration on 25 August 1982, and maximum values in 1976–78 (see Figure 8.2 for location of sampling points).

easier, and therefore are now generally used to determine algal biomass. Unfortunately, by measuring only Chl-a concentration, information on phytoplankton composition is lost, and it must also be noted that rather different Chl-a/biomass ratios have been published. In Lake Balaton Chl-a content was determined in parallel with the biomass at Szemes in 1976–77, at Tihany in 1977, and at Keszthely in 1978–79 (Vörös 1984). The correlation between Chl-a content and biomass was strong in all three basins. The mean Chl-a contents as a percentage of fresh weight were 0.30% at Tihany, 0.36% at Szemes, and 0.42% at Keszthely. The extremes were 0.09 and 0.68%.

Chlorophyll data for the lake are quite numerous. Since 1971 VITUKI has measured Chl-a levels at nine stations along the lake and the Transdanubian Water Authority at 16 offshore and several onshore stations at monthly intervals. The results from the two laboratories are in good agreement; both show increasing values toward the Zala River, and in recent years a significant increase has been detected at all stations.

The OECD Synthesis Report on Eutrophication Control (Vollenweider and Kerekes 1980) used a classification based on the annual maximum Chl-a content. Lakes with concentrations between 25 and 75 mg/m^3 are regarded as eutrophic, and above this level they are hypertrophic. If we select the maximal Chl-a values from 1976–78 data, it turns out that at this time only basin I fell into the hypertrophic category, basins II and III showed eutrophic values, and basin IV could be still regarded as mesotrophic (Figure 8.4). During the 1982 algal bloom the Chl-a content was two or three times higher than the maximal 1976–78 values, and so the whole lake became hypertrophic, except basin IV, which changed from mesotrophic to eutrophic.

8.3. Primary Production

Primary production is one of the best indicators of trophic state (Rodhe 1969). The first primary production measurements using the ^{14}C method were carried out in 1962–63, at which time there were no differences in the productivities of the four basins (Böszörményi *et al.* 1962). According to our present terms they corresponded to the mesotrophic level. A more detailed study started in 1972, when primary production at four depths was measured every two weeks using the ^{14}C method in basin IV (1972–73), basin I (1973–74), basin II (1974–75), basin III (1976–77), and again in basin IV (1977) (Herodek and Tamás 1975a, b, 1978; Herodek *et al.* 1982).

In 1972 the maximal primary production in basin IV was not much higher than in 1962, at 0.6 g C/m^{2}d. The vertical production profile varied according to weather conditions, since the transparency of the shallow water is basically determined by the amount of the sediment swirled up by wave action. During long calms the highest production was measured in the deepest sample, while strong storms restricted the euphotic layer to the uppermost 1 m. Typically, there was photoinhibition in the surface sample, maximum productivity was at 1 or 2 m, and at 3 m photosynthesis was reduced due to inadequate illumination.

In 1973, however, a completely different picture was found in basin I. Owing to the self-shading of phytoplankton, even on calm days the maximum was at the surface, at 1 m photosynthesis dropped to half the surface value, and below 2 m there was practically no photosynthesis at all. The maximal production (13.6 g C/m^{2}d) was one of the highest values measured in a European lake. In basin II the vertical photosynthesis profiles were similar to those of basin I; production was 3–4 times higher than in 1963, and in July 1974 a maximal production of 3.1 g C/m^{2}d was found.

The autumn of 1974 was very rainy, and this increased the diffuse load, which can be an important nutrient source especially in basin II. The extra load probably caused the extremely strong water coloration by diatoms from October to February. That year the lake did not freeze and production remained high throughout the winter. In February there was a serious fish kill; its origin has not been unequivocally clarified, but it is possible that anaerobic conditions in deeper water layers contributed to it.

In basin III (1976–77) the algae did not alter the optical properties of the water significantly, but production was much higher than in basin IV, and attained 2.6 g C/m^{2}d. Primary production shows a definite seasonality; it has a small peak during the spring diatom outburst, diminishes in May, and increases again from mid-June. With increasing eutrophication the summer peak becomes more pronounced.

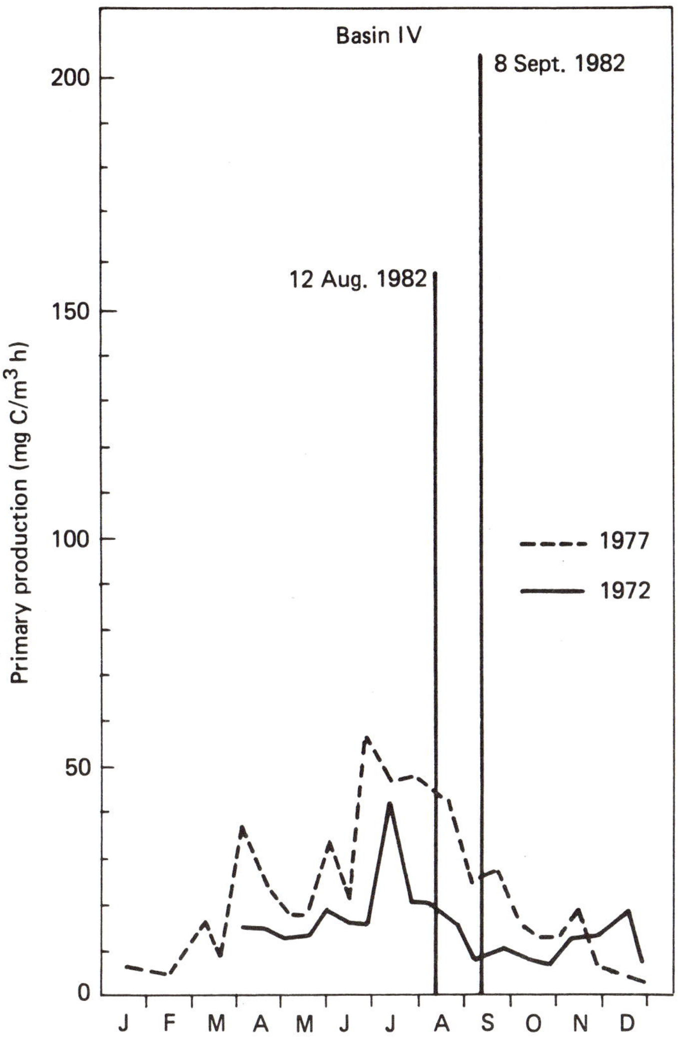

Figure 8.5. Primary production in the optimally illuminated layer in basin IV.

In 1977 the maximal primary production in basin IV was double that of 1972, at 1.7 g C/m^{2}d, proving that the whole lake was already in a state of rapid eutrophication. The annual productions in 1977 calculated from these measurements are as follows:

Basin I	830 g C/m^2
Basin II	301 g C/m^2
Basin III	274 g C/m^2
Basin IV	182 g C/m^2

With the given surface areas of the basins (Table 1.2), this means that about 26600, 36100, 39300, and 39700 tons of carbon were photosynthetically

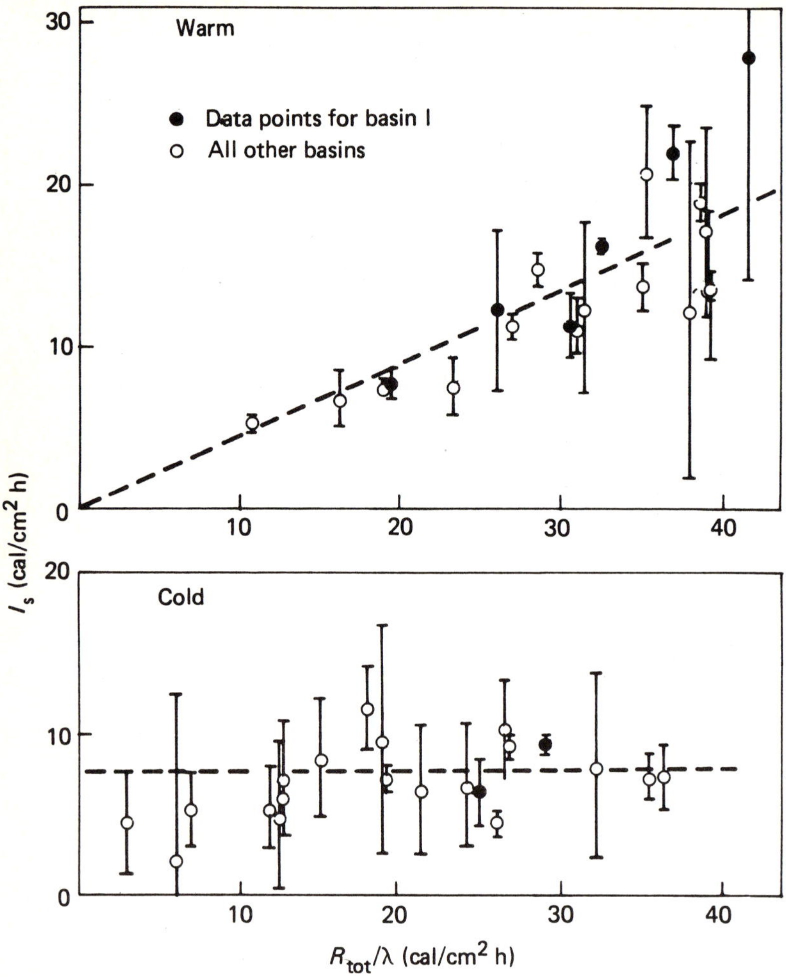

Figure 8.6. Saturation light intensity, I_s, as a function of total average irradiation, R_{tot}/λ (van Straten and Herodek 1982).

fixed in the four basins, and that total phytoplankton production was about 141 700 tons/yr in the mid-1970s.

In June–October 1979 measurements were repeated in basin I (Vörös *et al.* 1984). Production in the optimally illuminated layer was similar to that in 1973, although the production per surface area had diminished; the increased phytoplankton biomass restricted the euphotic layer to 50 cm. Production was measured simultaneously using ^{14}C and O_2 methods, which gave good agreement.

Systematic field measurements were started again in 1983. The 1982 algal bloom prompted some measurements to be taken in basins III and IV. At both stations the euphotic layer was restricted to the upper 2 m, since below that level illumination was less than 1% that at the surface. In the optimally illuminated layers of basins III and IV (Figure 8.5) production levels were four to five times higher than those found in 1972 and 1977.

In studies of the 1970s primary production, phytoplankton biomass, and light intensity were simultaneously measured at four depths. The light optima of the phytoplankton were estimated from several hundred observations using a nonlinear least square method (van Straten and Herodek 1982). In cold-water phytoplankton communities the light optimum was 8 cal/cm^{2}h, while for summer phytoplankton it was 10–20 cal/cm^{2}h, a linear increase with illumination (Figure 8.6) showing that there was definite light adaptation. Cold-water phytoplankton showed the fastest growth rates at 8°C. In summer the highest daily production/biomass ratios were found at 24–26°C. This ratio (if both production and biomass are expressed in terms of C content) attained 4.7 d^{-1} in the optimally illuminated layer in basin III, but in basin I values above 10 d^{-1} were occasionally found, showing that in Lake Balaton the phytoplankton biomass turnover is very rapid.

8.4. The Limiting Nutrient Problem

It is generally supposed that lakes in the temperate zone are basically P-limited, because the ratios of other nutrients to P are much higher in lake water than in living cells. In sewage effluent, on the other hand, the N/P ratio is lower than in organic matter, mainly due to the high P content of detergents, so that polluted lakes can change from being P- to N-limited. In such waters N-fixing blue-green algae have a special advantage and become the dominant plankton species. Beyond this reasonable simplification there are many theoretical and methodological problems concerning limiting factors. It should be realized that the term "limiting nutrient" is used in at least two different ways. One is the factor that limits the instantaneous growth rate of plankton, and the other is the nutrient that determines the maximal biomass attained during the year. At present there is no absolute method to determine either of these factors, but the problem can be approached with a combination of different techniques.

The simplest method is the chemical determination of the N/P ratio in the water. It is supposed that above a weight ratio of 15 the water is P limited, and below 7 it is N limited, and between these two values there is double limitation. According to data from VITUKI, in the period 1976–78 the average total N/total P ratio was about 20, suggesting frequent P limitation, but with large seasonal variations. In 1981 this ratio was generally lower than 15, and in basin I values below 5 were also found (Istvánovics 1982). Dissolved inorganic N compounds and orthophosphates are certainly utilized by algae, but at present chemical analysis cannot tell us what other P and N forms are biologically available; N/P ratios contain important information, but they do not solve the problem unequivocally.

Algal bottle tests are also used, in which lake water inoculated with a test alga (usually *Selenastrum capricornutum*) is enriched with different nutrients. The nutrient that causes the largest biomass increase during the test is regarded as the limiting factor (Miller *et al.* 1978). Dobolyi and Ördög (1981), using water samples from basins I and IV, found P limitation, but they refrained from generalizations on the basis of such a few measurements. The interpretation of these experiments is complicated since the water is filtered prior to the biotest to remove the original phyto- and bacterio-plankton. This filtration also retains other particulate nutrients, and their availability during such tests may differ from that under natural conditions due to the absence of bacteria. However, despite this inherent methodological problem it seems worthwhile to continue these assays with more frequent sampling of lake water.

The measurement of primary production in water samples enriched with different nutrients (Goldman 1961) has become a widely used technique in studies of the nutrient limitation of natural phytoplankton. This seemed to be an ideal method, revealing the immediate reaction of natural phytoplankton, but during preliminary experiments we obtained no immediate increase in C fixation. In accordance with the experiments of Lean and Pick (1981) the addition of orthophosphate decreased production for a time. It seems that P-deficient cells use their energy first for phosphate uptake, and that photosynthetic C-fixation increases only after the P depletion from the water, or when P attains a certain level within the cells. Therefore, after the addition of nutrients water samples must be incubated for a period before the C fixation can be measured. It is true that with this method information about instantaneous limitation is lost, but it has the advantage that it works with natural plankton.

Nutrient limitation was studied using this procedure for basins I and IV from August 1980 to August 1981 (Istvánovics 1982). Water samples of 200 cm^3 were enriched either with 30 mg/m^3 phosphate-P, with 140 mg/m^3 nitrate-N, or both. The samples and unenriched controls were incubated at ambient lake water temperature and at 4000 lux illumination. After four days, $Na_2{}^{14}CO_3$ was added to the flasks and they were further incubated for 3 h. The samples were then filtered and C fixation was determined from the radioactivity of the algae and the total carbonic acid content.

It was found that in basin IV samples production was increased in the early spring by P, in late spring and summer by N, in autumn again by P, and in winter by none of the nutrients. In basin I no such regular seasonality was found; in both spring and summer the P and N limitations alternated. In addition, in basin IV production was increased by only one nutrient at a time, whereas in basin I either P or N could increase production in the same sample. In basin IV the two nutrients given together increased the C uptake much more than the limiting nutrient alone, while in basin I this difference was not so pronounced.

In basin I blue–green algae were found that could fix N from the atmosphere, whose growth is therefore P limited, whereas, on the other hand, there are algae that are unable to fix N, whose growth is therefore N limited. This may explain why both nutrients alone increased C fixation in basin I samples. In basin IV there is usually a high nitrate concentration in early spring, and the phytoplankton is correspondingly P limited. Nitrate is exhausted from the water during diatom outbursts, resulting in N limitation in late spring. The N limitation in summer 1981 showed that even this part of the lake became overloaded with P, and in 1982 there was indeed a blue-green invasion here also.

8.5. P Metabolism

In water chemistry phosphate content is determined by measuring the blue color intensity after reaction of orthophosphate with molybdenic acid (Murphy and Riley 1962). Other P compounds must first be transformed into orthophosphate by acid destruction. In a routine analysis part of the water sample is filtered through a membrane filter of 0.45 μm pore size. From one aliquot of the filtrate the dissolved reactive P is determined directly, without previous destruction, and is usually regarded as representing the original orthophosphate content of the water. Another aliquot is digested prior to the coloring reaction with molybdenic acid to give the amount of total dissolved P. Total P is determined by applying the same treatment to the unfiltered part of the sample. Particulate P cannot be measured directly, but is calculated as the difference between total P and total dissolved P.

These P fractions were determined by L. Tóth from monthly samples taken at nine stations in 1971 (Tóth 1974). The total P concentration changes along the lake according to a gradient similar to those of phytoplankton biomass and production, but the slope of the P curve is somewhat flatter. In 1976–78 the mean total P concentrations were 78, 54, 38, and 29 mg/m^3 in basins I–IV, respectively. The total dissolved P amounts to about half of the total P. The dissolved reactive P level is low, with mean values of 4–6 mg/m^3.

Dissolved unreactive P, calculated by subtracting dissolved reactive P from total dissolved P, is generally believed to represent dissolved organic P compounds, although Dobolyi (1980) found that in ultraviolet (u.v.)-treated filtered Balaton water a large part of the dissolved P remained unreactive. U.v. treatment is believed to oxidize all organic compounds, so that the unreactive P must have been present in inorganic form, i.e., in condensed phosphates. In the central lake basin these condensed phosphates amounted to 2–67 mg/m^3, except in winter, when under the ice the concentration was only 0.2 mg/m^3. To examine this matter further, condensed phosphates were then identified by gel chromatography. First, bivalent cations were removed by passing the water through a cation exchanger resin to prevent lime precipitation, which could have caused P removal during the subsequent concentration of the samples. The water was then u.v.-treated to oxidize all organic compounds, and the solution was concentrated by freeze-drying. Chromatographic examination of such samples on Sephadex-25 fine columns revealed P compounds up to a condensation degree of 10. The condensed phosphate contents of the lake water, of the Zala River and a smaller tributary, the Tetves creek, of the effluent of a sewage treatment plant, and of a photooxidized algal suspension were analyzed in the same way. The elution curve of the lake water sample resembled that of the photooxidized algal suspension, while higher polyphosphates were absent from the inflowing waters (Figure 8.7), suggesting that the polyphosphates in the lake are produced by phytoplankton. Algae are known to contain significant polyphosphate reserves (Rhee 1973), and it may be possible that these are released into the water during cell degradation. It is known that algae, and especially bacteria, can utilize dissolved polyphosphates, but the rate of such processes under natural conditions is not known. A better understanding of the rôle of polyphosphates awaits further kinetic studies.

In the P metabolism of lakes orthophosphate is of great importance. It was suggested by Oláh *et al.* (1977) that orthophosphate levels in Lake Balaton are controlled primarily by adsorption in suspended sediment or in freshly formed biogenic lime. On the other hand, experiments by Dobolyi and Herodek (1980) indicate that active uptake by algae determines the orthophosphate level in the lake. First a plexiglass box, open at the bottom, was placed in shallow water (the water within this enclosure was stirred by an electrically driven paddle). By adding phosphate the P orthophosphate concentration was raised from the original 2.4–25 mg/m^3 and the dissolved reactive phosphate concentration was then measured at various times. After 1 h it dropped to the original 2.5 mg/m^3, but if the algae were killed by NaOCl, the phosphate-P concentration remained constant at 25 mg/m^3, confirming that P elimination is connected with life processes. This result did not, however, exclude the possibility that P is removed by biogenic lime precipitation induced by algal photosynthesis. In the third experiment, therefore, EDTA-Na_2 was added to the water in the box to keep Ca in solution in a complex form. It was found also that in this experiment, where biogenic lime formation was prevented, the added phosphate disappeared at the same rate as in the parallel experiment without added EDTA-Na_2. Thus the algae themselves, rather than the suspended sediment or biogenic lime, appear primarily to take up the phosphate from the water.

The same conclusion has been drawn from *in vitro* experiments. Carrier-free $H_3{}^{32}PO_4$ was added to water samples, which were filtered through membrane filters of 0.45 μm pore size after 30 and 60 min. The radioactivity of the filtrate was measured directly to determine the amount of orthophosphate remaining in solution after incubation. The membrane filters with the retained particulate material were photooxidized, resuspended in fresh water, and filtered again. The radioactivity remaining on the filter after this u.v. treatment represents the P incorporated into inorganic material, while the fraction taken into solution corresponds to the P utilized by algae. After 30 min 72% of the added labeled orthophosphate was incorporated into the particulate organic fraction, and only 3% into particulate inorganic matter (Figure 8.8).

The turnover time of the orthophosphate was studied by an isotope technique applied to water samples taken from basins I and IV in 1980–81 (Istvánovics 1982). Carrier-free $H_3{}^{32}PO_4$ was added to water samples of 200 cm^3, and at different times (usually after 2, 4, 8, 16, and 32 min) 5-cm^3 subsamples were passed through membrane filters and the radioactivity in the

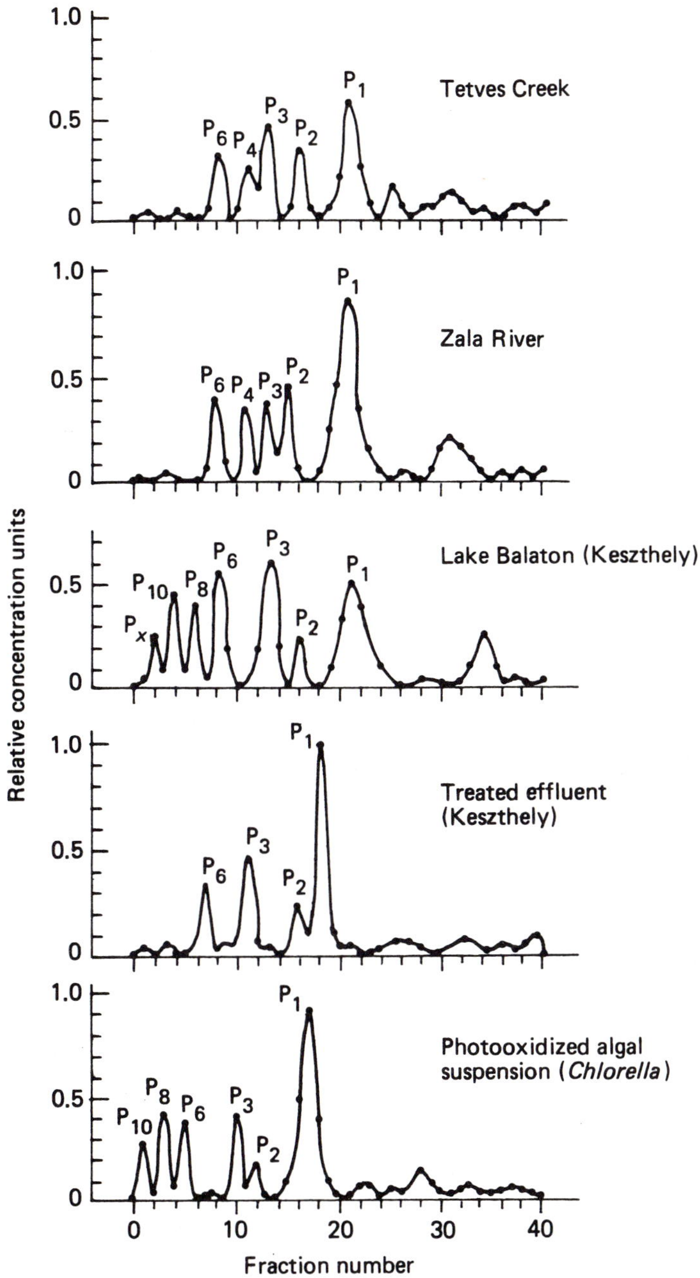

Figure 8.7. Condensed phosphate concentrations in Tetves Creek, in the Zala River, in the lake, in treated sewage effluent, and in a photooxidized algal suspension (Dobolyi 1980).

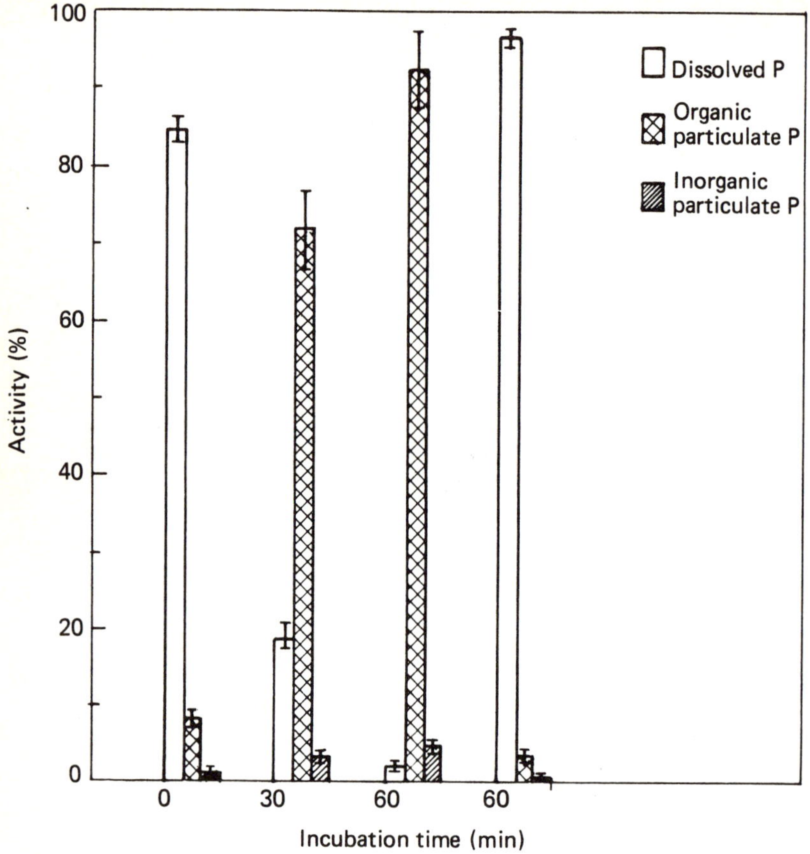

Figure 8.8. Distribution of radioactivity among various forms of P after *in vivo* incubation of lake water samples with $H_3{}^{32}PO_4$, without (0, 30, and 60 left) and with (60 right) suppression of algal metabolism (Dobolyi and Herodek 1980).

filtrate measured. Radioactivity decreased rapidly over time, and then became constant (Figure 8.9). This uptake behavior can be described by a simple two-compartment model (Rigler 1968) which predicts isotope equilibrium between particulate and dissolved phases and first-order kinetics for the decrease in activity in the filtrate:

$$\ln \frac{I_t - I_a}{I_0 - I_a} = -kt \quad ,$$

where I_0 is the percentage isotope activity of the filtrate at $t = 0$, I_t is the activity at any given time, I_a is that in the isotope equilibrium, and k is a kinetic constant. The reciprocal of the absolute value of k is the orthophosphate turnover time, which proved to be very short in both basins and changed seasonally. Under ice turnover times were relatively long (100 min in basin I, and 401 min in basin IV), but in early spring values below 10 min were

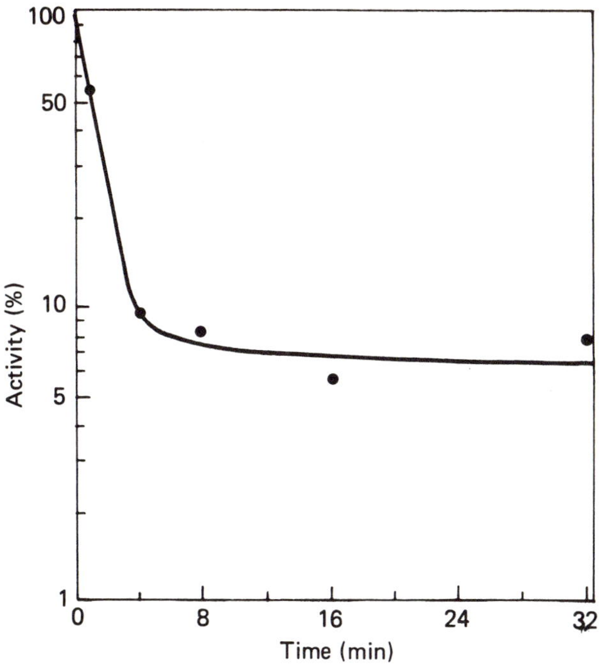

Figure 8.9. Decrease in radioactivity in solution after addition of $^{32}PO_4^{3-}$ to a sample of lake water.

found at both stations. In late spring there was a second maximum (335 min in basin I and 97 min in basin IV); in summer the turnover times were 1–4 min in basin I, and 4–8 min in basin IV, while in October they rose again. The uptake rate is found by dividing the orthophosphate concentration by the turnover time. If we calculate uptake rates from the observed turnover times, unrealistically high values are obtained that indicate that the algae would take up more P than C, which is not very likely. However, it was demonstrated by Rigler (1966) that the molybdate technique overestimates orthophosphate, especially at low concentrations. He added varying amounts of unlabeled phosphoric acid and carrier-free labeled phosphoric acid to lake water samples, measured the phosphate turnover times, and calculated the uptake rate as the sum of the original and added phosphate per turnover time. The curves showed decreasing uptake with increasing concentrations, which contradicts all biological expectations. Rigler obtained Michaelis–Menten type curves only by assuming that the original phosphate concentration in the water was much lower than that measured with the molybdate technique. This method cannot give us real values, but it can prove that the measured ones are overestimates. Similar experiments with Lake Balaton water showed that in most cases the real concentration must be much lower than the chemically determined one. To evaluate these results the turnover times were plotted as a function of the amount of added phosphate. According to the

Michaelis–Menten formula

$$v = v_{max} \frac{S_0 + S_a}{K_s + S_0 + S_a} ,$$

where

v = uptake rate
v_{max} = uptake rate at substrate saturation
S_0 = original substrate concentration
S_a = concentration of the added substrate
K_s = half-saturation constant.

Since the uptake velocity is defined as

$$v = \frac{S_0 + S_a}{T} ,$$

where T is the turnover time of the substrate in the water,

$$T = \frac{S_a}{v_{max}} + \frac{K_s + S_0}{v_{max}} .$$

The straight lines obtained proved that phosphate uptake follows Michaelis–Menten kinetics. From such graphs (Figure 8.10) we can determine the sum of S_0 and K_s and the value of v_{max}. Unfortunately, even with this technique we cannot obtain the original orthophosphate concentration, but it must be lower than the $S_0 + K_s$ value. In 10 of the 15 samples studied, $S_0 + K$ proved to be lower than the S_0 measured by the molybdate method.

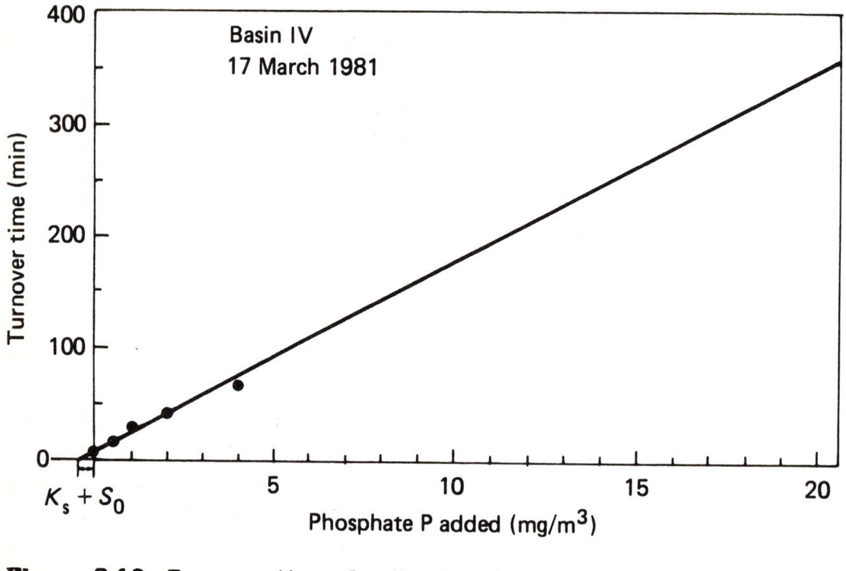

Figure 8.10. Turnover time of orthophosphate in lake water samples as a function of the added phosphate P.

The $S_0 + K_s$ values are higher in the hypertrophic basin I than in basin IV. It is very likely that S_0 and K_s are positively correlated due to the adaptation of algae to external nutrient concentrations, and in this case the orthophosphate level is usually higher in basin I than in basin IV.

The maximal uptake rates (actually attained only at nutrient saturation) are also much higher in basin I than in basin IV. These values are highest in late summer during the blooms of blue–green algae when the biomass is also maximal, and N-fixation renders the system P-deficient.

8.6. N Metabolism

In addition to P, forms of N were also analyzed by Tóth in samples collected monthly at the nine stations along the lake from 1975 on. N forms were studied in the course of N fixation (Oláh *et al.* 1981) and limiting nutrient studies (Istvánovics 1982).

In 1976–78 the mean total N concentrations were 1.4, 1.2, 0.9, and 0.7 g/m^3 in basins I–IV, respectively. The total dissolved N amounts to about two thirds of the total N. The ammonia-N concentration in basin I sometimes reaches 50 mg/m^3,and shows strong seasonal and diurnal changes, while in basin IV it is at the limit of detectability. The nitrite concentration is only a few mg/m^3. Among the inorganic N compounds, nitrate has the highest concentrations, but it varies seasonally, with the highest values in early spring after ice break, when the nitrate concentration reaches several hundred mg/m^3. This reserve is then utilized in the diatom outburst. In summer the nitrate concentration remains low, but in autumn it increases again. For ammonia-N and nitrate uptake by phytoplankton we have no data at present, although experiments with ^{15}N-labeled compounds are now under way.

N fixation, on the other hand, was studied in detail from 1977 to 1980 (Oláh *et al.* 1981) using the acetylene reduction technique. This method, introduced into limnology by Stewart *et al.* (1967), is based on the ability of the nitrogenase enzyme, which is responsible for N fixation, to reduce acetylene to ethylene. This method is more sensitive and, above all, much cheaper and less time-consuming than direct measurement with ^{15}N, and has therefore become widely used.

Acetylene is injected into bottles containing 150 cm^3 of lake water, and after thorough shaking the bottles are suspended in the lake at the depths from which the water was collected. After 3 h *in situ* exposure the bottles are again shaken vigorously. A gas sample was analyzed by a Pye Unicam GCV gas chromatograph with a glass column 2.1 m long and 4 mm in diameter, filled with Poropack-N. The temperature was 80°C, the carrier gas was N_2, with a flow rate of 40 cm^3/min. A flame ionization detector was used.

N fixation was found to be very high in basin I, in basin II it was one order and in basins III and IV two orders of magnitude lower than in basin I. These values corresponded well with the distribution of the blue–green algae at the time of the experiments. In later years, blue–green algae appeared in all parts of the lake (for example, summer 1982), and it is likely that the horizontal distribution of N fixation has changed accordingly since then. The seasonality of N fixation corresponds equally well to the occurrence of blue–green algae. It is low in June, increases in July, has its maximum in August when the blue–green algal population is in the log phase, in September it is still high, and then it drops rapidly in parallel with the destruction of the blue–green algae.

N fixation is usually highest in the morning, but a second peak sometimes appears in the late afternoon. Activity during the night is also appreciable, i.e., the algae can use energy from dark respiration in addition to that from sunlight. In the well-illuminated layer N fixation is usually greater than near the bottom, and these vertical differences are more pronounced in August–September (the time of the maximal activity) than in June. From measurements taken between 1977 and 1980, N fixation in the summer in basin I is estimated at 7–13 g N/m^2, whereas in winter it is much lower.

N fixation in sediment has also been measured (Oláh *et al.* 1981). Intact cores were taken in tubes 5 cm in diameter, and acetylene was injected into the sediment through perforations in the tube walls. After incubation the sediment was shaken, and the gases collected and analyzed as in the experiments with lake water. According to these studies N fixation in the sediment of basin I is about 4 g N/m^{2}yr, which is higher than in most other lakes. In sediment the N is fixed by bacteria.

The Zala River transports about 1000–1500 tons N/yr into basin I (Joó 1980; Chapter 14 in this volume), corresponding to an areal load of about 35 g N/m^{2}yr. Thus, on an annual basis, two thirds of the total N input to basin I is external in origin, but in the critical summer period N fixation is the main source of N.

8.7. Conclusions

The biomass, cell number, Chl-a concentration, and primary production of phytoplankton has increased in all basins of Lake Balaton, but eutrophication has been more rapid in western basins. It seems reasonable to model the lake as a sequence of four basins with different inputs, and to connect them with a hydrological submodel.

Distinct phytoplankton communities are dominant in different seasons, so that the use of more phytoplankton compartments with different parameters (maximum growth rate, light and temperature optima, half saturation constants for nutrients, etc.) appears appropriate.

Both primary production and phytoplankton biomass have maxima in spring, and again in summer. The spring biomass peak is higher in the less eutrophicated areas, while the summer peak is higher in hypertrophic areas. Primary production is higher in all parts of the lake in summer. Models should be expected to simulate these dynamics. Production/biomass ratios are high in Lake Balaton, so that high maximal growth rates should be used in models.

Lakes in their original states are usually P limited in the temperate zone, but they may become N limited due to increasing P loads. The total N/P ratios for Lake Balaton suggest a transient state. Biotests with natural phytoplankton indicate that in some periods N is the limiting nutrient, while in others it is P. In order to describe in detail the seasonal dynamics of phytoplankton, both P and N should be included in models.

In deep lakes there is a spatial separation between regions of maximum algal photosynthesis and maximum orthophosphate concentrations during most of the year, whereas in shallow lakes, such as Balaton, phosphate and algae are permanently mixed in the same water layer, and orthophosphate cannot accumulate in water in larger amounts. Experiments suggest that orthophosphate levels in the lake are controlled by microorganisms that compete for the limiting nutrients rather than by absorption by inorganic material.

Lake models are generally calibrated to orthophosphate concentrations, determined by the molybdate technique, although isotope studies demonstrate that chemical measurements overestimate real orthophosphate concentrations, particularly at low levels. The use of a half saturation constant appreciably lower than that used in other models may be appropriate.

Intensive exchange has been found between dissolved orthophosphate and particulate organic P. Gross uptake by microorganisms may be several times higher than net uptake. Little is known about other dissolved P compounds. Polyphosphates are present in significant concentrations in Lake Balaton water, probably released from the phytoplankton by metabolism or decay.

Among N compounds, nitrate shows a regular annual cycle. Its concentration is about 200–300 mg/m^3 at ice break, and this drops to 10 mg/m^3 during the spring diatom outburst. It is suggested that N limitation halts the mass development of phytoplankton in April.

The appearance of N-fixing blue–green algae in summer also indicates N deficiency. N fixation usually does not greatly influence the annual N budget of the lake, but in summer in the southwestern part of Lake Balaton it is the main source of N. These algae, which represent a serious nuisance in many lakes, should be incorporated into eutrophication models as a separate group.

References

Böszörményi, Z., Cseh, E., Felföldy, L., and Szabó, E. (1962) Some methodical questions on photosynthetic measurements with a radioactive tracer ^{14}C technique in Lake Balaton. *Ann. Biol. Tihany* 29:39–63 (in Hungarian with English summary).

Dobolyi, E. (1980) Identification of different phosphorus forms and their role in the eutrophication process of Lake Balaton, in J. Barica and L.R. Mur (Eds.) *Hypertrophic Ecosystems* (The Hague: Junk) pp 13–22.

Dobolyi, E. and Herodek, S. (1980) On the mechanism reducing the phosphate concentration in the water of Lake Balaton. *Int. Rev. Hydrobiol.* 65:339–43.

Dobolyi, E. and Ördög, V. (1981) Determination of the limiting plant nutrient in the water of Lake Balaton by algal assay procedure. *Int. Rev. Hydrobiol.* 66:810–5.

Goldman, C.R. (1961) Primary productivity and limiting factors in Brooks Lake, Alaska. *Verh. Int. Verein Limnol.* 14:120–4.

Herodek, S. and Tamás, G. (1975a) Phytoplankton production in Lake Balaton. *Symp. Biol. Hung.* 15:29–34.

Herodek, S. and Tamás, G. (1975b) The primary production of phytoplankton in the Keszthely basin of Lake Balaton in 1973–1974. *Ann. Biol. Tihany* 42:175–190.

Herodek, S. and Tamás, G. (1976) The mass and production of phytoplankton and eutrophication in Lake Balaton. *Hidrol. Közl.* 56:219–28 (in Hungarian with English summary).

Herodek, S. and Tamás, G. (1978) The mass and production of phytoplankton and eutrophication in Lake Balaton II: The Szigliget basin in 1974–1975. *Hidrol. Közl.* 58:384–91 (in Hungarian with English summary).

Herodek, S., Vörös, L., and Tóth, F. (1982) The mass and production of phytoplankton and eutrophication in Lake Balaton III: The Balatonszemes basin in 1976–1977 and the Siófok basin in 1977. *Hidrol. Közl.* 62:220–9 (in Hungarian with English summary).

Istvánovics, V. (1982) *Phosphorus Uptake and Limiting Nutrients of Phytoplankton in Lake Balaton.* Doctoral Thesis (University of Natural Sciences, Budapest) (in Hungarian).

Joó, O. (1980) Data for the eutrophication of Lake Balaton and considerations related to control activities, in G. van Straten, S. Herodek, J. Fischer, and I. Kovács (Eds.) *Proc. 2nd Joint MTA/IIASA Task Force Meeting on Lake Balaton Modeling* Vol. II (Veszprém: MTA VEAB), pp 139–66.

Lean, D.R.S. and Pick, F.R. (1981) Photosynthetic response of lake plankton to nutrient enrichment: A test for nutrient limitation. *Limnol. Oceanogr.* 26:1001–19.

Miller, W.E., Greene, J.C., and Shiroyama, T. (1978) *The* Selenastrum capricornutum Printz *algal assay bottle test.* Report No. EPA-600/9-78-018 (Washington, DC: EPA).

Murphy, J. and Riley, J.P. (1962) A modified single solution method for the determination of phosphate in natural waters. *Analytica Chim. Acta* 27:31–6.

Oláh, J., Tóth, E.O., and Tóth, L. (1977) Phosphorus metabolism in Lake Balaton. *MTA Biol. Oszt. Közl.* 20:111–39.

Oláh, J., El Samra, M.I., Moneim, M.A.A., Tóth, L., and Vörös, L. (1981) Nitrogen fixation in fish producing agroecosystems. *A Halhústermelés Fejlesztése* 10:1–208 (in Hungarian with English summary).

Rhee, G.Y. (1973) A continuous culture study of phosphate uptake, growth rate and polyphosphate in *Scenedesmus* sp. *J. Phycol.* 9:495–506.

Rigler, F.H. (1966) Radiobiological analysis of inorganic phosphorus in lake water. *Verh. Int. Verein. Limnol.* 16:465–70.

Rigler, F.H. (1968) Further observations inconsistent with the hypothesis that molybdenum blue method measures orthophosphate in lake water. *Limnol. Oceanogr.* 13:7–13.

Rodhe, W. (1969) Crystallization of eutrophication concepts in Northern Europe, in *Eutrophication: Causes, Consequences, Correctives* (Washington, DC: National Academy of Sciences), pp 50–4.

Sebestyén, O. (1953) Quantitative plankton studies in Lake Balaton II: Decennial changes. *Ann. Biol. Tihany* 21:63–89 (in Hungarian with English summary).

Sebestyén, O. (1954) Quantitative plankton studies in Lake Balaton III: The biomass of the pelagic *Dinoflagellatae*. *Ann. Biol. Tihany* 22:185–97 (in Hungarian with English summary).

Stewart, W.D.P., Fitzgerald, G.P., and Burris, R.H. (1967) *In situ* studies on N_2 fixation using the acetylene reduction technique. *Proc. Nat. Acad. Sci. USA* 58:2071–8.

Tamás, G. (1955) Quantitative plankton studies on Lake Balaton VI: Biomass of phytoplankton in the 1940s. *Ann. Biol. Tihany* 23:95–110 (in Hungarian with English summary).

Tamás, G. (1974) Biomass changes of phytoplankton in Lake Balaton during the 1960s. *Ann. Biol. Tihany* 41:323–42.

Tamás, G. (1975) Horizontally occurring quantitative phytoplankton investigations in Lake Balaton, 1974. *Ann. Biol. Tihany* 42:219–79.

Tóth, L. (1974) Examination of the water quality of Lake Balaton. *VITUKI Report No.* 7782/4/8 (Budapest: VITUKI), p 48 (in Hungarian).

van Straten, G. and Herodek, S. (1982) Estimation of algal growth parameters from vertical primary production profiles. *Ecological Modeling* 15:287–311.

Vollenweider, R.A. and Kerekes, J.J. (1980) *OECD Cooperative Program on Monitoring Inland Waters (Eutrophication Control).* Synthesis Report (Paris: OECD).

Vörös, L. (1980) The biomass, composition and diversity of phytoplankton in Lake Balaton in 1976. *Botan. Közl.* 67:25–33 (in Hungarian).

Vörös, L. (1982) Quantitative and structural changes of phytoplankton in Lake Balaton between 1965–1978. *Aquacultura Hungarica* 3:137–44 (in Hungarian).

Vörös, L. (1984) The biomass of phytoplankton as a water quality index. *Hidrol. Közl.* (in Hungarian with English summary).

Vörös, L. and Kiss, N. (1985) The seasonal periodicity of phytoplankton and its relationship to eutrophication, in G. Fekete (Ed) *Problems of Cenological Succession* (Budapest: Akadémiai Kiadó), pp 121–34 (in Hungarian).

Vörös, L., Vízkelety, É., Tóth, F., and Németh, J. (1984) Trophic conditions in the Keszthely basin in 1979. *Hidrol. Közl.* (in Hungarian with English summary).

CHAPTER 9

Wind-Induced Water Motion

P. Shanahan, D.R.F. Harleman, and L. Somlyódy

9.1. Mathematical Modeling of Lake Circulation

The circulation of water within a lake or reservoir is an important determinant of the lake's water quality behavior. The two major classes of motion, horizontal and vertical, significantly influence mass transport and thus water quality. Horizontal circulations are caused by the travel of water between the inflow and the outflow of the lake and the force of wind upon the water surface. Vertical circulations interact with the differences in water density in a stratified lake and are produced when various agents disrupt the normally stable stratification. Turbulence due to wind, inflows of high- or low-density water, and heating or cooling at the water surface are typical agents.

Excluded from this chapter are a number of types of water motion that do not involve the large-scale travel of water masses, such as surface waves. The discussion is general, however, in the sense that models appropriate to both deep and shallow lakes are presented.

Mathematical formulation of lake circulation

The equations of fluid motion in a lake are the departure point from which all mathematical circulation models must begin. These include the equation of conservation of mass (or the continuity equation), and the equation of conservation of momentum in each of the three coordinate directions. For an incompressible fluid, the continuity equation is:

$$\frac{\partial u}{\partial x} + \frac{\partial v}{\partial y} + \frac{\partial w}{\partial z} = 0 \quad , \tag{9.1}$$

where x and y are the horizontal components, as shown in Figure 9.1, z is the vertical component, measured downwards from the mean water surface

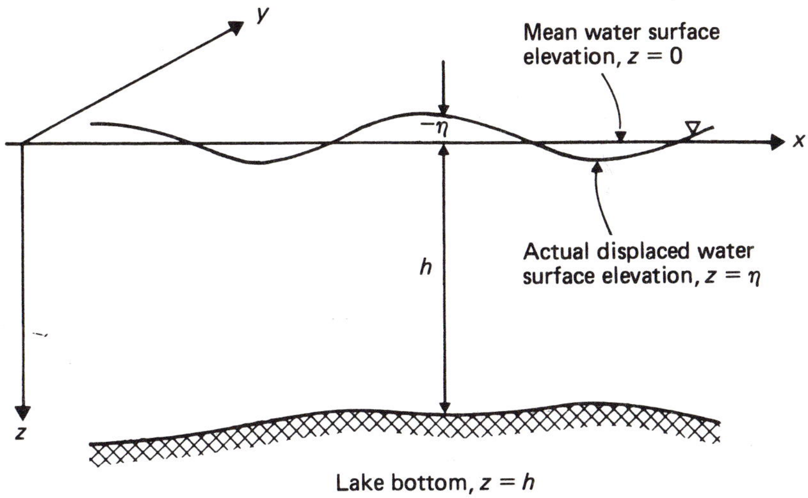

Figure 9.1. Definition sketch for mathematical formulation.

elevation, and u, v, w are the fluid velocity components in the x, y, and z directions, respectively. The momentum equations express the acceleration of the fluid resulting from various forces. The equations in the horizontal plane are given as equations (9.2) and (9.3). For clarity, the correspondence between the terms in the equations and the physical accelerations and forces they represent are also given:

$$\underset{(1)}{\frac{\partial u}{\partial t}} + \underset{(2)}{\left\{u \frac{\partial u}{\partial x} + v \frac{\partial u}{\partial y} + w \frac{\partial u}{\partial z}\right\}} = \underset{(3)}{fv} \underset{(4)}{- \frac{1}{\rho} \frac{\partial p}{\partial x}} + \underset{(5)}{A_H \left\{\frac{\partial^2 u}{\partial x^2} + \frac{\partial^2 u}{\partial y^2}\right\}} + \underset{(6)}{\frac{\partial}{\partial z}\left[A_V \frac{\partial u}{\partial z}\right]}, \tag{9.2}$$

$$\underset{(1)}{\frac{\partial v}{\partial t}} + \underset{(2)}{\left\{u \frac{\partial v}{\partial x} + v \frac{\partial v}{\partial y} + w \frac{\partial v}{\partial z}\right\}} = \underset{(3)}{-fu} \underset{(4)}{- \frac{1}{\rho} \frac{\partial p}{\partial y}} + \underset{(5)}{A_H \left\{\frac{\partial^2 v}{\partial x^2} + \frac{\partial^2 v}{\partial x^2}\right\}} + \underset{(6)}{\frac{\partial}{\partial z}\left[A_V \frac{\partial v}{\partial z}\right]}, \tag{9.3}$$

where t is time, f is the Coriolis parameter, p is the fluid pressure, ρ is the fluid density, A_H is the horizontal eddy viscosity, and A_V is the vertical eddy viscosity. The terms of these equations have the following meanings:

(1) The instantaneous or local acceleration of the fluid at a point.
(2) The convective acceleration, caused when fluid is transported from one point to another of different fluid velocity.
(3) The Coriolis force due to the Earth's rotation.
(4) The horizontal pressure force.
(5) The horizontal transport of momentum due to shear stresses.
(6) The vertical transport of momentum due to shear stresses.

Equations (9.2) and (9.3) incorporate the assumption that the turbulent momentum flux due to the Reynolds shear stresses can be represented by the product of an eddy viscosity and the first spatial derivative of the velocity. This problem of turbulence closure can also be solved by other means, including complete submodels for turbulent energy transport based on turbulence theory (see e.g., Vreugdenhil 1973). Unfortunately, these more sophisticated models are far more complicated and expensive for simulation. Further, turbulence models remain somewhat experimental and are less thoroughly proved for practical problems. Thus the great majority of circulation models employ the eddy viscosity assumption.

The momentum equation in the vertical direction is entirely similar to those above, but includes an additional term on the right-hand side to represent the force due to gravitational acceleration, g. The equation is considerably simplified by the realization that the pressure and gravitational forces dominate all others. Neglect of the lesser terms is known as the hydrostatic approximation, and leads to equation (9.4):

$$\frac{1}{\rho}\frac{\partial p}{\partial z} = -g \quad . \tag{9.4}$$

The boundary conditions for these equations are specified at the free surface, the lake bottom, and the lake shore. At the free surface, the kinematic boundary condition specifies that continuity be maintained:

$$-\frac{\partial \eta}{\partial t} - u\frac{\partial \eta}{\partial x} - v\frac{\partial \eta}{\partial y} = w \qquad \text{at } z = -\eta \tag{9.5}$$

where $-\eta$ is the free surface displacement. An additional condition at the free surface represents the shear stress due to the wind:

$$-\rho A_\mathrm{v}\frac{\partial u}{\partial z} = \tau_\mathrm{s}^x \quad \text{and} \quad -\rho A_\mathrm{v}\frac{\partial v}{\partial z} = \tau_\mathrm{s}^y \qquad \text{at } z = -\eta \quad , \tag{9.6}$$

where τ_s^x is the x-component of the shear stress on the surface, and τ_s^y is the y-component.

At the lake bottom, a no-slip boundary condition specifies that the fluid in direct contact with the rough bottom cannot move:

$$u = v = 0 \qquad \text{at } z = h \quad , \tag{9.7}$$

where h is the lake depth. Alternatively, a shear stress condition similar to that at the surface may be specified instead:

$$-\rho A_\mathrm{v}\frac{\partial u}{\partial z} = \tau_\mathrm{b}^x \quad \text{and} \quad -\rho A_\mathrm{v}\frac{\partial v}{\partial z} = \tau_\mathrm{b}^y \qquad \text{at } z = h \quad , \tag{9.8}$$

where τ_b^x is the x-component of the bottom shear stress, and τ_b^y is the y-component.

At the lake perimeter, a no-flow, no-slip boundary condition applies:

$$u = v = 0 \quad \text{at the } x \text{ and } y \text{ boundaries} \quad . \tag{9.9}$$

These equations and boundary conditions are complex and their solution is difficult, the nonlinear convective terms and boundary conditions being particularly troublesome. As a consequence, most solution methods depend upon simplification by averaging to reduce the problem dimensions, or by neglecting the less important terms in the equations. Such simplifications must be justified compromises of the physics of the lake and the mathematics of the equations. Assumptions made solely to simplify solutions may be unfounded and may lead to erroneous predictions.

Alternative model formulations

The choice of the lake circulation model is determined by its intended use and the physical characteristics of the water body to be modeled. Here we consider the range of available models as classified by spatial representation. Two general types of models may be defined: circulation models, which simulate two- or three-dimensional flow, and simplified models with fewer dimensions. The following discussion briefly describes the possibilities within these two broad classes, with reference to Figure 9.2.

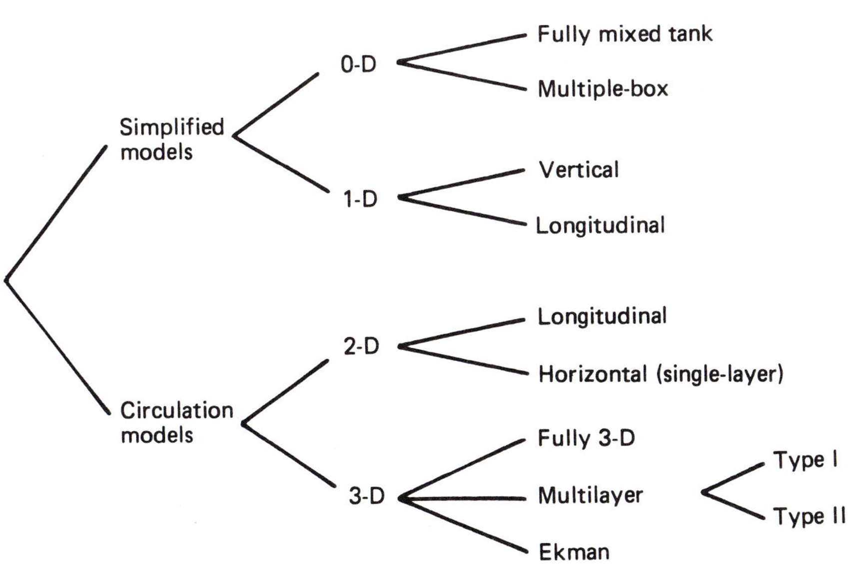

Figure 9.2. Lake model alternatives.

Simplified Models

The most simplified model is the zero-dimensional, in which the lake is assumed to act as if it were entirely homogeneouos. This representation is referred to as the fully mixed tank or continuous-flow stirred tank reactor. A variant on the fully mixed tank is the multiple-box model in which different lake sectors are represented as connected fully mixed tanks. Zero-dimensional models satisfy equation (9.1), the equation of mass conservation, but make no attempt to satisfy conservation of momentum, equations (9.2), (9.3), and (9.4). Thus, they cannot be considered to be true circulation or hydrodynamic models.

One-dimensional (1-D) models, although not generally called circulation models, satisfy the broad definition used in this chapter. Two major types of 1-D models exist: vertical and longitudinal (Brown 1978). Vertical 1-D models consider the lake to be horizontally homogeneous, but with a distinct vertical density structure. These are successful in modeling temperature changes in deep stratified lakes, but are not appropriate for shallow lakes such as Balaton.

Longitudinal 1-D models are less commonly applied to lakes than their vertical counterparts owing to their narrower applicability, but they have frequently been applied to rivers and estuaries. These models are applicable to long, narrow lakes in which vertical variations due to stratification are negligible. The majority of lakes which lend themselves to these models are characterized by large throughflows, to the point that Brown (1978) describes them as essentially sluggish rivers. Another class of lakes for which these models may be useful are shallow lakes which are long and narrow, without necessarily large throughflows. In these applications, 1-D models include most of the characteristics of circulation models. In particular, wind and bottom friction forces are considered.

Circulation Models

Two- and three-dimensional models are generally classified as true circulation models that consider the forces of wind and bottom friction, as well as the influence of inflows and outflows, to predict the motion within the lake. In reviewing circulation models, we draw upon our own search of the literature as well as published reviews by Cheng *et al.* (1976), Lindijer (1976, 1979), and Simons (1979) to outline the major classes of models. Three classes are defined by Cheng *et al.* (1976): single-layer, multilayer, and Ekman-type models. To this group we add some less common variations.

Single-layer models proceed from the assumption that the lake is vertically homogeneous (unstratified) to eliminate consideration of vertical variations in currents and other parameters. The vertical variation is removed by integrating the continuity and momentum equations from the free surface to the lake bottom, reducing the 3-D problem to one of only two dimensions. The integration process transforms the problem variables from velocities to horizontal mass transports, defined as:

$$q_x = \int_h^{-\eta} u\,dz \qquad q_y = \int_h^{-\eta} v\,dz \quad . \tag{9.10}$$

The integration also incorporates the surface and bottom boundary conditions into the resultant equations.

Single-layer models simulate mass flux and free surface motion well, but omit all detail concerning the vertical circulation structure. Their major use has been in storm surge studies in both deep and shallow lakes. Their applicability is more general in shallow lakes, where the assumption of vertical homogeneity holds well.

Multilayer models extend the single-layer methodology to stratified water bodies. Basically, the process applied to the entire water column in the single-layer models is applied piecewise to a number of layers through the lake. A different density may exist in each layer, and the vertical eddy viscosity may vary from layer to layer as well. The equations of continuity and momentum are vertically integrated over the depth of each layer, incorporating the free surface boundary condition into the top layer equation, and the bottom condition into the equation for the lowest layer. Interlayer conditions must also be specified, and become part of the layer equations as well. The final result of this procedure is a series of equations that represent the motion within each individual layer, with the layers coupled via the interlayer conditions.

There are two approaches to the construction of the layers. In the type I approach, the position of the layers is fixed in space and vertical transports occur between layers to maintain continuity. These transports also transfer momentum between the layers. In type II models, the layers are considered distinct, as if separated by thin membranes. No mass transport occurs between the layers; rather, the layers displace vertically to maintain continuity. The layers communicate via momentum transport due to interfacial stress.

Multilayer models correct the deficiencies of single-layer models, and type I models especially predict both free surface elevation and current well. The type II approach is less common than type I, and is most appropriate for distinctly stratified lakes with a clearly developed thermocline.

Ekman-type models simplify the equations of motion considerably more than the methods above. Based on the assumption that inertial forces are much less than the Coriolis force, the horizontal momentum equations are linearized by dropping the convective acceleration terms. Also, a linear friction law is assumed for bottom friction. These important simplifications permit the form of the vertical distribution of the horizontal velocity to be determined analytically. Once the form of the vertical structure is known, completion of the solution requires only that the variation in horizontal space be defined. This information is supplied by the solution of the vertically integrated conservation equations. The Ekman-type model solution specifies the 3-D variation of the horizontal currents only. The smaller vertical velocity component is not determined.

The Ekman-type solution, owing to the simplification of the equations, is the easiest method for computation, but the assumptions made reduce the model's applicability and require that the model's suitability be evaluated for each application. The model remains useful for a wide range of lakes nevertheless, having been applied to Lakes Superior (Lien and Hoopes 1978), Ontario (Bonham-Carter and Thomas 1973), Erie (Gedney and Lick 1972), Okeechobee (Su *et al.* 1976), and Velen (Bengtsson 1973).

A rarer type of 2-D model is that which computes motion in a vertical plane along the lake (e.g., Edinger and Buchak 1979). These models, which are used for long, deep but relatively narrow lakes, ignore lateral flow variations. Invariably, lakes that satisfy these criteria are impounded streams whose hydrodynamics are dominated at the upper end by inflow currents and at the lower end by temperature stratification. Such reservoirs typically exhibit a distinct 2-D temperature structure characterized by tilted isotherms. Models of this type combine the features of the two versions of 1-D model discussed above.

The final, and clearly most complex, modeling alternative is the fully 3-D model, which attempts to determine the lake's flow field in its full complexity, often with simultaneous consideration of the vertical density structure. Fully 3-D models are generally similar to 2-D circulation models, except that the vertical momentum equation is retained and the horizontal equations are not vertically integrated. Such models are not common, however, owing to the complexity and expense inherent in a 3-D grid. Thus, while examples do exist (e.g., Liggett 1970), the general state of the art for these models is not advanced to the point of practical application.

9.2. Application to Lake Balaton

Circulation in Lake Balaton

Although the existing data base is far from complete, sufficient information exists to construct an approximate description of Lake Balaton's circulation. The circulation is a composite produced by hydrologic flow through the lake, wind-induced currents, seiching, and lesser influences. The major unknown aspects of water motion in the lake are the character of spatial variations in the current over both horizontal and vertical axes, and the response of the circulation to changes in wind force over short time periods.

Hydrologic Flow

We define hydrologic flow as that produced by the inflow of water to the lake from streams, runoff, and rainfall, and the outflow from the lake by evaporation and stream discharge. Various components of hydrologic flow are discussed in Chapter 1. The longitudinal transport velocities (directed from west to east) associated with these hydrologic flows are small, of the order of 0.05 cm/s.

Wind-driven Flows

The influence of the wind overwhelms the slow hydrologic flow in establishing the instantaneous pattern of flow in Lake Balaton. The shallowness of the lake permits a circulation response to even mild winds, producing currents one or two orders of magnitude greater than those due to the hydrologic flow.

Surrounding hills, and the geography of the lake itself, exert a major influence upon the circulation caused by the wind. The hills produce local effects by blocking and deflecting the wind, leading to a spatially nonuniform wind field. The circulation is further modified by the constraints imposed by the lake boundaries. Although comprehensive field observations of the lake circulation have not been made, a rough picture of the circulation is found in the work of Győrke (1975), who used a physical hydraulic model of the western part of the lake to model circulation and sediment transport under artificially steady winds. The model physical scales were distorted by a factor of ten: the vertical scale was 1:100 while the horizontal scale was 1:1000. To capture the effect of the nonuniform wind field over the lake, hills were also modeled and a battery of fans were arranged to blow in a nonuniform pattern on the model. The model results show a complex system of flow gyres greatly influenced by the lake geometry and spatial variation of the wind field (Figure 9.3). Flow patterns of a similar character are also apparent in satellite photographs of Lake Balaton.

The vertical structures of wind-induced currents in the lake were the subject of recent field measurements conducted by Shanahan *et al.* (1981). The theory of wind-driven circulation predicts, for steady conditions, a profile in which currents at the surface align with the wind, with an opposing return current along the bottom (Plate 1970, Liu and Perez 1971).

In the field studies, the actual vertical velocity profile in Balaton was contrasted with the profiles given by theory. The observations adhere to the theoretically predicted steady state profile only occasionally. More typical is a highly transient velocity structure, with increasing variability as the measurement depth increases. Apparently, the currents experience the conflicting influences of the lake-wide seiche motion and the local wind-driven motion, as well as the inherently transient process of turbulent momentum transport. One exception to this picture of transience and variability is in the Tihany Strait (Figure 9.4) where Shanahan *et al.*'s observations of strong, unidirectional currents in the upper 5 m corroborate Muszkalay's (1973) observations discussed below.

Seiches

A seiche is the pendulum-like motion of the lake water surface after the cessation of a force that has displaced the surface from its equilibrium level position. The most common forcing agent causing seiches is the shear force of a sustained wind. Such a wind will cause a set-up, the superelevation of water level on the downwind shore above the level, undisturbed position. When the wind stops blowing, the superelevated waters flow downward, initiating the

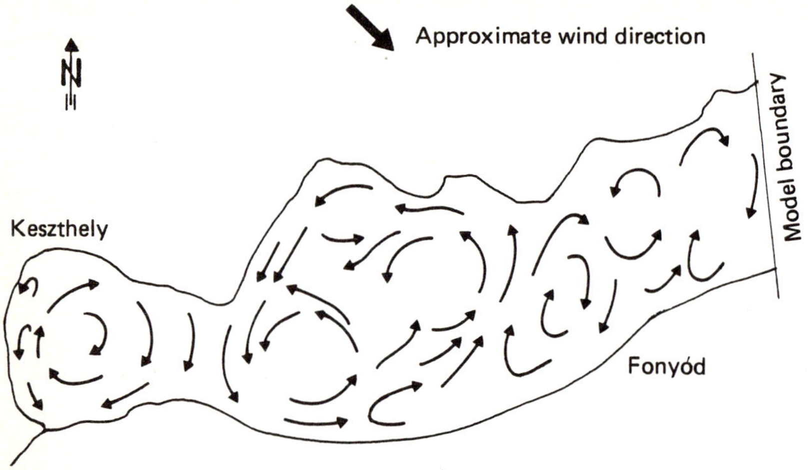

Figure 9.3. Qualitative flow pattern observed in Gyórke's physical model of western Lake Balaton (from van Straten *et al.* 1979).

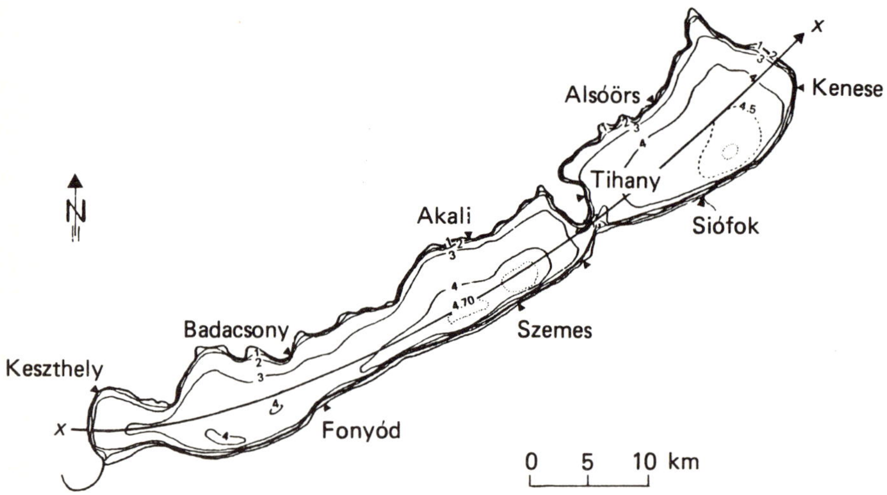

Figure 9.4. One-dimensional model coordinate system. Depth contours in meters. Small triangles indicate water surface elevation recording stations.

periodic seiche motion. Seiches may be especially significant in shallow lakes, since the magnitude of set-up increases as mean water depth decreases (Sibul 1955).

The seiche is a well observed phenomenon in Lake Balaton, with different seiche periods arising according to the direction and location within the lake. Hutchinson (1975) cites work done by Cholnoky at the turn of the century who found a longitudinal seiche period of between 10 and 11.5 h, while the transverse seiche was but 40 min. Other seiche periods have been distinguished for portions of the lake to the east and west of Tihany Strait (1 h and 2.5 h, respectively). In more recent work, Muszkalay (1973) found a range of seiche periods from 10 min to 1 d, with a mean of 5.5 h.

The most detailed studies of Lake Balaton's seiches are those of Muszkalay (1973), who collected nearly a full decade of water surface elevation observations at up to ten stations around the lake. Simultaneous measurements of wind speed at three stations and of water current at Tihany Strait complete his data base. The measurements show the lake to be in seemingly constant motion. A strong wind of only a few hours' duration can lead to observable seiches.

Using these observations, Muszkalay determined empirical formulas relating the wind strength, duration, and direction to the resulting set-up. His formula for longitudinal slope due to winds directed within 22.5° of the lake's long axis is:

$$I = 3.8 \times 10^{-7}\, T^{1/4}(W_x' - 2.8) \quad , \tag{9.11a}$$

where I is the dimensionless slope of the water surface (m/m), T is wind duration (h), and W_x' is the longitudinal component of the maximum instantaneous wind speed (m/s). Somlyódy and Virtanen (1982) statistically analyzed historical wind records for Balaton and found that for hour-long periods, the instantaneous peak wind speed exceeded the hourly average wind speed by a factor of 1.2–1.3. This information can be used to restate equation (9.11a) in terms of averaged wind speeds:

$$I = 3.8 \times 10^{-7}\, T^{1/4}(f_1 W_x - 2.8) \quad , \tag{9.11b}$$

where W_x is the average longitudinal wind speed (m/s), and f_1 is an adjustment factor ($1.3 \geq f_1 \geq 1.2$). In these relations, I is determined from the difference in the extreme stages at Keszthely and Kenese (Figure 9.4). It is a fictitious quantity in the sense that these stages may not in fact occur at precisely the same time, although the time lag is not large. The maximum observed longitudinal denivellation (that is, the net difference in water surface elevation from one end of the lake to the other) is roughly 1 m. In the transverse, which is the more common direction for strong winds, a denivellation of 0.4 m has been observed.

The creation of set-up, and subsequent seiche oscillation, is accompanied by the transport of considerable quantities of water, particularly where the lake narrows at Tihany. Muszkalay took advantage of this geometry and deployed four current meters along a single vertical mooring line in Tihany Strait. The maximum velocity observed by Muszkalay was 1.4 m/s. Figure 9.5 shows a typical set of measurements relating wind, water surface motion, and velocity at Tihany (see also Figure 9.4). The event of Figure 9.5 is caused by a wind transverse to the lake, the predominant direction for storm winds and the type of event comprising most of Muszkalay's published examples.

Muszkalay also employed his observations of current to define empirical relations giving the peak velocity in Tihany Strait during a storm. He specified the velocity at 1 m below the water surface as a function of i (m/m), the

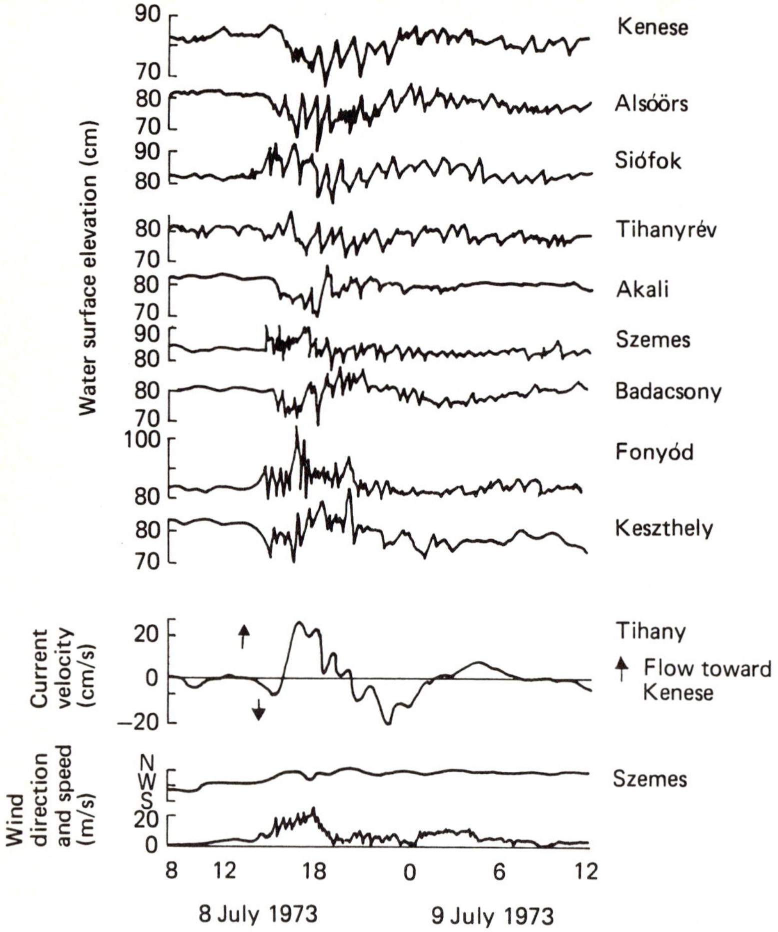

Figure 9.5. Example of Muszkalay's (1973) seiche observations (see Figure 9.4 for measurement station locations).

longitudinal water surface slope between Alsóörs and Szemes (Figure 9.4) as:

$$v = 500\sqrt{i} - 0.5 \quad , \tag{9.12}$$

where v is velocity (m/s). Muszkalay (1973) determined from current meter records that the current at Tihany Strait usually flows in a single direction throughout the water depth.

A significant factor in seiche behavior is the force of friction, an influence magnified by the shallowness of Lake Balaton. The effects of friction are to lengthen the observed oscillation period to greater than that predicted by frictionless theory, and to attenuate quickly the seiche amplitude (Hutchinson 1975). Frictionless theory predicts the period, T, to be $T = 2L / \sqrt{gH}$, where L is the lake length, g is gravitational acceleration, and H is the mean lake depth. This computes to 7.3 h in Balaton, well below the observed uninodal seiche period. According to Hutchinson, such a marked decrease in the period is a phenomenon unique to shallow lakes, with Balaton and Lake Okeechobee in Florida the only observed examples.

Wind Climate

The discussions above make clear the importance of the wind to the resultant motion in the lake. Thus a discussion of Balaton's circulation is incomplete without an accompanying description of the general characteristics of wind in the Balaton region. The following description is based on Béll and Takács (1974).

The prevailing wind direction is from the north–northwest. This trend is even more pronounced if events combining strong winds (greater than 8 m/s) and summer periods are considered. The monthly average wind speed ranges from 2 to 5 m/s, but the maximum may reach 30 m/s. The hourly average wind speed exceeds 8 m/s at Siófok during approximately 15% of the year.

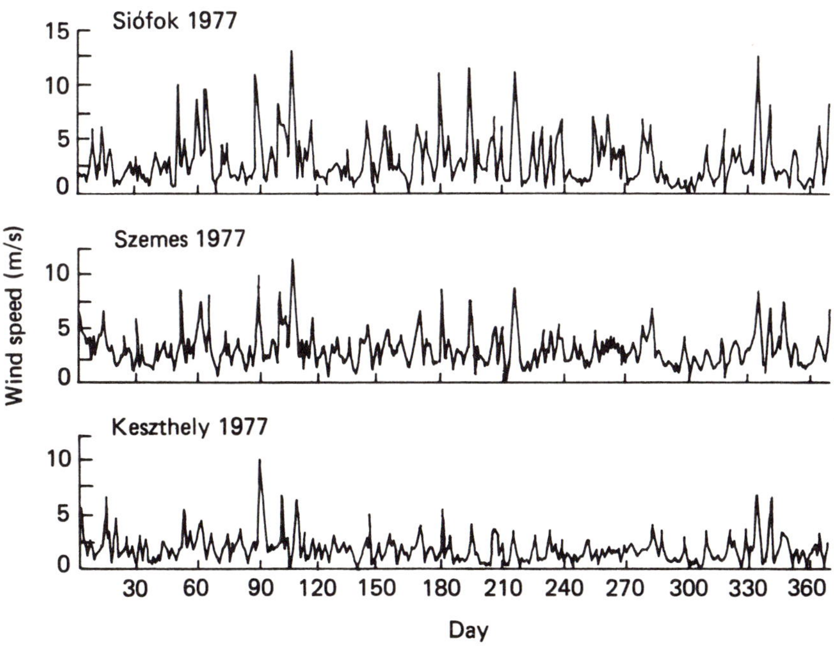

Figure 9.6. Comparison of observed daily average wind speeds at Siófok, Szemes, and Keszthely, 1977 (see Figure 9.4 for measurement station locations).

The variation of the wind in both space and time is strongly influenced by the hills immediately to the north of the lake. The sequence of hills acts to shelter, channel, and otherwise modify the wind field, leading to a highly variable wind field over the entire lake. The average wind characteristics show this clearly: at the eastern end of the lake the prevailing wind direction is from the northwest, at the western end it is from the north, while at the middle of the lake on the south shore the direction is northeast. The wind speed is also strongly affected. For example, the annual average wind speed is higher by 40–60% at Siófok than at Keszthely, a trend observable in nearly all storms (see Figure 9.6). Béll and Takács (1974) also report indications of significant variations in the wind across the lake as well as along it. Unfortunately, the limited number of regular wind recording stations (Siófok, Szemes, and Keszthely, Figure 9.4) prevents an accurate and comprehensive characterization of the wind field. The lack of data and the highly variable wind field imply considerable uncertainty in the wind data available for circulation models of Lake Balaton. In a subsequent section, we address the nature of these uncertainties and examine the problems they create for circulation modeling.

Introduction to the Lake Balaton circulation model studies

At various times in the Lake Balaton case study, virtually all applicable circulation model methodologies have been examined and tested. These include the 1-D longitudinal model, the 2-D single-layer model, the type I multilayer model, and two Ekman-type models.

Three circulation model alternatives were developed and evaluated in detailed studies of Lake Balaton. The models share a common foundation in their dependence upon the equations of motion that include wind and bottom friction as forces. The models differ in their parameterization of bottom friction and in their dimensionality. Included in these detailed studies are the 1-D model by Somlyódy and Virtanen (1982) and Somlyódy (1983), the 2-D single-layer model by Shanahan and Harleman (1982), and the 3-D Ekman-type model by Shanahan *et al.* (1981). Results from these models are compared and discussed below. In addition to these models studied in detail, two other models were tested but found inadequate or computationally impractical.

An Ekman-type model developed by Young and Liggett (1977) was adapted by Fisher (1980) for application to Lake Balaton. This model has the attractive flexibility of the finite-element technique to represent horizontal space, although its temporal formulation is quite limiting. The circulation is solved through time in the Laplace domain and retransformed to the time domain by numerical inversion. This technique permits the model to accept only step functions as wind input. The model was successfully tested in steady-state simulations of Lake Balaton, but a workable approach to transient simulation was not developed. Without a transient model, verification runs based on historical events could not be performed. The steady-state simulations failed to duplicate the spatial character of observed currents in Balaton.

In other research at IIASA, a type I multilayer model developed by O. Vasiliev and V.I. Kvon (Institute of Hydrodynamics, Siberian Branch of USSR Academy of Sciences, Novosibirsk) was also tested for application to Lake Balaton (van Straten and Somlyódy 1980). The model uses a finite-difference method to solve the nonlinear equations of motion closed by a two-equation turbulence model. The application to Lake Balaton simulated 221 grids in five vertical layers, but computational constraints allowed only time step functions to be considered as wind input. Results showed realistic water surface elevation predictions, but a failure to capture the character of observed current patterns. This is probably due to the inability to simulate a spatially varying wind field and to the coarseness of the vertical grid. These computational limitations, as well as the inability to perform transient solutions, led to use of the simpler formulations, discussed in detail below.

9.3. One-Dimensional Model Studies

Model construction

Governing Equations

The 1-D model developed by Somlyódy and Virtanen (1982) and Somlyódy (1983) includes the single spatial coordinate, x, the distance along the lake oriented to conform to the plan of the lake (Figure 9.4). A 1-D system of equations is derived by integrating equations (9.1) and (9.2) over the lake cross-section. The resultant equations of conservation of mass and momentum are those often employed for river flow (Mahmood and Yevjevich 1975, Kozák 1977):

$$\frac{\partial A}{\partial t} = -\frac{\partial Q}{\partial x} \tag{9.13}$$

$$\frac{\partial U}{\partial t} = -g\frac{\partial h}{\partial x} - \frac{1}{2}\frac{\partial}{\partial x}(U^2) + \frac{1}{(h+\eta)\rho}(\tau_s^x + \tau_b^x) \quad . \tag{9.14}$$

Equation (9.13) may alternatively be written as

$$B\frac{\partial h}{\partial t} = -\frac{\partial}{\partial x}[UB(h+\eta)] \quad , \tag{9.15}$$

where $U = Q/A$ is the longitudinal flow velocity, averaged over the cross section, $A = B(h+\eta)$ is the cross-sectional area, B is the lake width, and Q is the flow rate.

The shear stresses are described by introducing the drag coefficient C_D and bottom friction coefficient λ (see Lick 1976, Virtanen 1978):

$$\tau_s^x = \rho_a C_D W_x |W| \tag{9.16}$$

$$\tau_b^x = -\rho\lambda U|U|^n = -\rho\lambda\frac{Q|Q|^n}{A^{1+n}} \quad , \tag{9.17}$$

where ρ_a is the air density, W and W_x are the wind speed and its longitudinal component, respectively, and n is the bottom friction exponent ($0 \leq n \leq 1$). Here, the quadratic law will be used, $n = 1$. As seen from equation (9.17), τ_b^x is related to the average cross-sectional velocity rather than to the local velocity at the bed, so that λ is a lumped parameter. With $n = 1$, the friction factor λ is related to the Chézy coefficient C as $\lambda = g/C^2$.

Before solving, the equations are rearranged with η and Q as unknown variables and dimensionless quantities are introduced (see Somlyódy and Virtanen 1982). Boundary conditions for one of the variables should be defined at the two ends of the lake, $x = 0$ and $x = X$, respectively. For lake problems $Q(t,0)$ and $Q(t,X)$ are generally given. If $Q(t,0) = Q(t,X) = 0$, no inflow and outflow take place, a situation that will be considered here.

Numerical Solution

The equations are solved by a four-point implicit finite-difference scheme (Mahmood and Yevjevich 1975) to which an efficient matrix sweep technique is coupled. Time derivatives are approximated by differences centered in space and time, while for space derivatives the differences are centered in space but weighted in time. For coefficients and nonderivative terms, space-centered, forward-time approximations are employed except for the bottom shear term [equation (9.17)] where a more detailed approach, centered in space but weighted in time, is used (see Somlyódy and Virtanen 1982).

The implicit solution scheme is unconditionally stable in time. Owing to the matrix sweep method employed for solution of the set of algebraic equations, the solution expense is proportional to the number of grid points, N, rather than the N^3 proportionality for conventional matrix inversion methods. However, because the method requires a stepwise "sweep" from one end of the lake to the other, spatial stability is conditional. Somlyódy and Virtanen (1982) have developed the governing stability conditions.

Application to Lake Balaton

The 1-D circulation model was first calibrated for application to Lake Balaton and then verified. Calibration focused on the determination of C_D and λ, and included extensive sensitivity testing. Following calibration, the model was used to simulate actual historical events. Recorded wind speed and direction were used to drive the model and the model results were compared with records of water surface elevation and current. Calibration was based on the empirical characterizations of lake behavior developed by Muszkalay (1973). Thus, the verification by historical events served as an entirely independent test of model parameters.

The model finite-difference scheme employed a spatial increment of $\Delta x = 2000$ m leading to 40 calculation sections along the lake. The cross-sectional geometry was determined from hydrographic survey data collected by VITUKI (1976). A computation time step of $\Delta t = 1800$ s was used.

Model Calibration

Calibration of C_D and λ began with the definition of realistic ranges based on published values. In these studies C_D was assumed to be constant and not a function of wind speed. The slight wind dependence employed by some researchers was considered insignificant here due to the uncertainty in

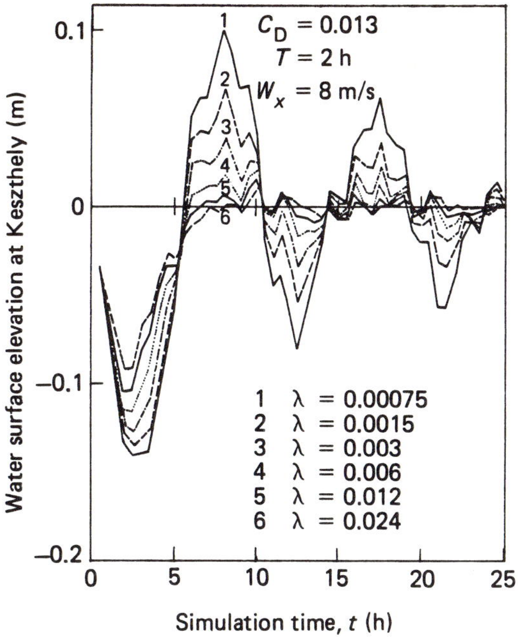

Figure 9.7. Histories of water surface elevation at Keszthely showing sensitivity to bottom friction coefficient, λ.

the wind data. C_D ranges from approximately 0.001 to roughly 0.0015 (Wu 1969, van Dorn 1953, Hicks *et al.* 1974, Bengtsson 1978, Graf and Prost 1980). For λ, no direct observations are available for lakes. The range observed in turbulent channel flow with small bottom roughness (0.007–0.03) served as a guideline for lake situations. Values at or near the lower bound are most likely in lakes.

The model was calibrated using a simple deterministic event at a standard simulation, which was designed to exercise the model under uncomplicated conditions to reveal its basic behavior. Uncertainty in the wind input data was not considered during calibration, although it was considered in a subsequent study described below. The standard calibration simulation was a simplified seiche event in the course of which a steady wind of speed W_x blows along the long axis of the lake, which decreases as a step function to zero at time T, the duration of the storm.

Model performance was evaluated using predicted water surface elevation at Keszthely and Kenese as an indicator of seiche behavior. These model results were examined for conformity with the observed characteristics of the lake, including the period and damping of the longitudinal seiche and the wind set-up quantified by Muszkalay (1973). Typical calibration simulation results are given in Figure 9.7, which shows the oscillation of the water level at the western end of the lake. As can be seen, the dynamics of the system are very fast, characterized by a seiche period of around 10 h. The bottom

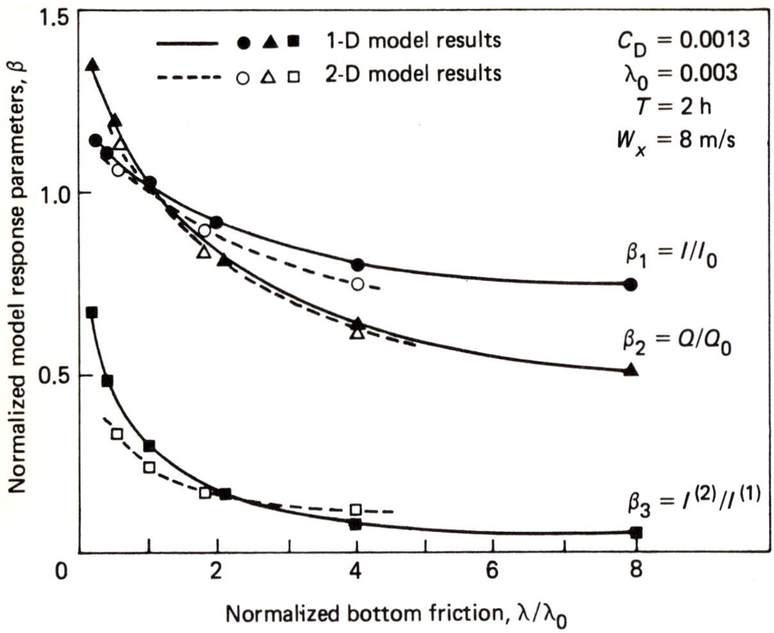

Figure 9.8. Summary of model sensitivity to the bottom friction coefficient, λ.

shear coefficient influences the peak amplitude but particularly the damping, an important feature that we refer to later on.

With increasing wind input duration, the duration of a negative (or positive) denivellation also increases. In the case of a step-like input, the water level approaches a steady state through several small oscillations. The flow at a given cross section shows a pattern similar to that in Figure 9.7, but the dynamics are even more rapid and the oscillation obviously decays for both short and long duration.

Figure 9.8 summarizes the sensitivity to λ in terms of three parameters, β_1, β_2, and β_3. β_1 is the ratio of the predicted surface slope, I, to a reference slope, $I_0 = 0.3 \times 10^{-5}$; β_2 is the similar ratio for the flow at Tihany, Q, where the reference flow $Q_0 = 2100$ m^3/s (the reference quantities I_0 and Q_0 are the slope and flow found in the calibration simulation); and β_3 evaluates seiche damping. β_3 is defined as the ratio of the maximum surface slope at the second denivellation, $I^{(2)}$, to that at the first, $I^{(1)}$. The variation in λ is also represented in Figure 9.8 in terms of a reference quantity, $\lambda_0 = 0.003$. It is apparent from Figure 9.8 that the maximum water level difference, β_1, is quite insensitive to λ. Although λ varies 30-fold, β_1 varies only from +15% to −27%. For $\lambda/\lambda_0 > 8$, β_1 is practically constant. The flow at Tihany Strait, β_2, is only slightly more sensitive to λ. The most sensitive parameter is β_3, the damping ratio, a conclusion illustrated in both Figures 9.7 and 9.8.

Sensitivity to C_D is illustrated in Figure 9.9. A reference drag coefficient, $C_{D0} = 0.0013$, defines the plot ordinate. The model output is more

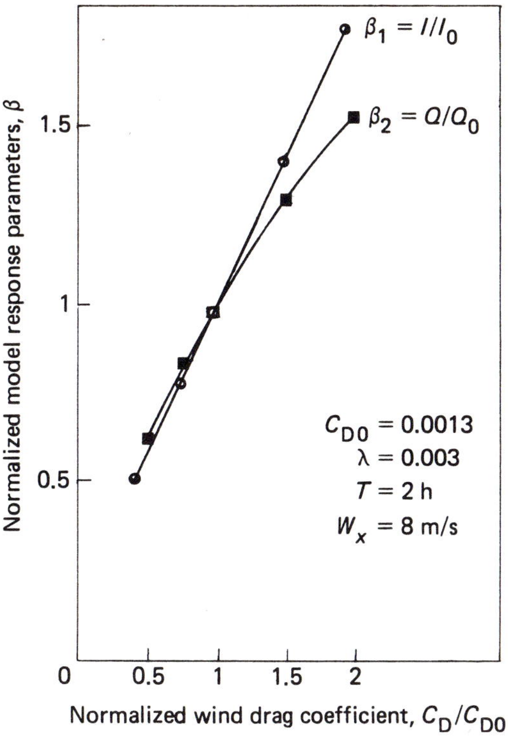

Figure 9.9. Summary of model sensitivity to the wind drag coefficient, C_D.

sensitive to C_D than to λ – an expected result since C_D directly determines the energy input to the system. The results show both β_1 and β_2 to depend approximately linearly on C_D.

Calibration of the 1-D model is based on the ability to match Muszkalay's empirical formulas for I and Q. Since λ and C_D have opposite influences on both I and Q, it follows that no unique, "best" parameter combination can be found for the model without further knowledge of the system. In the ranges $C_D = 0.0011–0.0014$ and $\lambda = 0.002–0.008$, multiple fittings of approximately the same quality can be arrived at for I and Q. At this stage the damping properties of the system can be utilized to supply further information. From the study of historical data it is apparent that the shallowness of the lake quickens damping; the second amplitude is around 30% of the first, while the fourth amplitude practically disappears. Proper damping characteristics are achieved with $\lambda = 0.003$ (or, equivalently, a Chézy coefficient of $C = 60$) and $C_D = 0.0013$. Both values are realistic for lake situations. In Figure 9.10 the predictions of the calibrated model are contrasted with Muszkalay's empirical formulas. The agreement is close, particularly in light of the considerable uncertainties in the field data used to drive the model.

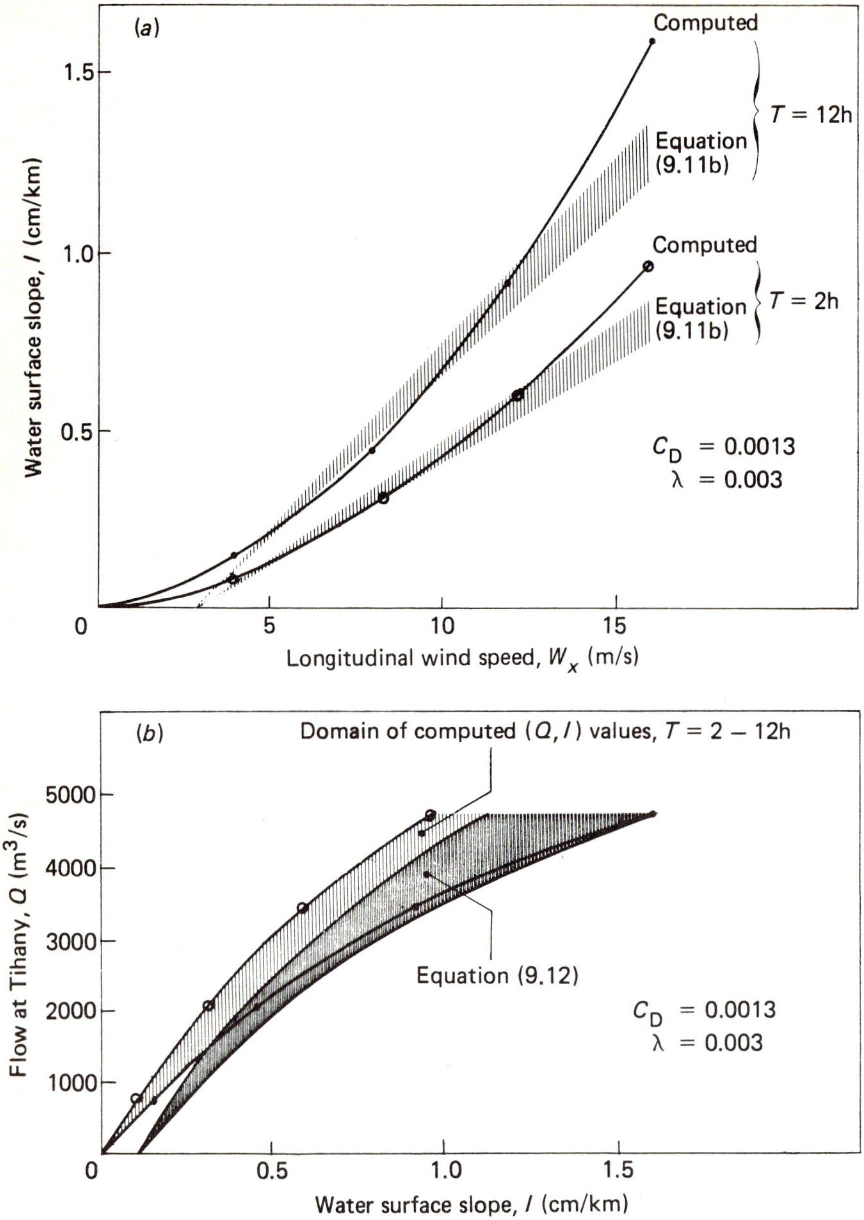

Figure 9.10. Comparison of model calibration with Muszkalay's empirical formulas. Comparison of computer predictions with (*a*) equation (9.11b), and (*b*) equation (9.12).

Model Verification

The calibrated model was tested against typical historical storm events selected from Muszkalay's (1973) observations. Altogether more than ten events of different natures were simulated without changing the parameter values. The complete simulation series is reported by Somlyódy and Virtanen (1982); here, three events are presented. The first event is characterized by winds along the lake, the second is produced by winds across the lake, and the third consists of a sequence of events of different character.

Input for the verification runs employed the hourly wind data recorded at Szemes. The wind strength was varied along the lake to account for the observed variations noted above. The model computation time step was set as 1 h. Model results were compared with water levels observed at the two ends of the lake (Keszthely and Kenese) and, if available, discharge derived from velocity measurements in Tihany Strait.

The storm of 16 November 1966 began at 8 a.m. and continued for over a day. The entire simulation included a first, relatively long-lasting storm with wind blowing from the east, followed by three smaller storms. Wind speed and direction for the entire period are plotted in Figure 9.11(*a*) and the resultant longitudinal component of the wind force in Figure 9.11(*b*). The shape of the water level versus time curve in Figure 9.11(*c*) is quite similar to that of the wind force, and from this example a linear relationship between the two could be hypothesized. The maximal denivellation reaches 0.7 m, approximately one fourth of the average lake depth, one of the highest values observed. No second amplitude can be observed, due mainly to the gradual decay in the wind shear. The agreement between simulated and observed water levels is excellent. Field observations of flow do not exist for this event. The model results [Figure 9.11(*d*)] show a highly fluctuating discharge that varies between –2000 and 3000 m^3/s.

The storm beginning 8 a.m. on 8 July 1963 is typical for the lake: a strong wind blew across the lake, perpendicular to the lake's long axis, resulting in a relatively small longitudinal wind force (Figure 9.12). The behavior of the water level is confused by small and random fluctuations that are not reproduced well in the simulated history. Also, the observed flow is much greater than that simulated. For this particular event, the model failed to duplicate realistically the observed behavior of the lake. As is shown below, this is a consequence of the large uncertainties in the wind direction.

The nine-day period beginning 18 April 1967 at 2 p.m. represents the most comprehensive historical event simulated (Figure 9.13). During the simulation period, more than five individual wind events may be distinguished, covering a range in wind speed from 0 to 25 m/s and a complete range in wind direction. The result is a variable and irregular history of longitudinal wind force, as shown in Figure 9.13(*b*). The simulated water levels [Figure 9.13(*c*)] are much noisier in character than those observed. In general, the predictions and observations at Kenese are in reasonable agreement, but at Keszthely the model overpredicts the elevation around t = 100 h when the wind blows from the north. As with the previously discussed event, the

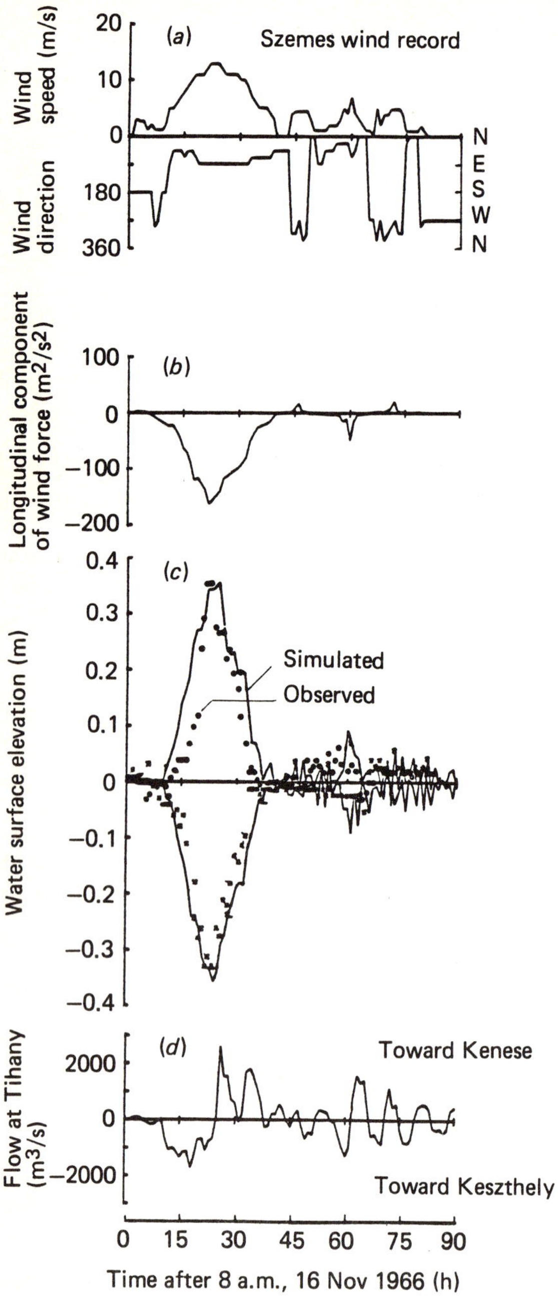

Figure 9.11. Comparison of simulation results with observations for the event of 16 November 1966. (*a*) Wind speed and direction at Szemes; (*b*) longitudinal component of wind force; (*c*) water surface elevation at Keszthely and Kenese; (*d*) flow at Tihany.

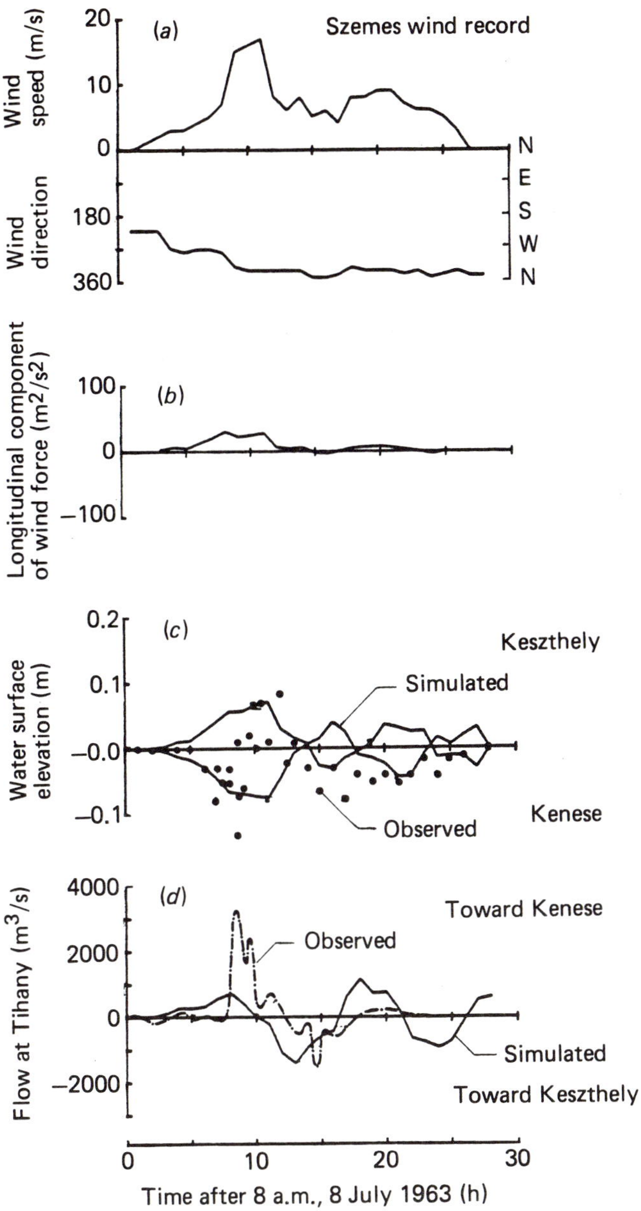

Figure 9.12. Comparison of simulation results with observations for the event of 8 July 1963. (*a*) Wind speed and direction at Szemes; (*b*) longitudinal component of wind force; (*c*) water surface elevation at Keszthely and Kenese; (*d*) flow at Tihany.

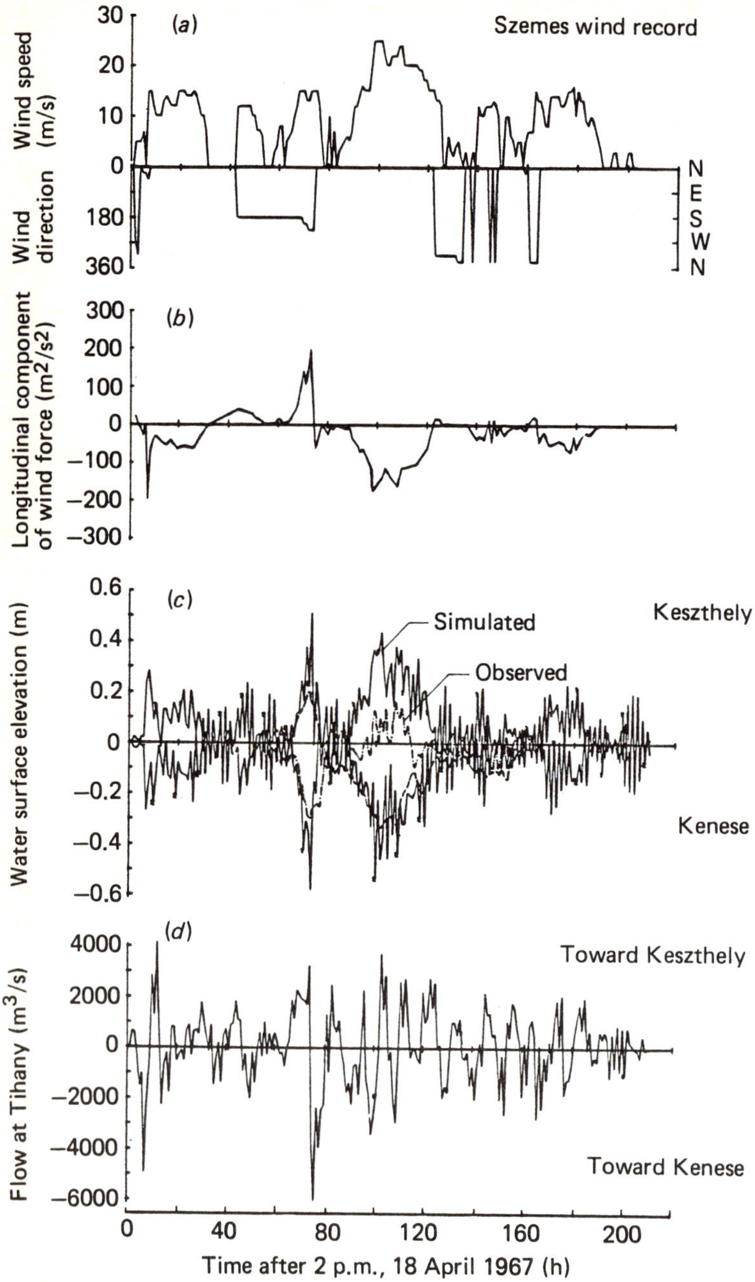

Figure 9.13. Comparison of simulation results with observations for the event of 18 April 1967. (*a*) Wind speed and direction at Szemes; (*b*) longitudinal component of wind force; (*c*) water surface elevation at Keszthely and Kenese; (*d*) flow at Tihany.

discrepancy is probably due to uncertainties in the wind field, in this case specifically due to spatial nonuniformities in wind.

In summary, comparison of the model results with historical observations shows that the model is satisfactorily calibrated. The model is verified for longitudinal wind conditions, but model performance is poor when the wind blows more or less perpendicularly to the lake's long axis. This problem is discussed further in the following section.

Effects of wind input uncertainty

Earlier in this chapter, the considerable uncertainties associated with the model wind input were raised. In this section we return to that problem to explore it in greater depth, particularly as it affects the model predictions.

The wind velocity vector is defined by its magnitude (speed) and direction. By far the greatest source of uncertainty is the direction. As shown in the model verification simulations, this is particularly so when the wind is nearly perpendicular to the lake. Unfortunately, this is the prevailing wind direction, from the north across the lake. The uncertainties in the wind input arise from a number of sources. The most important are:

(1) *Wind direction variability.* Owing to turbulent fluctuations, wind records often define direction as a band 40–60° wide rather than as a single line. Accordingly, the wind direction has a stochastic character that is lost when the wind data are time averaged or treated as a deterministic variable.

(2) *Measurement error.* The resolution of many wind recording instruments is discrete, in intervals of 22.5° or even 45°.

(3) *Spatial nonuniformity.* The wind field is spatially nonuniform, resulting in simultaneous direction measurements that can differ dramatically. The 90° difference between Keszthely and Siófok cited earlier is an example.

The effect of these errors was investigated in a series of model experiments, which we describe below.

Order of Magnitude Analysis

The importance of wind direction can be illustrated by a simple analysis. Here, we represent the wind direction by the variable α or its transform α^*. The definition of α is based upon the meteorological convention of measuring the direction from which the wind blows in degrees clockwise from true north. Its transform, $\alpha^* = \alpha - 247.5$, is the direction measured relative to the lake's long axis. Thus, if $\alpha^* = 180°$, the wind blows along the lake from east to west; if $\alpha^* = 90°$, as shown in Figure 9.14, the wind blows from north to south across the lake in the prevailing wind direction.

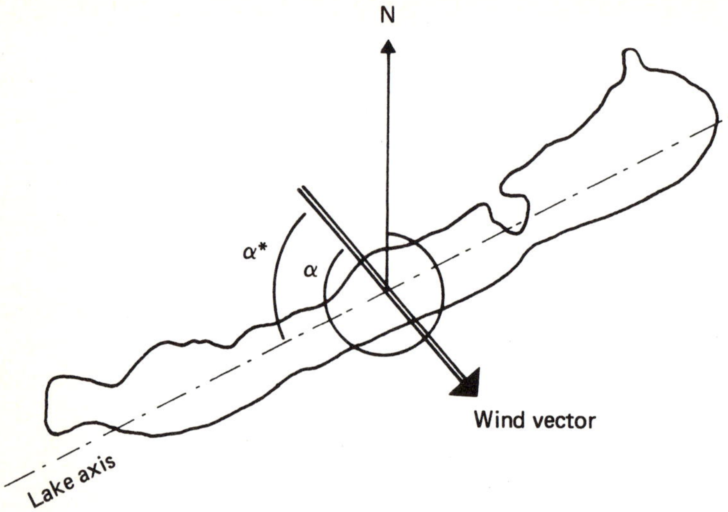

Figure 9.14. Definition sketch for wind direction measurements.

Based on the model calibration results in Figure 9.10 for a wind duration of 2 h, the maximum water level slope I and the discharge at Tihany Q can be approximated as functions of the longitudinal wind stress, τ_s^x. The approximate relations are (Somlyódy 1983):

$$I \simeq 2.8 \times 10^{-5} \rho_a C_D |W|^2 \cos \alpha^* \tag{9.18}$$

$$Q \simeq 1.1 \times 10^{-6} \sqrt{|I|} \quad , \tag{9.19}$$

where Q takes the same sign as I. Consider the influence of uncertainty in the wind direction by assuming $\alpha^* = \alpha_0^* + \Delta\alpha$, where $\Delta\alpha$ is a small error. Equation (9.18) thus becomes

$$I \simeq 2.8 \times 10^{-5} \rho_a C_D |W|^2 (\cos\alpha_0^* - \Delta\alpha \sin\alpha_0^*) \quad , \tag{9.18a}$$

with consequent changes in the value of Q computed from equation (9.19). Equation (9.18a) clearly shows the tendency of error propagation. The direction error has no influence under longitudinal winds, $\alpha_0^* = 0$ or 180°, although there is maximum effect during transverse winds, $\alpha_0^* = 90$ or 270°. Since α_0^* = 90° is the prevailing wind direction, direction error is of major importance.

To evaluate the importance of the error, consider a typical storm event: $W = 10$ m/s and $\alpha^* = 90°$ with an error of $\Delta\alpha = +22.5°$. The resultant changes in equations (9.18) and (9.19) lead to a change in denivellation of ±13 cm and in discharge of ±1500 m^3/s. The same results are found at higher wind speeds with smaller direction errors; e.g., an error of only ± 10° gives the same I and Q errors at a wind speed of 15 m/s. Clearly, errors in wind direction can significantly change the model predictions: recall that the predictions of I and Q in the historical verification events are of the same order as the I and Q prediction errors above.

Monte Carlo Simulations

Recalling the three sources of error in wind direction cited earlier, it is obvious that in the vicinity of $\alpha_0^* = 90°$, an error in a single direction datum can cause a drastic change in the simulation results. The most troublesome of the verification events, 8 July 1963 and 18 April 1967, are characterized by transverse winds, so that the lack of agreement between simulation and observation for these events is not surprising. A more accurate procedure, employing Monte Carlo simulation, was developed to account for wind data uncertainty.

The Monte Carlo simulation gives an estimate of the stochastic response of a deterministic system that is subject to stochastic inputs. In the present case, the system is the wind-driven circulation of Lake Balaton simulated by a deterministic model. The input to this system, the longitudinal wind force, is uncertain; thus the output, the predicted water motion, is uncertain. We cannot easily define the probability distribution of this uncertain output, but we can estimate the probability distribution of the wind input by accounting for the known sources of error and uncertainty.

The essence of the Monte Carlo procedure is to drive repeatedly the deterministic model with a distribution of inputs until a sufficient population of output observations is built to define a probability distribution. Each model simulation represents a single Monte Carlo trial. The stochastic model is constructed by adding a random direction error $\Delta\alpha_j$ to the wind direction α_j observed at time $t = j\,\Delta t$. Somlyódy (1983) analyzes separately three different distributions for the wind direction error – a Gaussian distribution, a discrete function (in which $\Delta\alpha_j$ takes values of 22.5°, 0°, or −22.5°), and a uniform distribution corresponding to the uncertainty sources discussed previously. Testing with the model shows 100 trials to be adequate to build a representative output distribution.

Somlyódy (1983) explored the implications of uncertainty in the wind force in Monte Carlo simulations of a number of events. Here we consider two events already discussed (Figures 9.11 and 9.12). For the first event (16 November 1966), because the winds are nearly longitudinal, we would expect small sensitivity to the wind direction, and this is confirmed by the Monte Carlo simulation results shown in Figure 9.15. The results show the mean and statistical character of the wind force, Keszthely water surface elevation, and Tihany flow distributions. Statistical information is shown by the mean, the simulated maximum and minimum, and the mean ±1 standard deviation. The results show very little variance in the predicted water surface level and somewhat greater, although still small, variance in the flow. The greater sensitivity of the flow to wind direction uncertainty indicates the greater difficulty in validating circulation-model flow predictions. The mean predictions for both surface level and flow agree reasonably well with the deterministic predictions in Figure 9.11. As a whole, uncertainty has a relatively minor influence on model results for this event, a major reason for the successful verification in the deterministic simulation.

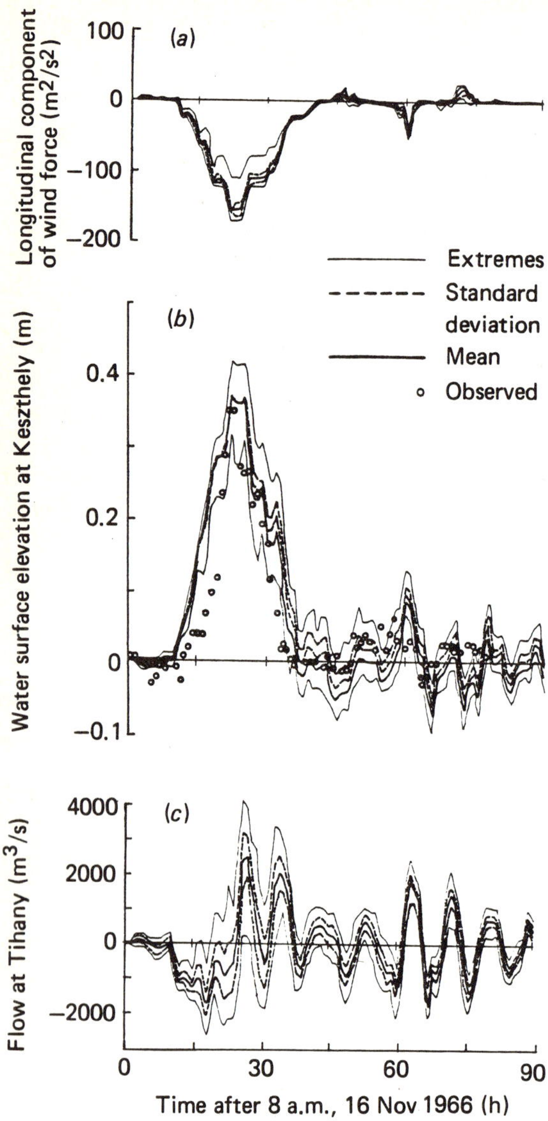

Figure 9.15. Monte Carlo simulation of the event of 16 November 1966. (*a*) Longitudinal component of wind force; (*b*) water surface elevation at Keszthely; (*c*) flow at Tihany.

The event of 8 July 1963 was also investigated in the verification simulation, but with far less success than the event just discussed. This event is the result of winds blowing nearly perpendicularly to the lake – a situation of greater uncertainty in the wind force. This is shown dramatically in the Monte Carlo simulation results (Figure 9.16). The wind force is surrounded by

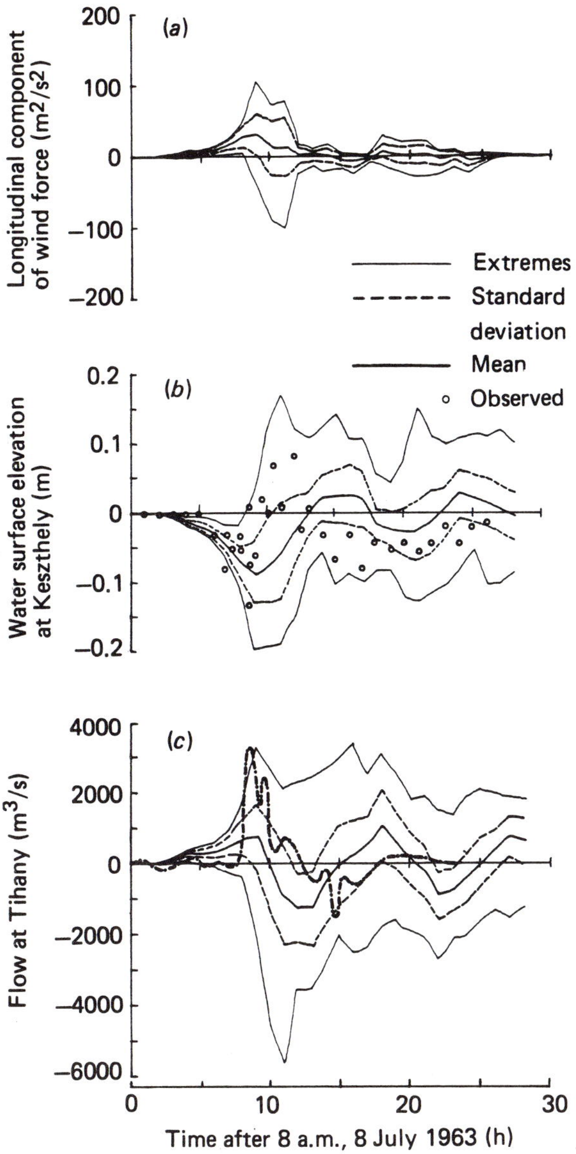

Figure 9.16. Monte Carlo simulation of the event of 8 July 1963. (*a*) Longitudinal component of wind force; (*b*) water surface elevation at Keszthely; (*c*) flow at Tihany.

wide error bands, leading to great uncertainty in the predicted water level and current. The water level predictions vary between +0.15 m and −0.15 m over most of the simulation period. The discharge varies over a very wide band as well. For both these predictions, the observed data fall within the range of Monte Carlo trials although the distribution means do not differ drastically from the deterministic simulation results. The results illustrate that the model verification can be called a failure only if uncertainty in the wind direction is not considered.

The Monte Carlo simulations supply considerable insight into the dynamics of the lake. From simulation results we can conclude that except when the wind blows along the lake, model results are more sensitive to wind direction errors than to the model parameters examined during model calibration. The results also lead to the conclusion that although the model is at best partially verified in the deterministic tests, the verification is successful in a stochastic fashion.

Long-Term Simulation

The longest simulation run with the 1-D model was the simulation of an entire year. The calendar year 1977 was selected for this test since that year served as the base case for the eutrophication models developed as a part of the case study (Chapters 3, 11, 12, and 13). The wind record for Siófok, consisting of eight observations per day, was used as the model input. All model and computation parameters remained the same as those used in the calibration and verification simulations.

Simulation results are shown for January and July 1977 in Figure 9.17. The plots show the flow through Tihany Strait (i = 28), at sections in the lake corresponding to the two other interbasin boundaries (i = 5 and 15), and at the midpoint (i = 35) of the Siófok basin. The results for January, when winds were light, contrast with July, when there were several large storms.

The character of the results for 1977 is in accord with that observed in the previous shorter simulations. The system shows very dynamic behavior in time, with rapid and wide fluctuations in flow. Flow also varies spatially, from a low near Keszthely (model section i = 5) to its highest at Tihany (section 28). The most dramatic variations in flow occur during July, with oscillations between ±3000 m^3/s over much of the lake.

Figure 9.18 shows a statistical summary of the results for the entire year. Shown are the daily average flow and standard deviation of the flow at Tihany. Comparison with Figure 9.17 shows that the daily average flow is roughly two orders of magnitude less than instantaneous values, although the underlying transience is reflected in the large standard deviation. Daily average flows can be ten times larger than the monthly hydrologic throughflow. However, if the averaging period is increased further, the average of the dynamic flow approaches zero.

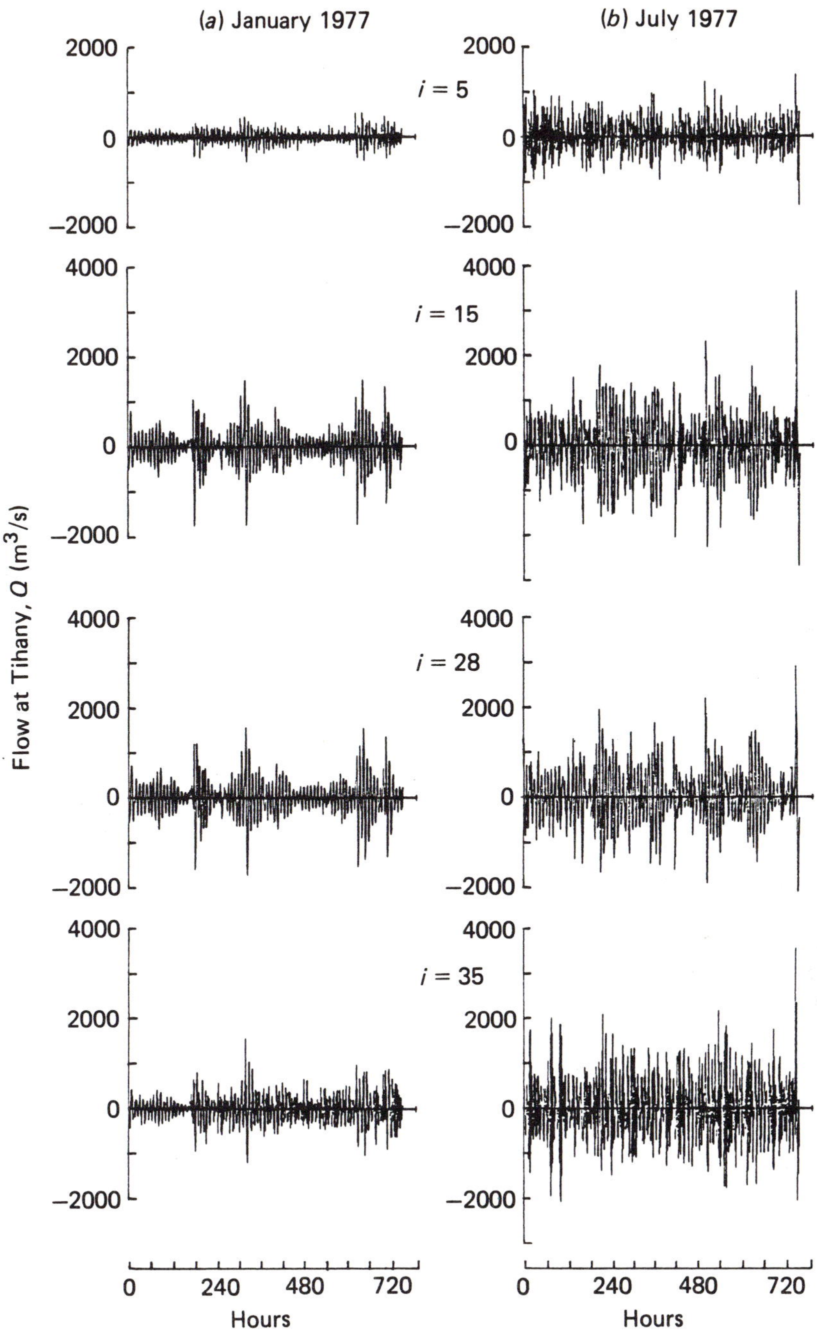

Figure 9.17. Results of long-term simulation.

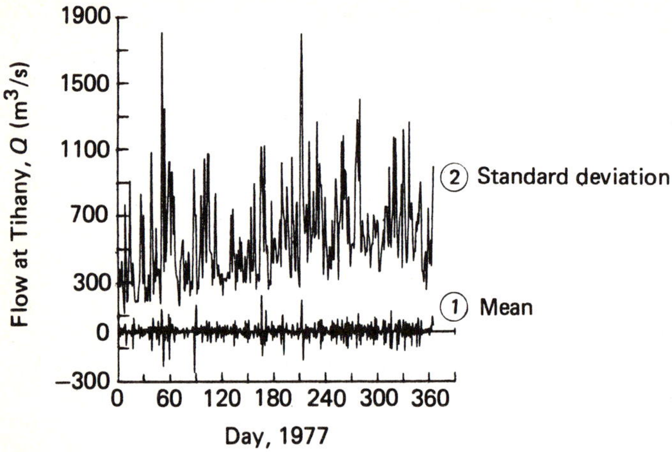

Figure 9.18. Summary statistics of long-term simulation.

9.4. Two-Dimensional Model Studies

Model construction

The 2-D model developed by Shanahan and Harleman (1982) is a fully transient model based on the linearized equations of motion. The model solves for mass transport in the two coordinate directions and for displacement of water surface as functions of time and horizontal space. The model permits a time-varying wind field to be specified for determination of the transient forces on the water surface, and also accounts for the nonlinear force of friction at the lake bottom. An explicit finite-difference technique is employed to solve the equations.

Governing Equations

The equations of motion upon which the 2-D model is based employ the shallow-water approximation and omit both the convective acceleration terms and the horizontal shear terms. Neglecting these terms is justified by the smallness of inertial and frictional forces relative to the Coriolis force in Lake Balaton. The simplified conservation of momentum equations, after integration over the lake depth, are thus [see equations (9.2) and (9.3)]:

$$\frac{\partial q_x}{\partial t} = -g(h + \eta)\frac{\partial \eta}{\partial x} + f q_y - \frac{\tau_b^x}{\rho} + \frac{\tau_s^x}{\rho} \tag{9.20a}$$

$$\frac{\partial q_y}{\partial t} = -g(h + \eta)\frac{\partial \eta}{\partial y} - f q_x - \frac{\tau_b^y}{\rho} + \frac{\tau_s^y}{\rho} \quad . \tag{9.20b}$$

The vertically integrated mass conservation equation is employed:

$$\frac{\partial q_x}{\partial x} + \frac{\partial q_y}{\partial y} = \frac{\partial \eta}{\partial t} \quad . \tag{9.21}$$

In these equations, the mass transport variables q_x and q_y represent water motion as the discharge per unit of horizontal length, as defined in equation (9.10).

The water of the lake is subject to shear forces at the water surface and lake bottom, which appear as the surface and bottom shear stresses, τ_s and τ_b, in equations (9.20). The wind-induced surface stresses are calculated as [for comparison see equation (9.16)]:

$$\tau_s^x = \rho_a C_D |W| W_x \qquad \tau_s^y = \rho_a C_D |W| W_y \quad , \tag{9.22}$$

where the notation is the same as above except that we have broken the wind vector, W, into its x and y components, W_x and W_y. The 2-m wind drag coefficient, C_D, is either fixed as a constant or calculated from published formula, such as those of Wu (1969), van Dorn (1953), or Hicks *et al.* (1974).

At the lake bottom, the shear stress is taken as a nonlinear function of the depth-average velocity in an equation similar to (9.17):

$$\tau_b^x = -\rho\lambda\bar{u}\sqrt{\bar{u}^2 + \bar{v}^2} \qquad \tau_b^y = -\rho\lambda\bar{v}\sqrt{\bar{u}^2 + \bar{v}^2} \quad . \tag{9.23}$$

The depth-average velocities are defined as:

$$\bar{u} = \frac{q_x}{h + \eta} \qquad \bar{v} = \frac{q_y}{h + \eta} \quad . \tag{9.24}$$

Numerical Solution

Simulation of transient wind-driven circulation events requires solution of equations (9.20) and (9.21) in time and horizontal space. This is accomplished in the 2-D model using an explicit finite-difference scheme, alternately solving the momentum and the continuity equations.

Application of the finite-difference scheme requires the division of the lake into a grid of discrete rectangular areas. A square grid or, more precisely, two square grids are used in the model. The two grids arise from the use of a staggered grid scheme in which velocity components are determined on the middle of the grid sides, while water surface displacements are found at the grid center points (Figure 9.19).

An explicit solution technique is employed, first solving the momentum equations for the current velocities at full time steps, then using those velocities in the continuity equation to solve for the water surface elevation at half time steps. From surface elevations, surface slopes can be determined for substitution into the momentum equation, which is solved at the next time step. This cycle is repeated until the simulation period is complete.

In solving the equations, the solution proceeds along the rows of the finite-difference grid from left to right, solving for all columns in a row before moving up to the next row. This sequence is followed until all rows have been completed, and is repeated for first the momentum and then the continuity equation at each time step. The particular sequence employed implies that the solution at a grid depends upon a mixture of variable values determined at

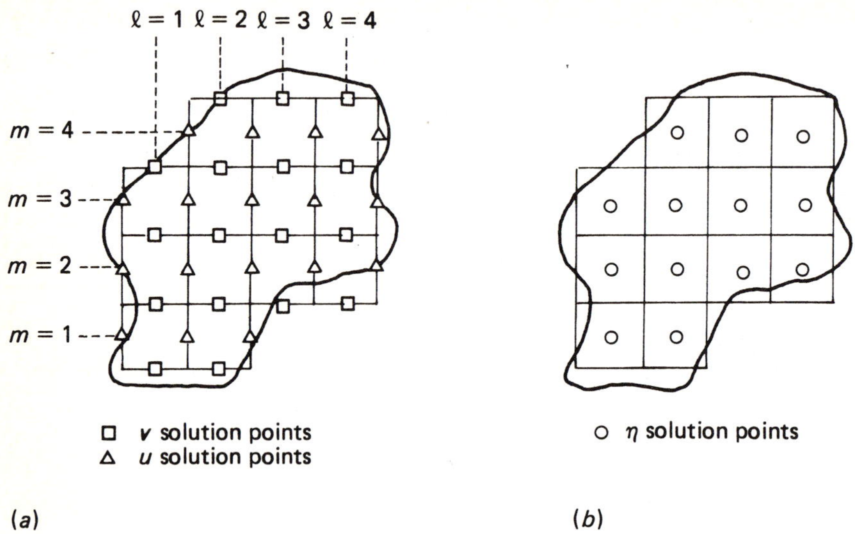

Figure 9.19. Finite-difference solution procedure. (*a*) Full time steps, solution for velocity; (*b*) half time steps, solution for water surface elevation.

the current and preceding time steps. This is permitted by a short computation time step, as limited by stability conditions.

The explicit solution scheme employed for the hydrodynamic model is conditionally stable, requiring that the well known Courant condition (which limits the time step to the time required for a surface gravity wave to traverse a grid square) be met in order to achieve a solution.

Application to Lake Balaton

Application of the 2-D model to Lake Balaton proceeded in two phases, similarly to that of the 1-D model: calibration and subsequent verification.

The finite-difference solution requires the specification of the computation time step, Δt, and grid spacing, $\Delta x = \Delta y$. The grid size selected was 1900 m, determined to be the largest able to approximate the geometry of Tihany Strait. A finer grid size (950 m) was also examined by Shanahan and Harleman (1982) but was found not to produce materially different results. The time step employed was 180 s.

Model Calibration

Calibration of the 2-D model followed the same procedures as used for the 1-D model. As illustrated in Figures 9.8 and 9.9, model sensitivity is essentially the same as that of the 1-D approach and the same parameter set (C_D = 0.0013 and λ = 0.003) was found to give the best fit with respect to Muszkalay's formulas.

Model Verification

The 2-D wind-driven circulation model was verified by the simulation of three historical events [observed by Muszkalay (1973)] selected to characterize different conditions in the wind forcing function.

The event of 4 and 5 July 1961 was produced by winds with a significant component along the lake's long axis. The consequent strong longitudinal seiche was well captured in the simulation, as shown by the comparison of stages at the two ends of the lake given in Figure 9.20. Flow data were not collected during this event.

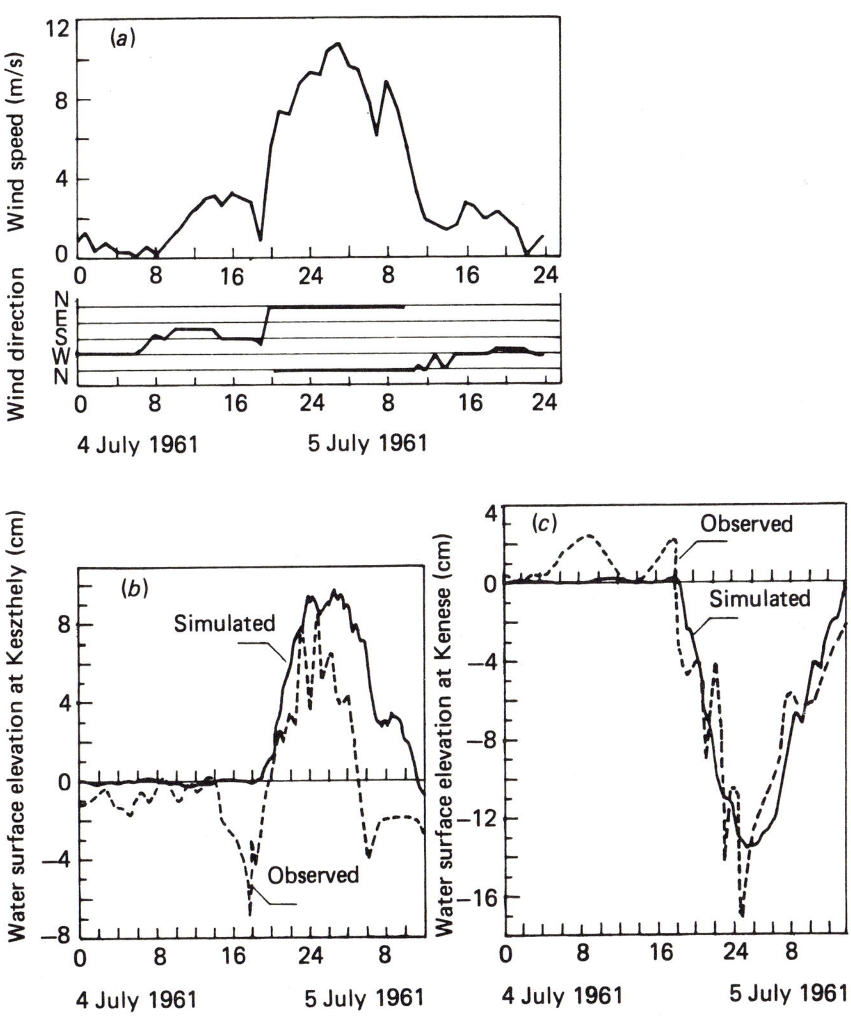

Figure 9.20. Comparison of simulation results and observations for the event of 4 and 5 July 1961. (*a*) Wind speed and direction at Keszthely; (*b*) water surface elevation at Keszthely; (*c*) water surface elevation at Kenese.

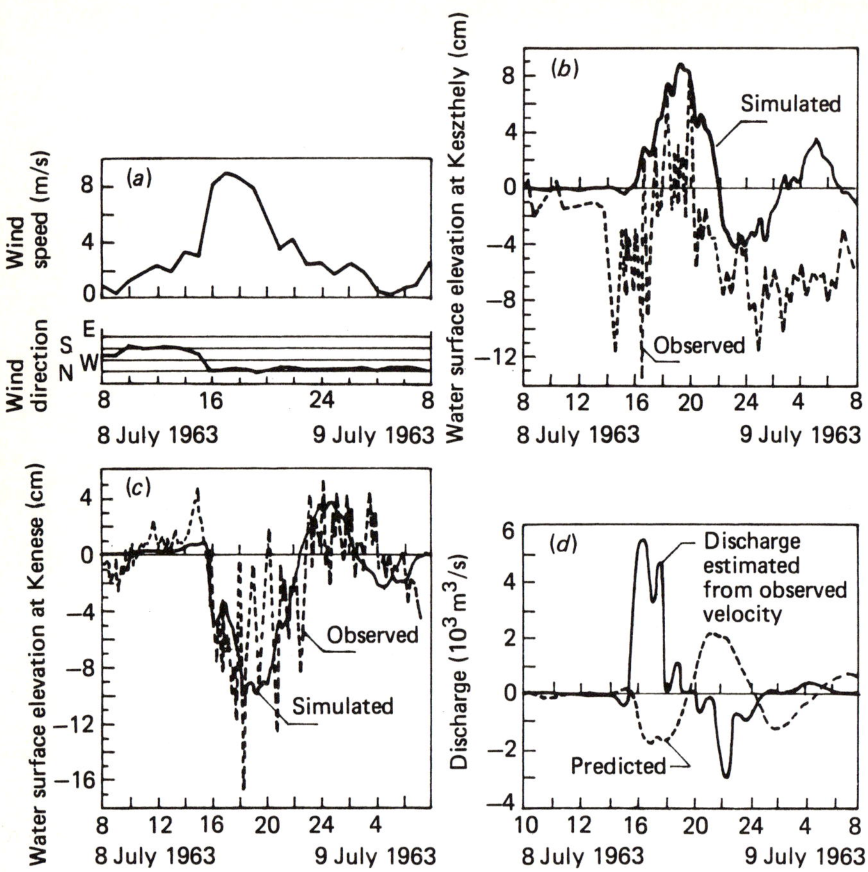

Figure 9.21. Comparison of simulation results and observations for the event of 8 and 9 July 1963. (*a*) Wind speed and direction at Keszthely; (*b*) water surface elevation at Keszthely; (*c*) water surface elevation at Kenese; (*d*) current at Tihany.

The event of 8 and 9 July 1963 produced a longitudinal seiche with comparable displacement to that of July 1961, but with much stronger simultaneous transverse seiching. Model results are compared with observations in Figure 9.21. Prediction of the stage at Keszthely is poor, while that at Kenese is generally good. Missing from the model predictions are most of the high-frequency oscillations seen in the observation record as a result of transverse seiching, probably due to the smooth wind record (hourly averages) relative to the 40-min transverse seiche period. Also contrasted in Figure 9.21 is the observed and predicted discharge through Tihany Strait: the agreement is quite poor. In general, the results from the 2-D model are of the same general character as the results of the deterministic simulation by the 1-D model (Figure 9.11). As with the 1-D model, the poor agreement may be attributed to uncertainties in the wind force due to the nearly transverse wind direction.

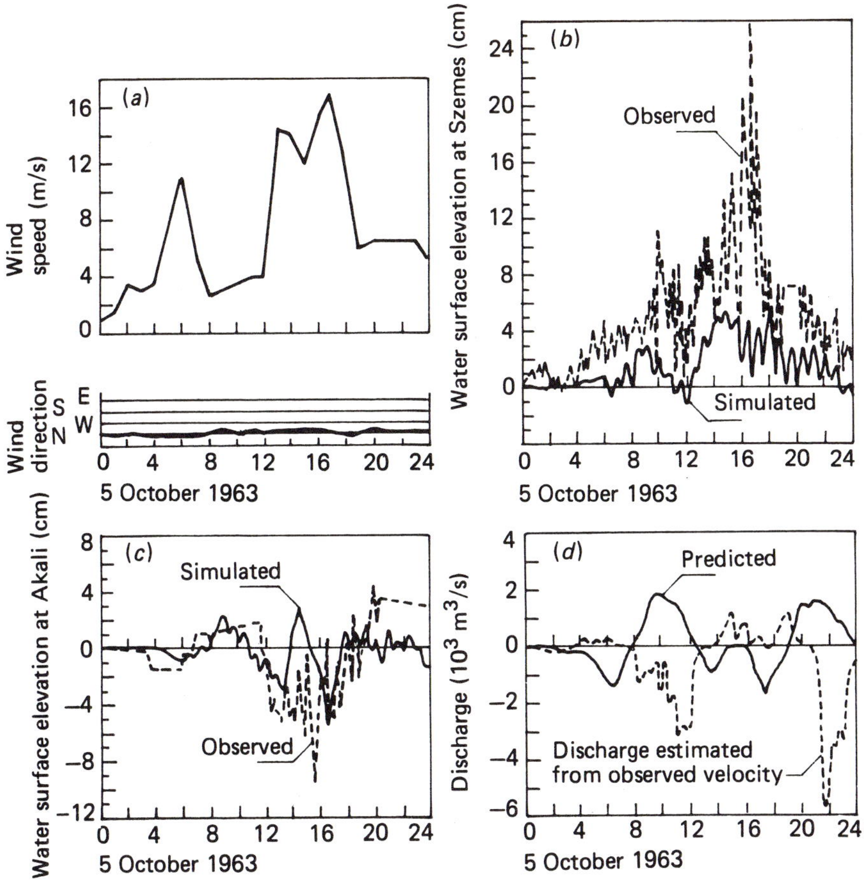

Figure 9.22. Comparison of simulation results and observations for the event of 5 October 1963. (*a*) Wind speed and direction at Szemes; (*b*) water surface elevation at Szemes; (*c*) water surface elevation at Akali; (*d*) current at Tihany.

The final event considered is that of 5 October 1963. Unlike the two previous events, which began with longitudinal winds, the storm of October 1963 brought only transverse winds. In fact, during the entire event the wind was nearly perpendicular to the lake. Under this situation, small changes in the wind direction lead to great changes in the magnitude of the longitudinal wind component including frequent reversals in longitudinal direction. With such conditions, local modifications of the wind field and errors in the wind data critically influence the model results making accurate predictions virtually impossible. This is confirmed by comparisons of predicted and observed transverse seiches at Akali and Szemes and discharge at Tihany (Figure 9.22).

The horizontal circulation pattern predicted by the 2-D model supplied additional, albeit qualitative, verification. Physical model studies and satellite photographs indicate the presence of horizontal circulation gyres in Lake

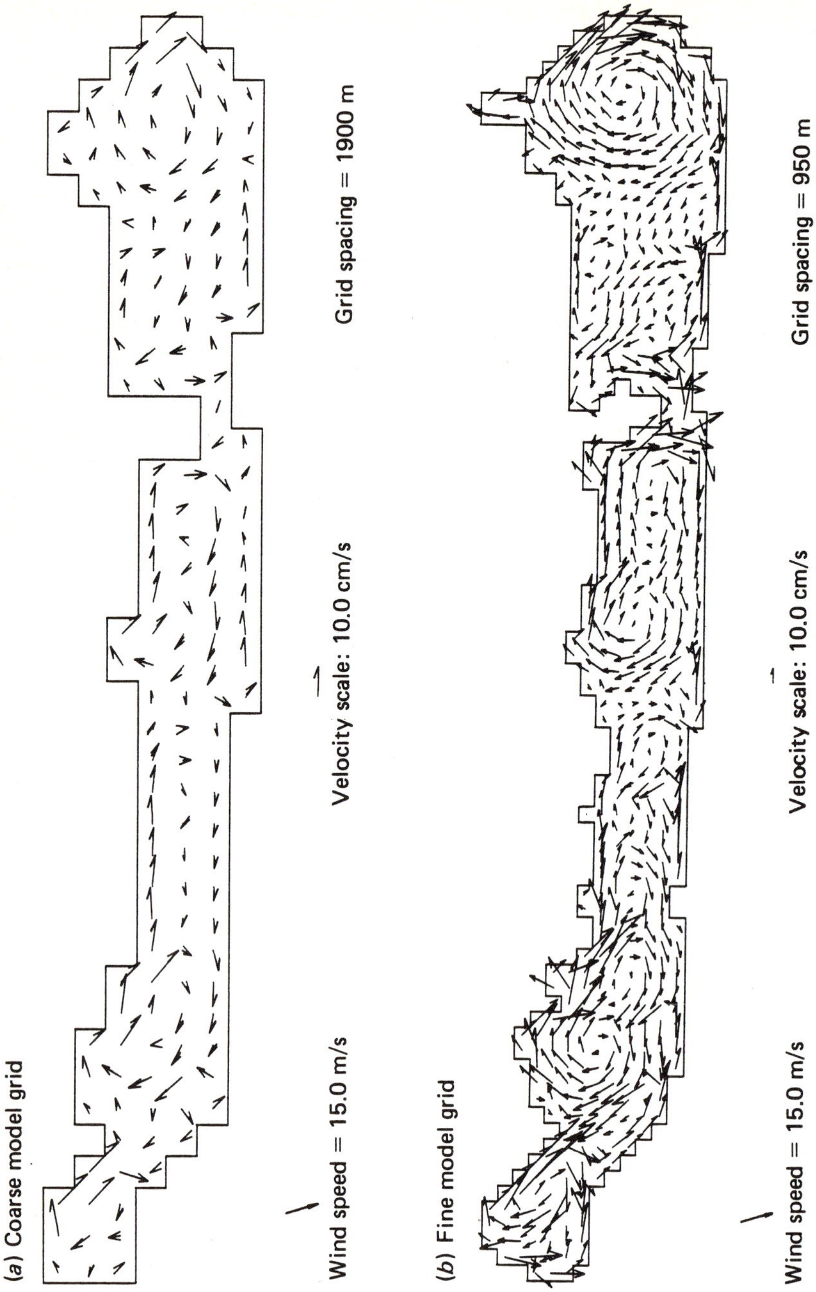

Figure 9.23. Simulated steady-state horizontal circulation.

Balaton. The steady-state model results given in Figure 9.23 are in qualitative agreement with the flow patterns in Figure 9.3. Figure 9.23 also shows the agreement between the coarse- and fine-grid 2-D model results. The two grid sizes result in essentially the same circulation, although with less detail in the coarse-grid results.

The preceding comparisons of the 2-D model results with observations of actual historical events show a general ability to capture the character of water motion in Lake Balaton – particularly longitudinal flow – but an inability to duplicate the fine structure of seiche behavior and the flow at Tihany. The model's ability to reproduce the character of large-scale motion is adequate as input for 1-D water quality simulation. The model's failures can be ascribed to uncertainties in the data available to drive and test the model, particularly in the wind data.

Effects of Wind Input Uncertainty

The influence of wind data uncertainty on 2-D model results was tested, though not in a Monte Carlo fashion, but with a deterministic sensitivity analysis: the wind direction scenario was incremented by +22.5° and –22.5°. For the sake of comparison with the 1-D model the event of 8 and 9 July 1963 was selected. The results of the simulations are shown in Figure 9.24 for the stage at Keszthely and the mean flow at Tihany. The results are not directly comparable with the extremes of the 1-D model result distributions shown in Figure 9.16, since those are the extremes observed at each time in the event as compiled from all simulations. Nevertheless, the sensitivity of the 2-D model results to wind direction uncertainty dramatically confirms the 1-D model findings.

Analysis of dispersion

It is shown in Chapter 10 that for the purposes of water quality, Lake Balaton may be reasonably approximated as a 1-D system. The long and narrow geometry of the lake supports this model, as does the distribution of inflow and outflow to the lake. However, development of a 1-D water quality model also requires the determination of 1-D transport information, which in turn requires a longitudinal velocity to be obtained by integrating the velocity profile over the width and depth of the lake cross section. This process of creating a 1-D advective velocity also causes all vertical and lateral variations in velocity to be eliminated. The dispersive effects of lateral velocity variations must then be captured through a new parameter, the longitudinal dispersion coefficient.

The importance of dispersion to 1-D water quality modeling underscores an essential difference between the 1-D and 2-D circulation models: the ability of the 2-D model to describe lateral velocity variations and therefore to indicate the character of dispersion in the lake. As part of the inquiry into the circulation of Lake Balaton, we now analyze the 2-D model's description of

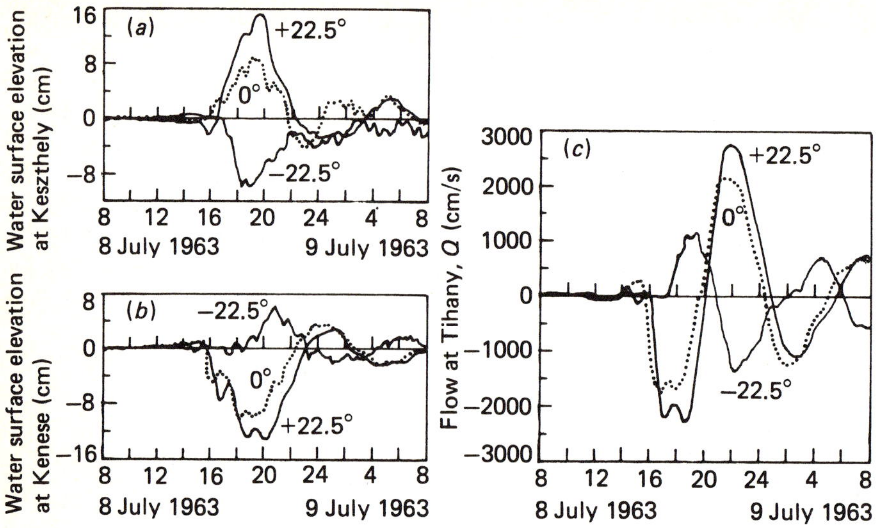

Figure 9.24. Effect of wind direction uncertainty on 2-D model simulation results for the event of 8 and 9 July 1963. (*a*) Water surface elevation at Keszthely; (*b*) water surface elevation at Kenese; (*c*) flow at Tihany.

dispersion and lateral mixing, using selected results from our simulation of the 8 July 1963 event to examine the spatial and temporal character of water motion.

Computation of Dispersion

Shanahan and Harleman (1982) propose a procedure to determine the dispersion coefficient from the 2-D model results. To compensate for the effect of seiche motion, they follow a method of computation developed by Holley *et al.* (1970) for oscillatory flow. However, whereas Holley *et al.* take lateral diffusion to be the dominant lateral transport mechanism, Shanahan and Harleman assume that lateral advection due to wind-driven currents dominates. The formula developed from these assumptions is:

$$D = \frac{1}{A} \int_0^B \mathrm{d}y q'' \int_0^y \mathrm{d}y' \frac{q''}{q_y} , \tag{9.25}$$

where D is the longitudinal dispersion coefficient, y is the lateral distance across the lake, B is the cross-sectional width, $q''(y)$ is the flow per unit width at y, less the mean flow per unit width in the cross section, q_y is the lateral flow per unit width, and A is the cross-sectional area. Following Fischer (1969), equation (9.25) may be approximated as:

$$D = \frac{KB\overline{(u'')^2}}{\bar{v}} , \tag{9.26}$$

where K is a constant of the order of 0.1, u'' is the local deviation of the depth-averaged velocity from the cross-sectional mean $(= q''/h)$, $\overline{(u'')^2}$ is the cross-sectional average of $(u'')^2$, and v is the mean lateral velocity in the cross section. From equation (9.26) D must be reduced to account for the effect of oscillatory flow as described by Holley ***et al.*** (1970). In oscillatory flow, dispersion is reduced if the period of oscillation is less than the time necessary for the dispersant to mix throughout the cross section. A reduction factor, $f(T')$, is defined by Holley ***et al.*** as a function of a characteristic time ratio T'. Here T' is defined as the dimensionless ratio of the seiche oscillation period to the lateral travel time:

$$T' = \frac{T_l}{B/2\bar{v}} , \tag{9.27}$$

where T_l is the longitudinal seiche period. The reduction factor $f(T')$ determines the ratio of the average dispersion over an oscillation period to the equivalent steady-flow dispersion from equation (9.25), i.e., $D_{osc}/D_{steady} = f(T')$, where D_{osc} is the average dispersion coefficient over an oscillation period, and D_{steady} is the integral over the oscillation period of the instantaneous steady dispersion coefficient. The instantaneous steady dispersion coefficient is the coefficient that would be computed from the velocity distribution at any instant, treating the distribution as though it were steady.

The method developed to compute dispersion from the lake circulation model results is somewhat speculative in the absence of comprehensive field or laboratory verification studies. In particular, the reduction factor to account for oscillatory flow was not determined for lake conditions. Nevertheless, it has been applied to prismatic and natural open-channel conditions as well as to pipe flow, finding variation within an order of magnitude (Holley ***et al.*** 1970, Chatwin 1975). The robustness of the theory over this range of conditions supports its extension to the lake environment. Nevertheless, use of Holley ***et al.***'s method is clearly an extension beyond the purpose for which it was originally derived.

Analysis of Two-Dimensional Model Results

The motion within the lake during one fourth of a seiche cycle is shown in Figure 9.25. This sequence of vector plots shows the predicted depth-average current at 1-h intervals from the time of maximum flow towards Keszthely to the time immediately after the seiche current reverses direction. In the central portion of the lake, strong unidirectional currents dominate the motion, while flow gyres are more evident in the eastern Siófok basin and in the western Keszthely and Szigliget basins. The character of the bulk motion with time is indicated in Figure 9.26 in time plots of the discharge at three sections (enumerated from west to east in Figure 9.25) along the lake. The flows are shown to be large, exceeding the hydrologic throughflow by two orders of magnitude.

The transports shown in Figure 9.26, being net transports, ignore all variations in the current across the lake section. Such variations are, of

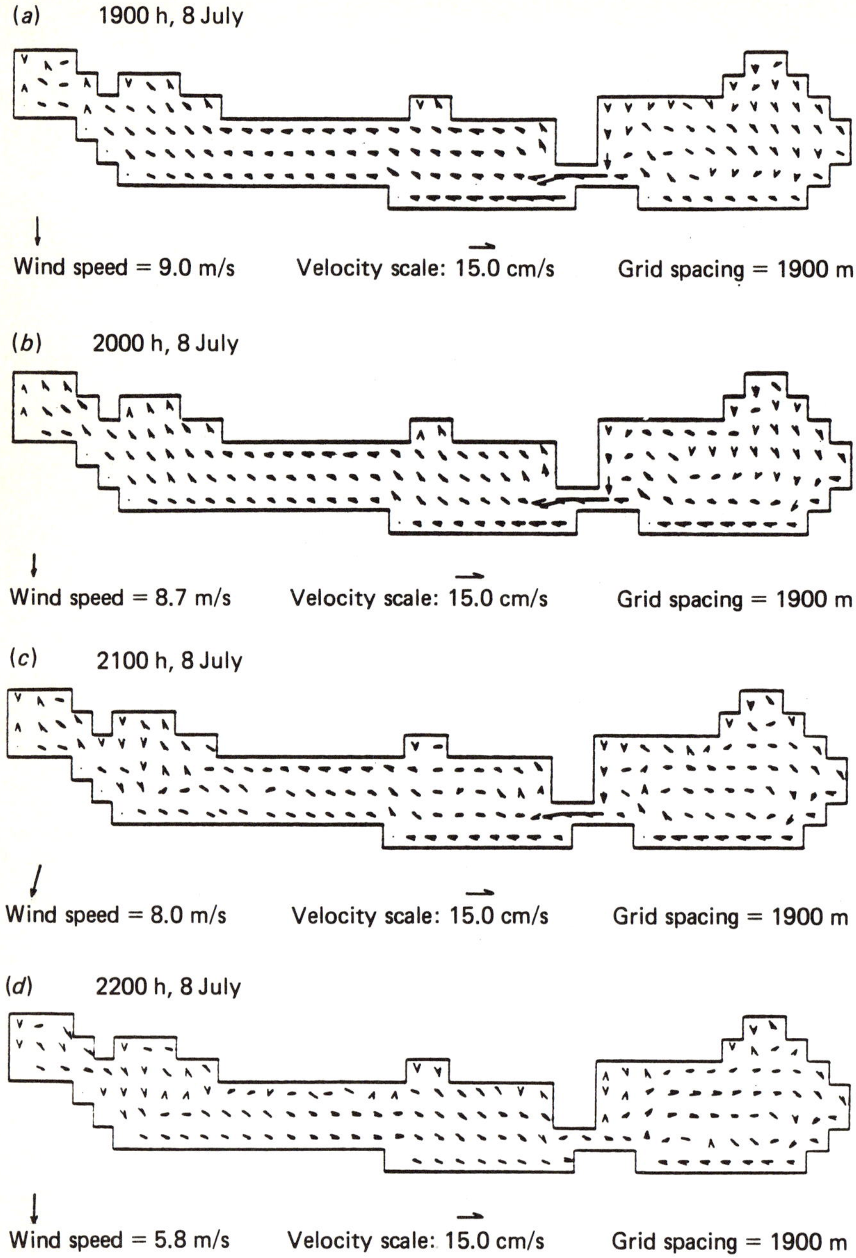

Figure 9.25. Wind-driven circulation predicted in simulation of the event of 8 and 9 July 1963.

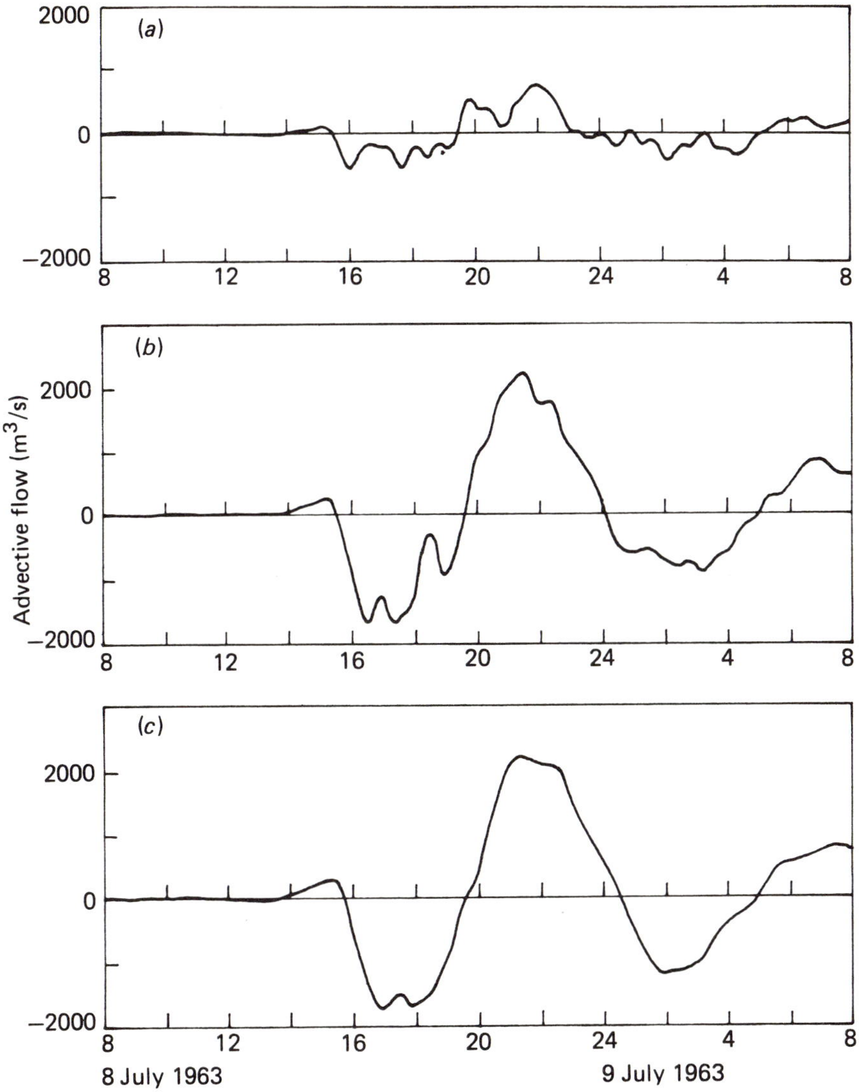

Figure 9.26. Advective flow predicted in simulation of the event of 8 and 9 July 1963. (*a*) Model section 6 (between basins I and II); (*b*) model section 18 (between basinsn II and III); (*c*) model section 29 (between basins III and IV).

course, responsible for dispersion and are thus very important in water quality predictions. The magnitudes of such current variations are shown in Figure 9.27 for two locations along the lake. Shown are the lateral profiles of the depth-average velocity at intervals of 1 h for section 18, typical of the middle of the lake where unidirectional flows dominate, and for section 36, midway in the Siófok basin, where gyre motion is more evident. The velocity profiles at this latter section are considerably less uniform than those at the mid-lake

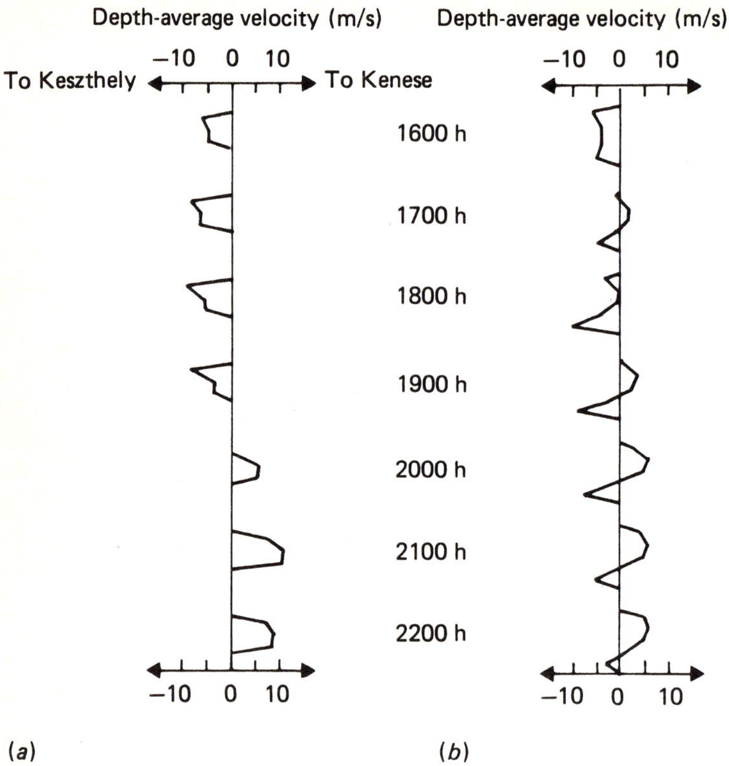

Figure 9.27. Lateral velocity profiles predicted in simulation of the event of 8 and 9 July 1963. (*a*) Model section 18 (between basins II and III); (*b*) model section 36 (mid-basin IV).

location. At either section, flow variations in time are greater than those in the lateral direction.

The results shown in Figure 9.27 indicate significant lateral velocity variations and hence the potential for significant dispersion. As discussed above, a method to compute the longitudinal dispersion coefficient from the circulation model results has been developed, but this determines only the average coefficient during a seiche period – results which are not particularly interesting during a single short event. Time histories of dispersion derived from long-term simulations are presented below, illustrating that significant lateral nonuniformities occur in the predicted wind-induced currents with consequent longitudinal dispersion.

Long-Term Simulation

The 2-D circulation model was used to perform a continuous simulation of events in July and August 1977. For wind data, a record of 3-h average wind speeds and directions measured at Keszthely was used, since Shanahan *et al.*

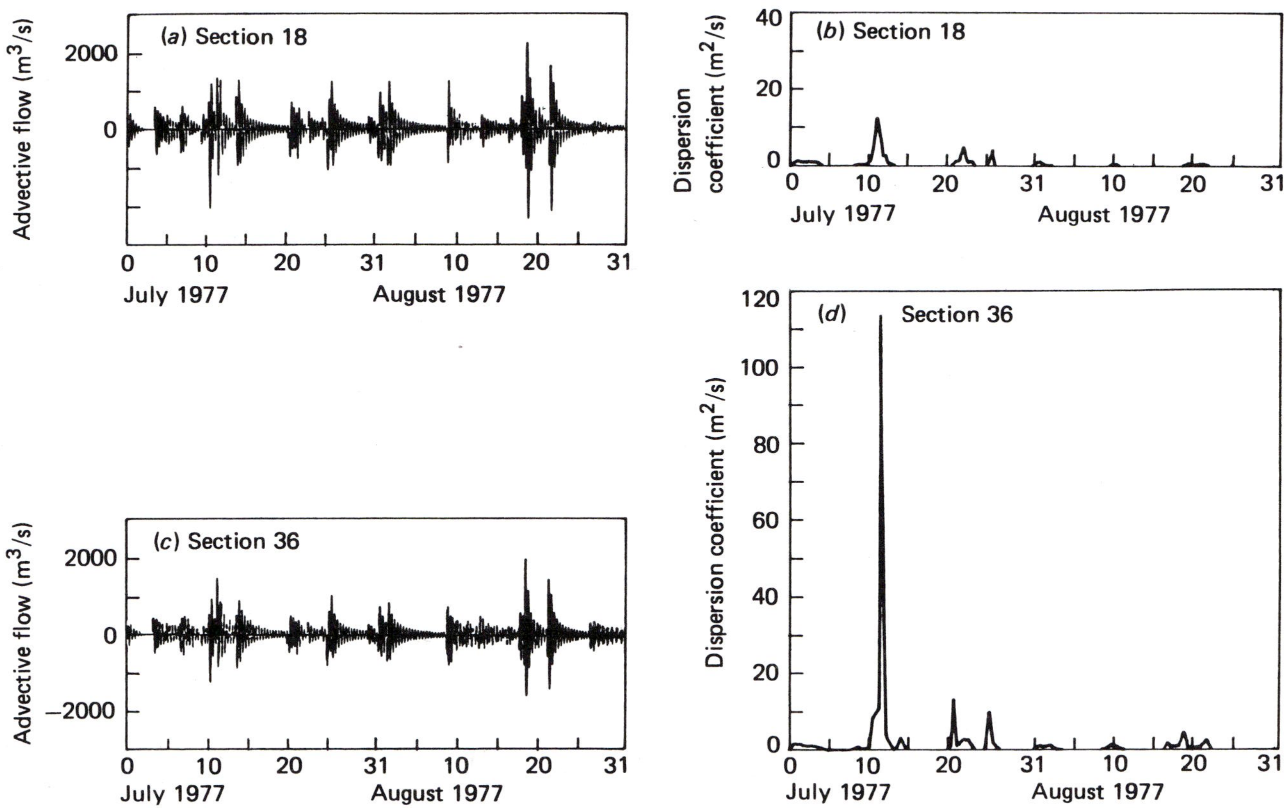

Figure 9.28. Advective flow and dispersion predicted in simulation of July and August 1977. (*a*) Advective flow and (*b*) dispersion coefficient at model section 18; (*c*) advective flow and (*d*) dispersion coefficient at model section 36.

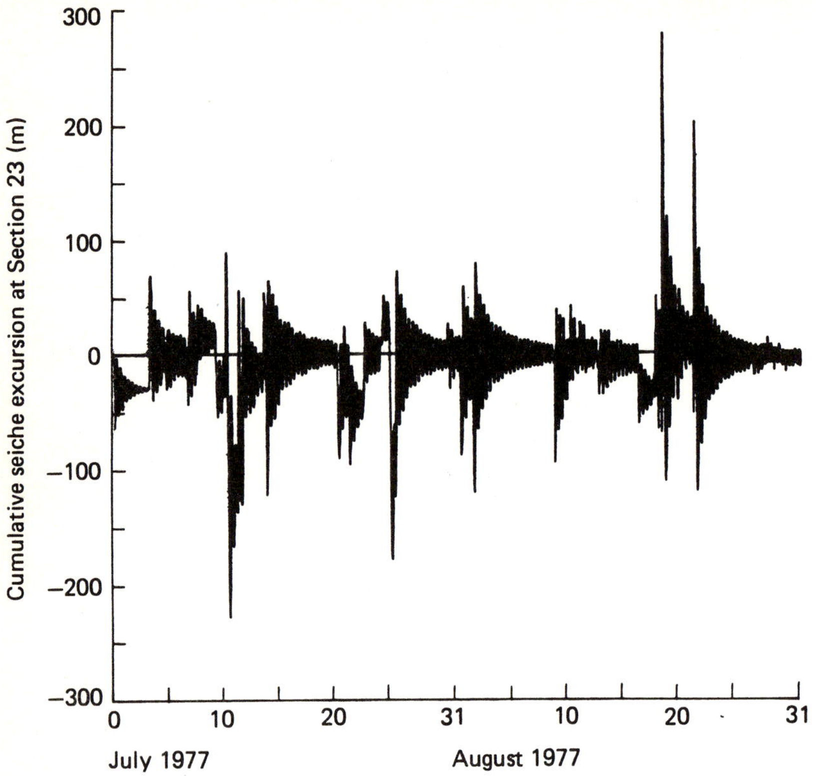

Figure 9.29. Longitudinal translation due to seiche advection predicted in the simulation of July and August 1977.

(1981) had achieved somewhat better results with that record compared with the one from Szemes. The wind speed was adjusted spatially to compensate for the generally lower wind speeds observed at Keszthely. The adjustment procedure followed that described by Shanahan *et al.* (1981).

Typical results from the circulation simulations for 1977 are plotted in Figure 9.28, which shows the longitudinal advection and computed dispersion coefficients at two sections plotted against time through July and August. Both advection and dispersion are highly transient, responding to episodic wind events. The advective motion's oscillatory character is striking, but it should be clear that there is no net translation of water over long time periods due to this component (see also Figure 9.18). This can be seen in Figure 9.29, which shows the cumulative seiche excursion determined from the predicted advective flow as a function of time. The seiche excursion is the average distance water travels during a half period of flow in one direction. The results in Figure 9.29 are for a location midway through basin III, roughly the nodal point of the uninodal seiche. The nodal point is the location along the lake at which the discharge due to the seiche is greatest.

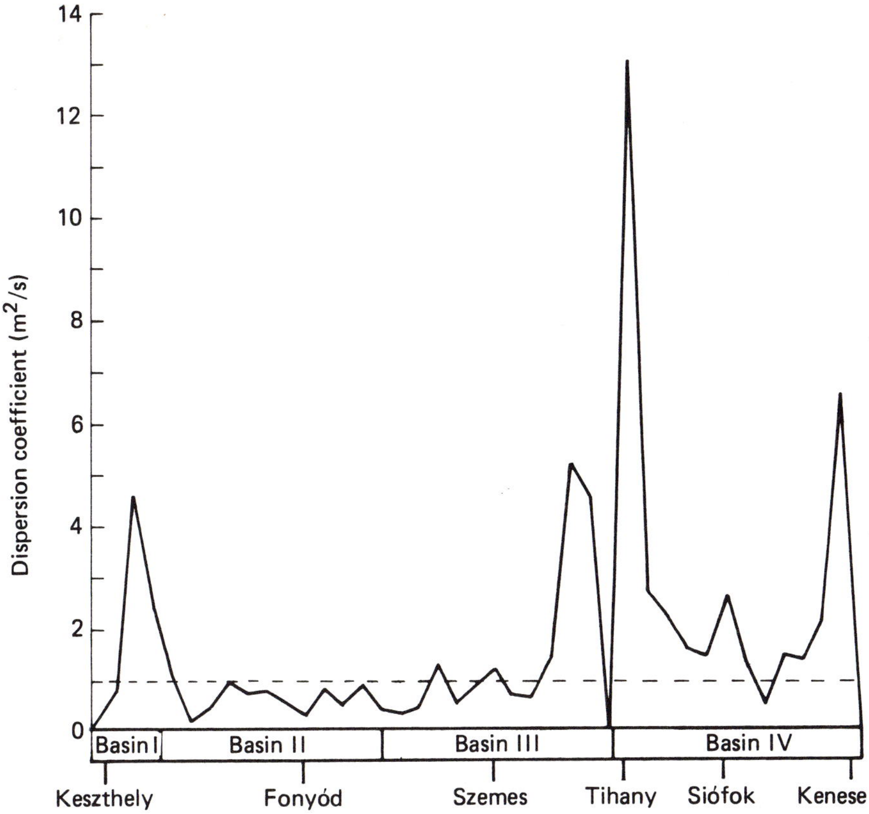

Figure 9.30. Spatial distribution of the computed dispersion coefficient averaged over the simulation of July and August 1977.

The transient character of dispersion is clearly illustrated in Figure 9.28. During most of the time, dispersion is low – strong mixing occurs on an occasional but fairly regular basis due to wind events. The spatial distribution of the average dispersion coefficient over the two-month simulation period is shown in Figure 9.30. A constant value of $D = 1.0$ m^2/s is also shown as a reference value. Dispersion is highest at the locations of greatest change in geometry due to the larger secondary currents at these sections. Also, a somewhat higher dispersion is maintained in the Keszthely basin and the Siófok basin (basin IV) where there is a greater tendency to gyre motion.

Despite the similarities in the calibration of 1-D and 2-D models, there are significant differences in their behavior. In particular, there is a much stronger tendency to dampen seiche oscillations in the 2-D model: (cf. the long-term simulations in Figures 9.17 and 9.28). Probably, the difference in the abilities of the models to resolve spatial detail is partially responsible for these differences in behavior. One can speculate that in the 2-D model significant flows in the shallow parts of the cross section lead to greater damping than can be modeled in the 1-D model.

9.5. Three-Dimensional Model Studies

Model construction

The 3-D model was developed by Shanahan *et al.* (1981) based on a previous one developed by Cooper and Pearce (1977) and later modified by Nelson (1979) for lake circulation studies. The Ekman-type model employs a mixed-solution technique, using a finite-difference formulation to capture horizontal variations in current velocity, but also using a Galerkin technique to determine the horizontal current as a continuous function in the vertical. The Galerkin model permits eddy viscosity to be a variable function of the depth in a piecewise linear fashion. The model derivation is based upon a linear bottom friction law, of the form

$$\tau_b^x = \rho\lambda_b u \qquad \text{and} \qquad \tau_b^y = \rho\lambda_b v \qquad \text{at } z = h \ , \tag{9.28}$$

where λ_b is the bottom friction coefficient. The Galerkin model is fully transient, determining the horizontal current velocities and the water surface elevation that change with time due to unsteady wind forces. For the numerical solution the reader is referred to Shanahan *et al.* (1981).

Application to Lake Balaton

Application of the 3-D model to Lake Balaton is described by Shanahan *et al.* (1981). The procedure follows essentially the same steps as described above for the 1-D and 2-D models. The results achieved were comparable to the 1-D and 2-D models in their ability to duplicate historical events.

Unfortunately, later studies revealed that the calibrations achieved with the 3-D model were empirical at best. The model computer program was found to contain a programming error that affected results when using particular forms of the eddy viscosity function, including the function found by calibration for Lake Balaton. The result of this error was a misaccounting of vertical momentum transport, creating an effective, but artificial, nonlinear bottom friction that could not be equated with any physical characteristics or phenomena.

Correction of the error eliminated the artificial nonlinear bottom friction, leaving only the linear friction relation. Unfortunately, linear bottom friction proved inadequate in Lake Balaton where friction is such an important influence: it was simply impossible to achieve simulation behavior with the damping characteristics of the actual lake. Further, the Galerkin formulation proved incompatible with a nonlinear friction law without creating an unreasonably expensive program to execute. The 3-D model was thus abandoned, the apparently successful calibration being considered fortuitous at best, without sufficient theoretical backing to be reliable.

The failure of the 3-D model supplies dramatic evidence of the importance of bottom friction, and particularly a nonlinear friction law, to the success of a circulation model for shallow lakes. It also illustrates the difficulty in software transfer and in adding the vertical dimension to circulation models of shallow lakes. Problems arise due to the very steep velocity gradients within the shallow water column and the parameterization of bottom friction. The similarity of the 1-D and 2-D model results suggests that the 3-D model would probably predict water surface elevations and net longitudinal transport similarly. The primary unanswered question concerns the ability of the 3-D model to predict vertical velocity structure, the confirmation of which would certainly also call for further detailed field studies.

9.6. Summary and Conclusions

Circulation dynamics in Lake Balaton

Hydrologic flows in to and out of Lake Balaton are minor compared with the dominating wind-driven flows. Physical model studies and satellite photographs reveal a pattern of wind-driven circulation characterized by two-dimensional gyres throughout the lake. Comprehensive field studies have defined the characteristics of Lake Balaton's seiche in some detail. The shallowness of the lake significantly influences the longitudinal seiche by lengthening its period and causing it to damp out after a few cycles.

The wind field over Lake Balaton is characterized by significant transience and spatial nonuniformity. In addition, the lake itself responds very quickly: storms of only 1–2 h duration induce considerable motion in the lake. The variability of the wind and the rapidity of the lake's response leads to the conclusion that the lake practically never reaches a steady state. This also implies a necessity to resolve the wind field in time and space to good accuracy – a task not feasible with the existing wind observation network.

Circulation modeling of Lake Balaton

Three different modeling options were tested in studies of Lake Balaton: one-, two-, and three-dimensional. The 1-D and 2-D models were calibrated and found generally able to replicate the motion within the lake. The 3-D Ekman-type model, with linear bottom friction, was found inadequate for shallow lakes.

Calibration of 1-D and 2-D models depends on the determination of two major parameters, the wind drag and bottom friction coefficients. These affect model performance oppositely, thus it is difficult to find a unique "best" parameter combination. To account for the damping characteristics of the seiche as well, a single-parameter combination is selected for both the 1-D and 2-D models. The selected parameters are a wind drag coefficient, C_D = 0.0013, and a bottom friction factor, λ_b = 0.003.

The 1-D and 2-D models have been successfully verified in comparisons with field observations of historical events when wind is directed longitudinally along the lake. For winds across the lake, the model cannot be validated in deterministic simulations. Analysis of the situation shows that the poor verification is largely the result of uncertainties in the wind direction. These errors are particularly serious when the wind is directed across the lake, as for the predominant northerly winds. Under cross-wind conditions, even small errors in direction lead to significantly different wind forces, including a reverse in longitudinal direction. Thus errors in wind direction can completely distort the time pattern of the simulation.

The impact of wind direction uncertainty has been assessed in Monte Carlo simulations performed with the 1-D model, which resulted in a successful verification, in a stochastic sense, even for cross-wind conditions (when the uncertainty bounds are the greatest). The analysis leads to general conclusions valid for both the 1-D and 2-D models – that input data sensitivity dominates parameter sensitivity in Lake Balaton circulation models.

Simulations performed with the 2-D model indicate the character of lateral velocity variation in the lake. Model results show that longitudinal fluxes due mainly to seiche motion are very strong and highly transient. The flows are generally unidirectional across the lake section, although they are occasionally interrupted by flow reversals along one shore or the other. Reversals are particularly important where the lateral sections change abruptly, as at either end of the lake and either side of the Tihany Peninsula, where horizontal flow gyres tend to occur. Lateral nonuniformities in velocity lead to the dispersion of dissolved or suspended material in the lake water and thus influence lake water quality.

One-dimensional versus two-dimensional model results

The capabilities of the 1-D and 2-D models overlap considerably, but differ in a few key respects. The areas of overlap allow direct comparison of the predictions made by the two models. With respect to the model calibrations, the wind drag and bottom friction parameter calibrations are identical. For model verification, there is a single event which was simulated with both models – 8 July 1963. Despite the fact that different wind records were used, the predictions by the two models are generally similar in character. Finally, both models were used to simulate the period of July and August 1977. Judging from the predicted advection (Figures 9.17 and 9.28) there are essential differences in the damping behavior of the two models. It is probable that the models' different abilities to resolve spatial variations in depth are partly responsible for the discrepancies. In general, the 1-D and 2-D models make predictions similar in character, except with respect to the damping of seiche motion.

Implications for eutrophication modeling

Mass transport is the link between lake circulation and water quality. Two types of mass transport information are required to effect this linkage in the framework of a 1-D water quality model. The first is advection, the net longitudinal flow determined over a lateral cross section. As is discussed in Chapter 10, this is effectively captured for the water quality model by the long-term hydrologic flow. The second component of mass transport is dispersion, the longitudinal mixing due to a nonuniform lateral velocity. This information is often assessed by estimation and calibration of the water quality model. However, 2-D or 3-D circulation model results can be used to determine velocity nonuniformity and thus dispersion directly. For the linked hydrodynamic–biochemical model developed for the Lake Balaton study, the 2-D circulation model was employed to compute longitudinal dispersion as a function of time. The development of the linked model – and the role of the 2-D circulation model in the linkage – is the topic of Chapter 10.

References

Béll, B. and Takács, L. (Eds.) (1974) ***The Climate of the Lake Balaton Region***, Vol. XL (Budapest: Hungarian Meteorological Institute) (in Hungarian).

Bengtsson, L. (1978) Mathematical models of wind induced circulation in a lake, in ***Proc. Hydrology of Lakes Symposium, Helsinki***. IAHS-AISH Publication 109, pp 123–32.

Bonham-Carter, G. and Thomas, J.H. (1973) Numerical calculation of steady wind-driven currents in Lake Ontario and Rochester embayment, in ***Proc. 16th Conf. on Great Lakes Research***, p 640.

Brown, R.T. (1978) ***Effects of Data Variability and Model Structure on Reservoir Ecosystem Simulation***. PhD Thesis, Dept. of Civil Engineering, MIT.

Chatwin, P.C. (1975) On the longitudinal dispersion of passive contaminant in oscillatory flow in tubes. ***J. Fluid Mechanics*** 71(3):513.

Cheng, R.T., Powell, T.M., and Dillon, T.M. (1976) Numerical models of wind-driven circulation in lakes. ***Appl. Math. Modeling*** 1:141.

Cooper, C.K. and Pearce, B.R. (1977) ***A Three-Dimensional Model to Calculate Currents in Coastal Waters Utilizing a Depth Varying Vertical Eddy Viscosity***. Report No. 226 (Cambridge, MA: Ralph M. Parsons Laboratory, Dept. of Civil Engineering, MIT).

Edinger, J.E. and Buchak, E.M. (1979) Reservoir longitudinal and vertical implicit hydrodynamics, in E.E. Driver and W.O. Wunderlich (Eds.) ***Environmental Effects of Hydraulic Engineering Works***. Proc. Int. Symposium, Knoxville, TN, September 1978 (Knoxville, TN: Tennessee Valley Authority).

Fischer, H.B. (1969) ***Cross-Sectional Time Scales and Dispersion in Estuaries***. Proc. of 13th Congress IAHR, Kyoto, Japan, IAHR, Vol. 3, p 173.

Fisher, I. (1980) ***Procedure for Obtaining Wind-Induced Circulation Patterns for a Lake Using the Laplace Transform Model of Young and Liggett (1977)***. IIASA Internal Report (Laxenburg, Austria: International Institute for Applied Systems Analysis).

Gedney, R.T. and Lick, W. (1972) Wind-driven currents in Lake Erie. ***J. Geophys. Research*** 77(15):2714.

Graf, W.H. and Prost, J.P. (1980) Aerodynamic drag and its relation to the sea state, with data from Lake Geneva. ***Archiv Meteorol., Geophis. und Bioklimatol.*** Series A, 29:67–87.

Gyórke, O. (1975) Studies into the factors affecting the morphological processes in shallow lakes by means of a hydraulic model, in ***Fundamental Tools to be Used in Environmental Problems***. 16th Congress IAHR, São Paulo, Brazil, Vol. 2, p 24.

Hicks, B.B., Drinkrow, R.L., and Grauze, G. (1974) Drag and bulk transfer coefficients associated with a shallow water surface. *Boundary Layer Meteorol.* 6:287.

Holley, E.R., Harleman, D.R.F., and Fischer, H.B. (1970) Dispersion in homogeneous estuary flow. *ASCE J. Hydraulics Div.* 96(HY8):1691.

Hutchinson, G.E. (1975) *A Treatise on Limnology*, Vol. 1, Part 1, *Geography and Physics of Lakes* (New York: Wiley).

Kozák, M. (1977) *Computation of Unsteady Flow in Open Channels* (Budapest: Akadémiai Kiadó) p 341 (in Hungarian).

Lick, W. (1976) Numerical modeling of lake currents. *Ann. Rev. Earth and Planetary Sciences* 4:49.

Lien, S.L. and Hoopes, J.A. (1978) Wind driven, steady flows in Lake Superior. *Limnol. and Oceanogr.* 23(1):91.

Liggett, J.A. (1970) Cell method for computing lake circulation. *ASCE J. Hydraulics Div.* 96(HY3):725.

Lindijer, G.J.H. (1976) *Three-Dimensional Circulation Models for Shallow Lakes and Seas; General Formulation and Stationary Models.* Report R 900-I (Delft, Netherlands: Delft Hydraulics Laboratory).

Lindijer, G.J.H. (1979) *Three-Dimensional Circulation Models for Shallow Lakes and Seas; Semi-Analytical Steady State Analysis of Wind-Driven Current Including the Effect of a Depth-Dependent Turbulent Exchange Coefficient.* Report R 900-II (Delft, Netherlands: Delft Hydraulics Laboratory).

Liu, H. and Perez, H.J. (1971) Wind-induced circulation in shallow water. *ASCE J. Hydraulics Div.* 97(HY7):923.

Mahmood, K. and Yevjevich, V. (Eds.) (1975) *Unsteady Flow in Open Channels* (Fort Collins, CO: Water Resources Publications).

Muszkalay, L. (1973) *Characteristic Water Motion in Lake Balaton* (Budapest: VITUKI) (in Hungarian).

Nelson, S.M. (1979) *Application of a Numerical Model for Three-Dimensional Sheared Flows.* MS Thesis, Dept. of Civil Engineering, MIT.

Plate, E.J. (1970) Water surface velocities induced by wind shear. *ASCE J. Eng. Mechanics Div.* 96(EM3):295.

Shanahan, P. and Harleman, D.R.F. (1982) *Linked Hydrodynamic and Biogeochemical Models of Water Quality in Shallow Lakes.* Report No. 268 (Cambridge, MA: Ralph M. Parsons Laboratory, Dept. of Civil Engineering, MIT).

Shanahan, P., Harleman, D.R.F., and Somlyódy, L. (1981) *Modeling Wind-Driven Circulation in Lake Balaton.* Collaborative Paper CP-81-7 (Laxenburg, Austria: International Institute for Applied Systems Analysis).

Sibul, O. (1955) *Laboratory Study of Wind Tides in Shallow Water.* Technical Memorandum No. 61, Beach Erosion Board, US Army Corps of Engineers.

Simons, T.J. (1979) *Hydrodynamic Models of Lakes and Shallow Seas.* Unpublished draft report, Canada Centre for Inland Waters, Burlington, Ontario.

Somlyódy, L. (1983) Input data uncertainty and parameter sensitivity in a lake hydrodynamic model, in M.B. Beck and G. van Straten (Eds.) *Uncertainty and Forecasting Water Quality* (New York: Springer) pp 129–55.

Somlyódy, L. and Virtanen, M. (1982) *Application of a One-Dimensional Hydrodynamic Model to Lake Balaton.* Collaborative Paper CP-82-47 (Laxenburg, Austria: International Institute for Applied Systems Analysis).

Su, C.L., Pohl, J.H., and Shih, G. (1976) Wind-driven circulation in Lake Okeechobee, in W.G. Gray, G.F. Pinder, and C.A. Brebbia (Eds.) *Finite Elements in Water Resources.* Proc. 1st Int. Conf. on Finite Elements in Water Resources, Princeton University (London: Pentech) p 4.113.

van Dorn, W.G. (1953) Wind stress on an artificial pond. *J. Marine Research* 12(3):249.

van Straten, G. and Somlyódy, L. (1980) ***Lake Balaton Eutrophication Study: Present Status and Future Program.*** Working Paper WP-80-187 (Laxenburg, Austria: International Institute for Applied Systems Analysis).

van Straten, G., Jolánkai, G., and Herodek, S. (1979) ***Review and Evaluation of Research on the Eutrophication of Lake Balaton – A Background Report for Modeling.*** Collaborative Paper CP-79-13 (Laxenburg, Austria: International Institute for Applied Systems Analysis).

Virtanen, M. (1978) ***Computation of Two-Dimensional Flow in Shallow Water Systems.*** Report prepared for the 9th Int. Post-Graduate Course on Hydrological Methods for Developing Water Resources Management (Budapest: UNESCO–VITUKI).

VITUKI (1976) ***Standard Cross-Sections of Lake Balaton*** (Budapest: VITUKI).

Vreugdenhil, C.B. (1973) ***Computational Methods for the Vertical Distribution of Flow in Shallow Water.*** Report W152 (Delft, Netherlands: Delft Hydraulics Laboratory).

Wu, J. (1969) Wind stress and surface roughness at air–sea interface. *J. **Geophys. Research*** 74(2):444–55.

Young, F.D.L. and Liggett, J.A. (1977) Transient finite element shallow lake circulation. *ASCE J. **Hydraulics Div.*** 103:109–21.

CHAPTER 10

Lake Eutrophication Model: A Coupled Hydrophysical–Ecological Model

P. Shanahan and D.R.F. Harleman

10.1. Introduction

Chapter 2 introduces the principle of decomposition and aggregation as an important basis of the approach to the Lake Balaton study. In Section 2.2, the principle is applied to the development of the lake eutrophication model (LEM), distinguishing between alternative levels of model complexity and detail. The four-box model, a "discrete" approach, and the one-dimensional (1-D) "continuous" model are first introduced in Section 2.2. The continuous model is capable of greater precision in representing the spatial differences caused by loading distribution, wind-induced circulation, and mass transport. Furthermore, the continuous model should more realistically reflect dynamic changes in water quality that are due to unsteady mass transport. Nevertheless, – as will be shown – the four-box model structure is an acceptable, albeit less accurate, modeling approach.

Four-box LEMs have been developed and applied to Lake Balaton (see Chapters 3, 11, 12, and 13). However, as stressed in Section 2.2, this simplified approach to modeling spatial structure is acceptable only if the four-box model results agree reasonably with results from the more spatially detailed continuous model. To assess the agreement between these modeling approaches, the development of a spatially detailed model that couples hydrophysical and ecological (or biochemical) submodels is presented in this chapter (see Figure 2.1). The results are then compared with a spatially simple four-box model that incorporates the same ecological submodel. Since the two models employ an identical ecological formulation, the differences in the model results over time and space can be attributed entirely to the effects of hydrodynamic transport and loading distribution.

This chapter offers not only conclusions specific to the Balaton study, but also general guidelines for the construction and spatial segmentation of a lake water quality model.

10.2 Analysis of Model Transport

Justification for investigating the coupled 1-D model is mainly provided by an analytical examination of transport in lake water quality models (Shanahan and Harleman 1984). The analysis is based upon conceptual reactor models that are analogous to the multiple-box 1-D lake models. One common reactor model, the tanks-in-series reactor, consists of sequential, fully mixed tanks and is an analog of the multiple-box model. Another, the dispersed flow reactor, approximates the 1-D lake model. Figure 10.1 illustrates these conceptual reactors schematically and defines the reactor parameters. The parameters include the Peclet number, which is the dimensionless ratio of advective to dispersive transport.

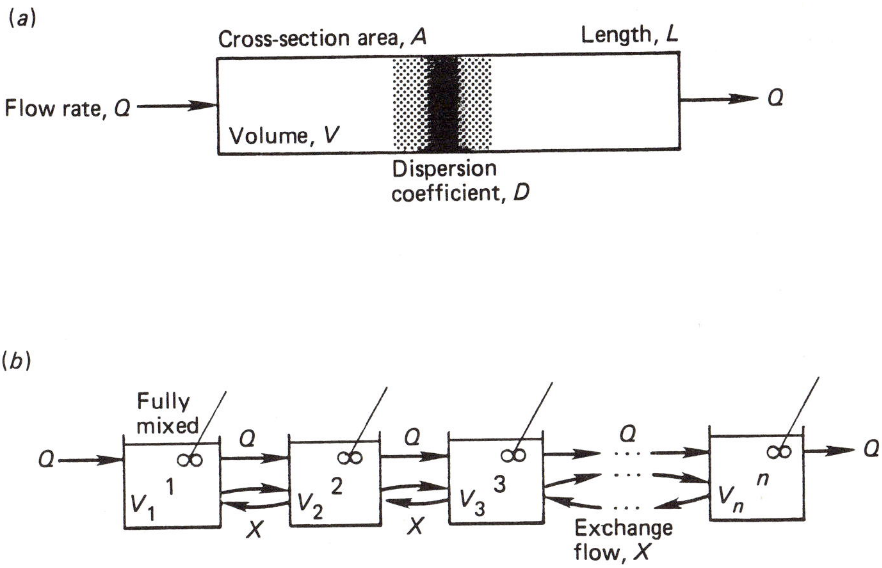

Figure 10.1. Schematic representation of conceptual reactors. (*a*) Dispersed flow reactor; (*b*) tanks-in-series reactor with exchange flow.

Modulation of the reactor parameters leads to a continuous variation in the flow and mixing characteristics of the tanks-in-series and dispersed flow reactors. These characteristics are conveniently evaluated by the concentration changes observed in the reactor outflow in response to pulse injection of a conservative, unreactive tracer at the inflow to the tank (Figure 10.2).

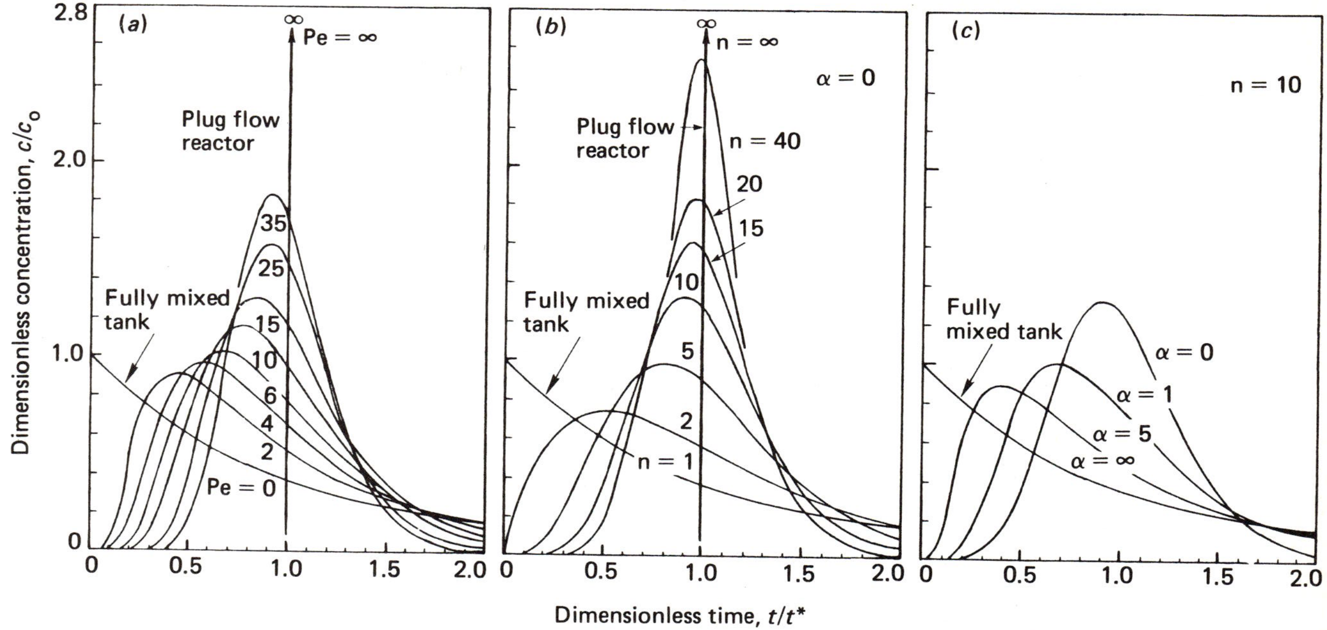

Figure 10.2. Conceptual reactor impulse responses. (*a*) Dispersed flow reactor (Thomas and McKee 1944); (*b*) tanks-in-series reactor without exchange flow; (*c*) 10 tanks-in-series reactor with exchange flow (Tuan *et al.* 1980).

The extremes of the response spectrum are the infinite dispersion of the fully mixed tank and the zero dispersion of the plug flow reactor. The degree of mixing is increased in the dispersed flow reactor as the dispersion coefficient increases from zero (the plug flow limit) to infinity (the fully mixed tank limit); the Peclet number correspondingly decreases from infinity to zero. In the tanks-in-series reactor, the degree of mixing increases as the number of tanks decreases [Figure 10.2(b)] or as the exchange flow increases [Figure 10.2(c)].

The response curves of the tanks-in-series reactor [Figures 10.2(b) and 10.2(c)] are similar in character and variation to those of the dispersed flow reactor [Figure 10.2(a)]. This similarity has prompted a number of researchers to relate their characteristics mathematically. Levenspiel and Bischoff (1963) give the approximate relation (10.1), valid for large n and Pe, for the tanks-in-series reactor without exchange flow:

$$\mathrm{Pe} = 2n - 1 \simeq 2n \quad . \tag{10.1}$$

Zvirin and Shinnar (1976a, 1976b) give relation (10.2) for the tanks-in-series reactor with exchange flow:

$$\mathrm{Pe} = \frac{2n}{1 + 2\alpha} \quad . \tag{10.2}$$

Both equations (10.1) and (10.2) presume that the volumes of all tanks in the tanks-in-series reactor are equal.

Equations (10.1) and (10.2) quantify the dispersive character of the tanks-in-series reactor as a function of the number of tanks and the exchange flow. Although the resultant dispersion appears to be similar in the tanks-in-series and dispersed flow reactor responses (Figure 10.2), there is a critical operational difference in the analogous lake model formulations. If the lake model is 1-D, the analog of the dispersed flow reactor, its mixing characteristics are determined by the specified dispersion coefficient. Some so-called numerical dispersion may also be present, but in a properly constructed finite-difference model it will contribute negligibly. On the hand, if the lake model is a multiple-box type, the exact analog of the tanks-in-series reactor, its dispersive character will be determined by the number of boxes. Since this dispersion is an implicit function of the model geometry, Shanahan and Harleman (1984) define it as "implicit dispersion."

In the development of most lake models, the number of boxes is chosen *a priori*, based on the geometry of the lake. The implicit dispersion of these models falls into one of three categories. If the boxes are too few, the model will be overly dispersive by virtue of its implicit dispersion, and so is inherently unable to accurately represent transport in the lake. If the number of boxes carries an implicit dispersion approximately that of the actual lake, then the model can accurately portray transport in the lake, but only if the exchange flow is zero. If the number of boxes is sufficiently large, then the implicit dispersion will be less than that of the lake. To compensate, exchange flows must be specified to increase the dispersive transport in the model. However, the exchange flow is not directly related to any observable characteristics of the lake; rather, it is a model-dependent property since it varies as the number of model boxes varies. It can be determined only by calibration. To maintain the separation of model transport from model biochemistry, calibration must be based upon an independent, and preferably conservative, constituent in the lake, and it should not be accomplished using the model state variables as a standard.

The above analysis can be applied to the Lake Balaton four-box model using a modified version of equation (10.1) to consider the effect of boxes of unequal volume. Zvirin and Shinnar (1976b) give equation (10.3) for the local Peclet number for the ith tank in a series of n unequal tanks.

$$\mathrm{Pe}_i = \frac{2V}{V_i} \frac{Q}{Q + X_i + X_{i+1}} \tag{10.3}$$

where $V = V_1 + V_2 + \cdots + V_n$ is the total reactor volume, and X_i is the exchange flow between tank $i + 1$ and tank i.

The local Peclet numbers for the Lake Balaton four-box models are tabulated in Table 10.1 and compared with average dispersion coefficients predicted by the 2-D circulation model (see Chapter 9). The results show reasonable agreement between the dispersion given by the circulation model and that implicit in the four-box model without exchange flow. Particularly fortunate in this respect is the presence of a small-volume box, and correspondingly low dispersion, at Keszthely. The higher dispersion in the remaining boxes is less critical since concentration gradients are more gentle east of Keszthely Bay. The steep concentration gradients associated with the Zala River nutrient load at Keszthely Bay, however, make the low dispersion in that box a critical factor in the success of the four-box model. The addition of minimal exchange flow affects the four-box model only slightly, but at its maximum value the exchange flow employed in SIMBAL (Chapters 3 and 11) is too great, and exaggerates dispersion in the lake.

Our conclusion is that the spatial discretization of the Lake Balaton four-box models creates an amount of implicit dispersion which is fortunately in rough agreement with that of the actual lake. Nonetheless, the preceding analysis has shown that this need not always be the case – indeed, inattentive selection of box boundaries may create a model intrinsically unable to portray dispersive transport in the lake. The remedy is careful evaluation of the box model's implicit dispersion and resegmentation of the model as warranted. The following step-by-step procedure is recommended for developing box models (Shanahan and Harleman 1984):

Step 1 Specify box boundaries based on lake geometry, locations of major loading sources or field sampling stations, and similarly important features in the lake.

Step 2 Determine the net (advective) flow between the boxes from available hydrologic data.

Step 3 Determine the implicit dispersion of the box model via equation (10.1) or (10.2).

Table 10.1. Dispersion in the Lake Balaton models.

	Box number						
	1		2		3		4
SIMBAL MODEL							
Model properties							
Volume (10^6 m^3)	82		413		600		802
Throughflow (m^3/s)	10.4		10.4		10.4		10.4
SIMBAL, maximum exchange flow							
Exchange flow (m^3/s)		37.0		41.5		10.3	
Local Peclet number	11.4		1.1		1.1		2.4
Equivalent dispersion coefficient (m^2/s)	5.9		37.7		27.2		9.3
SIMBAL, minimum exchange flow							
Exchange flow (m^3/s)		3.0		3.5		0.9	
Local Peclet number	35.9		5.7		4.4		4.4
Equivalent dispersion coefficient (m^2/s)	1.9		7.3		6.8		5.1
TWO-DIMENSIONAL CIRCULATION MODEL							
Dispersion coefficient computed by circulation model, averaged by basin and over time (m^2/s)	2.6		0.6		1.6		3.1

Step 4 Estimate the dispersion of the actual lake from field measurements, studies of similar lakes, or approximate relations such as those given by Liu (1977) or Murthy and Okubo (1977).

Step 5 Compare the dispersion of the multiple-box model with the estimated actual dispersion and determine which of the three possible tanks-in-series model regimes applies.

Step 6 If the box model is overdispersive, repeat the procedure beginning with Step 1 using smaller box sizes.

Step 7 If the box model is underdispersive, determine the correct value of exchange flows by calibration against a conservative constituent measured in the lake.

Step 8 Verify the advective and dispersive transports against measurements of conservative constituents. The verification field data must be independent of the calibration data employed in Step 7.

The above is a sound, albeit empirical, procedure for the development of lake water quality models. The procedure isolates the calibration of model transport from the calibration of water quality model kinetics, which both simplifies the construction of the model and strengthens its overall calibration.

In some cases, the relatively fine discretization of a finite-difference model may lead to a far more straightforward treatment of transport. The fine discretization approach is employed in the coupled hydrophysical–ecological model in order to contrast a direct treatment of transport with that of the four-box models. The following sections describe the coupled model further.

10.3 Construction of the Coupled Model

Analysis of mass conservation equation

In order to judge the relative importance of transport compared to biochemical reactions and load terms, an order of magnitude analysis is performed on equation (2.2), which for a single water quality component takes the form of

$$\frac{\partial C}{\partial t} = -(u + u')\frac{\partial C}{\partial x} + \frac{\partial}{\partial x}\left[D\frac{\partial C}{\partial x}\right] + R(C) + L_e + L_i(C) \quad . \tag{10.4}$$

Here, the notation is the same as in Chapter 2, except that velocity is separated into that of the hydrologic throughflow, u, and the wind induced circulation, u' (see Chapter 9), respectively.

First the influence of the term $u'\partial C/\partial x$ is discussed. As was shown in Chapter 9, the back-and-forth motion of the seiche creates zero net advection over the long-term. In the short term, particles are oscillating in the x direction around their original position (excursion) – similarly to tidal motion – but the changes are so fast that they have no influence on concentration and thus the daily average – as can also be demonstrated by a transport model – of C remains unchanged. The conclusion is that the term $u'\partial C/\partial x$ can be neglected in equation (10.4).

Next equation (10.4) is converted into a nondimensional form

$$\frac{\partial C^*}{\partial t} = \frac{U_0 t_0}{x_0} U^* \frac{\partial C^*}{\partial x^*} + \frac{D_0 t_0}{x_0^2}\frac{\partial}{\partial x^*}\left[D^*\frac{\partial C^*}{\partial x^*}\right] + \frac{L_{e0} t_0}{C_0}(R^* + L_e^* + L_i^*) \quad , \tag{10.5}$$

where

$$C^* = C/C_0 \ , \qquad t^* = t/t_0 \ , \qquad x^* = x/x_0 \ ,$$
$$U^* = U/U_0 \ , \qquad D^* = D/D_0 \ ,$$
$$L_e^* = L_e/L_{e0} \ , \qquad L_i^* = L_i/L_{e0} \ , \qquad R^* = R/L_{e0} \ .$$

In equation (10.5) the relative magnitude of the transport terms on the right-hand side is given by the dimensionless quantities $U_0 t_0/x_0$ for advection and $D_0 t_0/x_0^2$ for dispersion. The characteristic value for x_0 is logically chosen as the length of the lake, about 78 km; the characteristic time selected for t_0 is 2 years, approximately the filling time (see Chapter 1), while u_0 was taken as an average hydrologic flow of the order of 0.5 mm/s (40 m/d). Selection of D_0 is less straightforward since dispersion, as shown in Chapter 9, can reach instantaneous values as high as 100 m^2/s; here we assume a range between 0 and 14 m^2/s.

The order of magnitude analysis of equation (10.4) is completed by estimating the order of the loading and reaction term. We use the biologically available P load (465 kg/d, see Chapter 6) divided by the total lake volume for L_{eo}. With C_0 taken as a typical total P concentration of 0.05 g/m^3, t_0 again as the filling time, the magnitude of the loading and reaction term is $L_{eo}t_0/C_0 \simeq 3.6$. The magnitudes of the various terms of equation (10.4) are summarized in Table 10.2 (similar figures can be arrived separately, e.g., for the Keszthely Bay).

Table 10.2. Scaling analysis of mass conservation equation for Lake Balaton.

Process	Term in equation (10.4)	Dimensionless number	Value
Time rate of change	$\frac{\partial C}{\partial t}$	–	1
Advection (due to throughflow)	$U\frac{\partial C}{\partial x}$	$\frac{U_0 t_0}{x_0}$	0.8
Dispersion (due to seiche)	$\frac{\partial}{\partial x} D \frac{\partial C}{\partial x}$	$\frac{D_0 t_0}{x_0^2}$	0–0.3
Loading and reaction	$L_e + L_i + R$	$\frac{L_{eo} t_0}{C_0}$	3.6

The conclusion to be drawn from the analysis and Table 10.2 is that advection, dispersion, reaction, and loading must all be considered influential in the P dynamics of Lake Balaton. Accordingly, all must be included in a mathematical model of the lake's water quality.

Selection of model components

An impediment to constructing the coupled hydrophysical–ecological model is the necessity to cross disciplines, that is, to include expertise from ecology, hydrodynamics, mathematics, and other disparate fields. The Lake Balaton case study brought these disciplines together, thus making it possible to build the coupled model as a combination of already existing model components, as shown schematically in Figure 10.3.

The ecological component of the coupled model is the appropriately simple SIMBAL model, described in Chapters 3 and 11. Simpler models than SIMBAL would be unlikely to capture the essential character of eutrophication in the lake, yet more complex models might require too great a computational effort for the coupled model to become a practical tool.

The external nutrient loading component is supplied by the detailed loading estimates compiled by Jolánkai and Somlyódy (1981), as summarized in

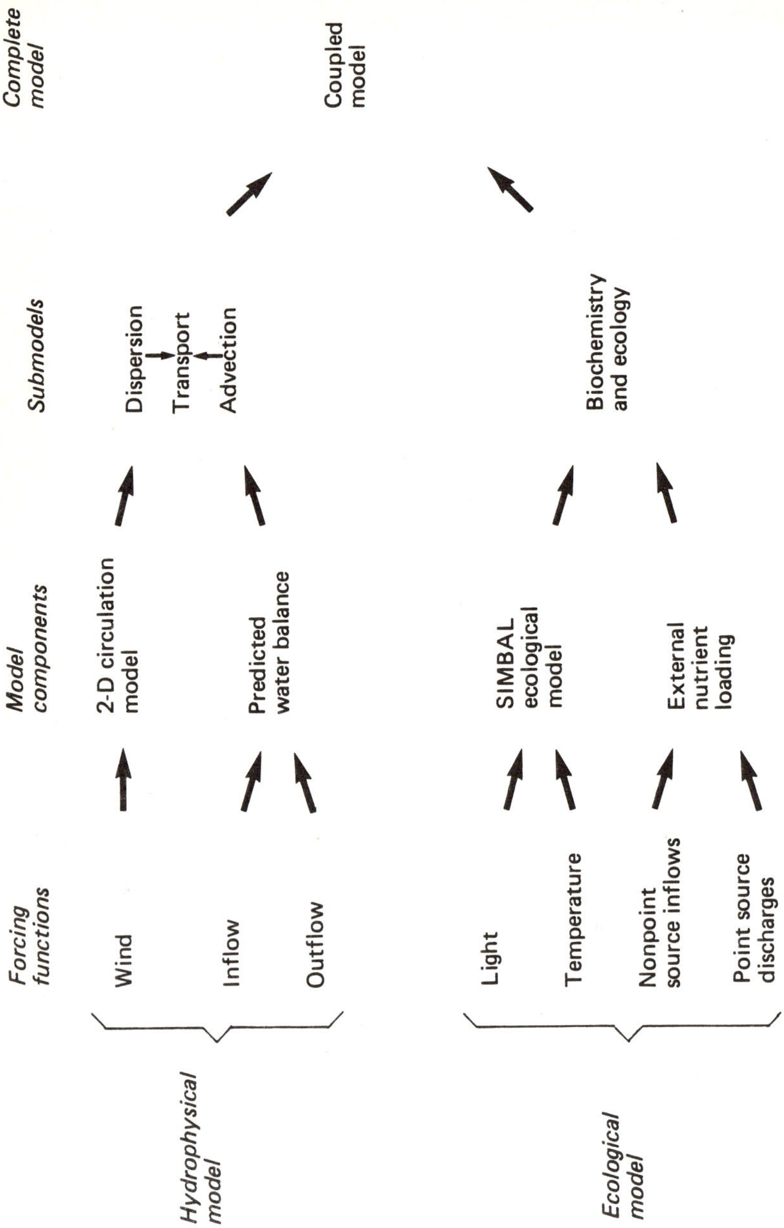

Figure 10.3. Model construction for coupled hydrophysical–ecological model.

Chapter 6. The loadings used in the coupled model are consistent with those used in the box models described in Chapters 3 and 11–13, with the exception that a finer spatial distribution is necessary in the coupled model with respect to the box models.

Spatial water balances are computed using the method given by Baranyi (1973), the essence of which is to determine the inflow and outflow to each model segment in such a way as to uniformly distribute the total net inflow or outflow to the lake. The balance is computed on a monthly basis using observed or calculated values of inflow, outflow, precipitation, and evaporation.

The hydrodynamic component is the 2-D circulation model developed by Shanahan and Harleman (1982). This model is used to compute the 1-D dispersion coefficient as a function of time and space and is described in Chapter 9. The inclusion of hydrodynamic inputs that vary over time and space is perhaps the single most important difference between the coupled model and its box model predecessors.

The inputs from the four model components are integrated (Figure 10.3) to form the coupled model, which determines the concentration of the four P components of the SIMBAL model. The solution assumes a 1-D spatial structure that is represented by finite difference techniques. The lake is divided into forty 1900-m grid cells: thus, the coupled model has a spatial resolution an order of magnitude greater than the four-box models of Chapter 3.

10.4 Model Solution

Finite difference formulation

The coupled model simultaneously solves the mass conservation equations [equation (10.4)] of four P components, using a fractional-step finite-difference method. The fractional-step method constructs the finite-difference approximation to equation (10.4) by breaking the solution into a series of three subproblems (Shanahan and Harleman (1982): advection, dispersion, and reaction and loading (see Tables 10.2 and 10.3).

The advantage of the fractional-step method lies primarily in its ability to separate approximately the nonlinear reaction solution from those for advection and dispersion, thus leading to a computationally economic procedure.

Simulation procedures

The coupled model was run in various simulations, all of which have some common features and were performed for a single algal growth period, 250 days beginning 28 February 1977. This particular year was selected for simulation because the necessary loading and forcing function data were available.

Table 10.3. Steps in numerical solution of mass conservation equation.

Advective step

$$C_i^{*} = C_i^t - \frac{\Delta t_a}{V_i}\left\{Q_{i+1}^t\left[\frac{1}{2}\left[\vartheta_a(C_i^t + C_{i+1}^t) + (1-\vartheta_a)(C_i^{*} + C_{i+1}^{*})\right]\right]\right.$$
$$- Q_i^t\left[\frac{1}{2}\left[\vartheta_a(C_{i-1}^t + C_i^t) + (1-\vartheta_a)(C_{i-1}^{*} + C_i^{*})\right]\right]$$
$$\left. - Q_{\text{out},i}^t\left[\vartheta_a C_i^t + (1-\vartheta_a)C_i^{*}\right]\right\}$$

Dispersive step

$$C_i^{**} = C_i^{*} + \frac{\Delta t_d}{V_i}\left\{X_{i+1}^t\left[\vartheta_d(C_{i+1}^{*} - C_i^{*}) + (1-\vartheta_d)(C_{i+1}^{**} - C_i^{**})\right]\right.$$
$$\left. - X_i^t\left[\vartheta_d(C_i^{*} - C_{i-1}^{*}) + (1-\vartheta_d)(C_i^{**} - C_{i-1}^{**})\right]\right\}$$

Reactive step

$$C_i^{t+\Delta t} = C_i^{**} + \Delta t_r R_i^t(C_i^{t+\Delta t}, C_i^{**}) + \frac{\Delta t_r}{V_i} L_i^t$$

Notation

C_i^t	concentration vector for grid i at time t
Δt	time step increment (varies for advective, dispersive, and reactive steps)
V_i	grid volume: $V_i = \frac{1}{2}(A_i + A_{i+1})\Delta x_i$
A_i	cross-sectional area at upstream face of grid i
Δx_i	length of grid i
Q_i^t	flow from grid $i-1$ to grid i at time t
X_i^t	dispersive flow between grid $i-1$ and grid i at time t: $X_i = \dfrac{D_i A_i}{\frac{1}{2}(\Delta x_{i-1} + \Delta x_i)}$
D_i	dispersion coefficient at upstream face of grid i
ϑ	implicit–explicit weighting parameter
R_i^t	reaction rate vector at grid i at time t
$Q_{\text{out},i}^t$	outflow from grid i at time t
L_i^t	mass loading to grid i at time t

An important assumption in the model simulations was the use of a typical, rather than actual, hydrodynamic dispersion history. The expense of running the 2-D circulation model precluded a complete simulation of the entire 250-day phytoplankton growth season. Instead, the model hydrodynamic component was run only for the months of July and August 1977, as a representative wind period. The character of seiche motion in the lake is sufficiently uniform to presume the dispersion history for this period to be typical. Thus, it was used repeatedly throughout the entire 250-day water quality simulation.

10.5 Simulation Results

Coupled model

A variety of simulations were made to study and evaluate the model. The base case for all simulations is that in which (1) the dispersion coefficient is computed from the circulation model by the method described in Chapter 9, (2) the biogeochemical model employs the average calibration parameters given by van Straten (1980) (Chapter 11), and (3) the loading distribution is that developed by Jolánkai and Somlyódy (1981) (Chapter 6).

Results from this simulation are depicted in Figure 10.4(*a*)–(*i*) as monthly profiles of the distribution along the lake of the total, algal, and dissolved inorganic P. Following an establishment period of one or two months, the general features of the P profile remain fairly constant throughout the growth season. The profile closely reflects the spatial distribution of the P sources, particularly the major influence of the Zala River. The Zala River nutrient inflows create high concentrations of total and algal P within Keszthely Bay, but these concentrations tend to drop east of the bay. This is explained by the relatively low mixing (dispersion) across the eastern end of the bay, as shown in Figure 9.30.

Another striking feature of the spatial P distribution is that of the Sió Basin. With the exception of local effects created by sewage discharges at Kenese and Füred, this basin is lightly loaded relative to the rest of the lake. As well, the narrow Tihany Strait allows very little dispersive exchange with the western part of the lake. Thus, with the exception of its eastern extremity near Kenese, the Sió Basin shows the lake's lowest concentration of total and algal P.

On a more detailed level, the spatial P distributions have local peaks at the locations of important nutrient load point sources. These persist throughout the phytoplankton growth season, despite advective and dispersive transport. The transient effect of a likely major load is apparent in the dissolved inorganic P distribution for November. A minor peak develops in the distribution at mid-lake (near Boglár) and is probably a consequence of the annual drainage of sewage ponds used for fish breeding. Since the nutrients are released toward the end of the growth season, they do not lead to an increase in algal P by increased algal growth.

The effect of the slow advection due to hydrologic throughflow cannot be discerned in the dynamics of the simulated P distribution. On the other hand, dispersion is more transient (Figure 9.28) and produces observable P dynamics at a point. For example, Figure 10.5 shows the changes in predicted P component concentrations over time in Keszthely Bay, east of the Zala River mouth (model grid number 2). The jagged curves are a consequence of transient mixing events that lead to exchange with the very concentrated waters

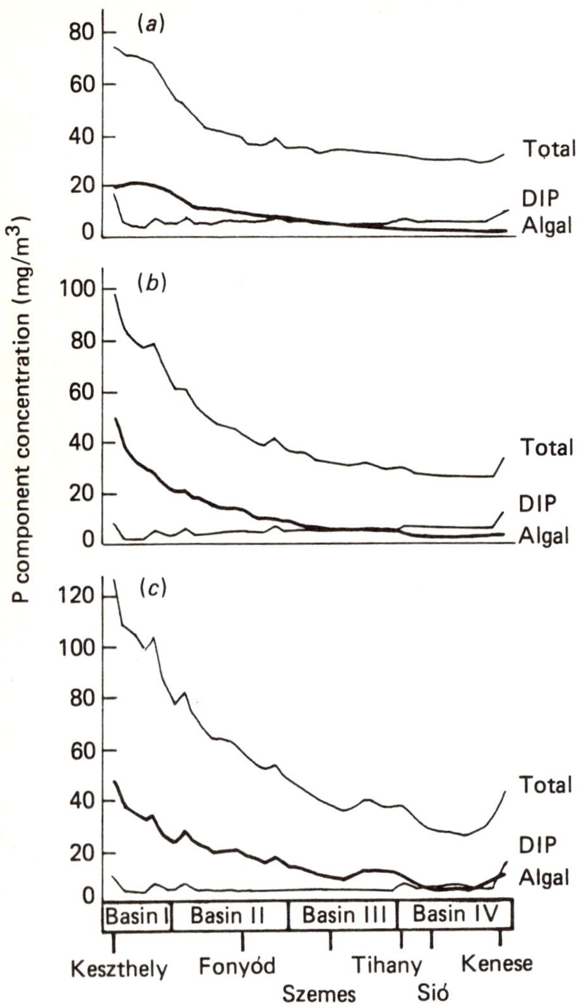

Figure 10.4. (*a*)–(*c*) Spatial distribution of P concentration through time, as predicted by the coupled model base case. (*a*) 6 March 1977; (*b*) 6 April 1977; (*c*) 6 May 1977. Total = total P; DIP = dissolved inorganic P; Algal = total algal P.

at the mouth of the Zala River (model grid number 1). The sudden increases in total P on 15 June and 15 August 1977 correspond to strong mixing events which lead to significant inflow of total P from grid 1 to grid 2. Concurrent decreases in total P occur at the Zala River mouth (grid 1), as shown in Figure 10.6. Similar dynamics occur along the entire lake in proportion to the steepness of the longitudinal concentration gradients. Generally, dispersion dynamics in the lake appear to be dominated by sporadic, short mixing events, separated by periods of relative calm. This is most clearly seen at the Zala River mouth where the rate at which nutrients are introduced is highest.

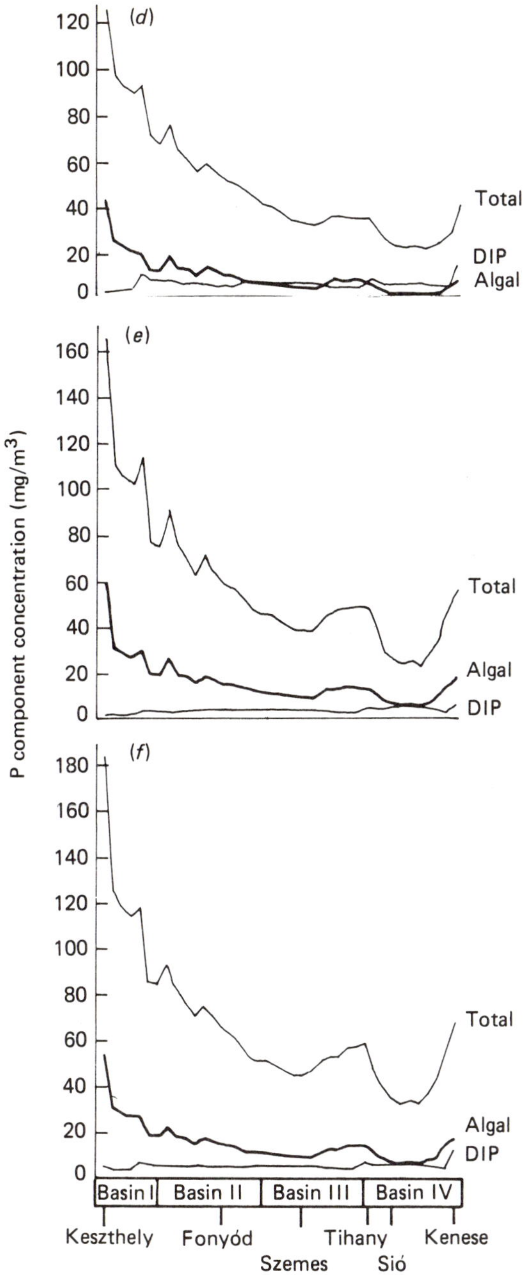

Figure 10.4 (*d*)–(*f*) Spatial distribution of P concentration through time, as predicted by the coupled model base case. (*d*) 5 June 1977; (*e*) 5 July 1977; (*f*) 4 August 1977. Total = total P; DIP = dissolved inorganic P; Algal = total algal P.

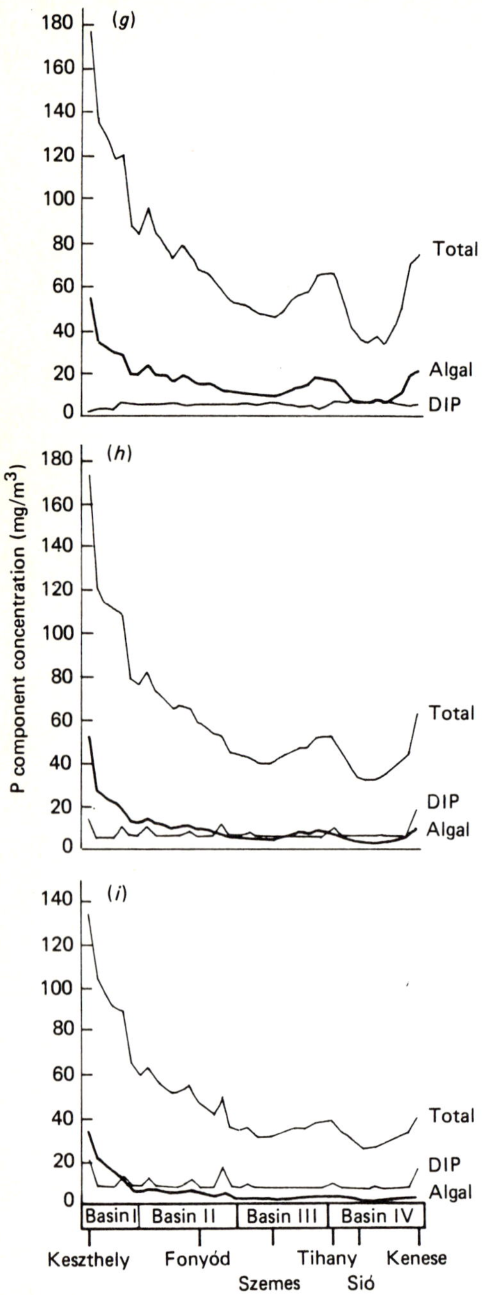

Figure 10.4 (*g*)–(*i*) Spatial distribution of P concentration through time, as predicted by the coupled model base case. (*g*) 3 September 1977; (*h*) 3 October 1977; (*i*) 6 November 1977. Total = total P; DIP = dissolved inorganic P; Algal = total algal P.

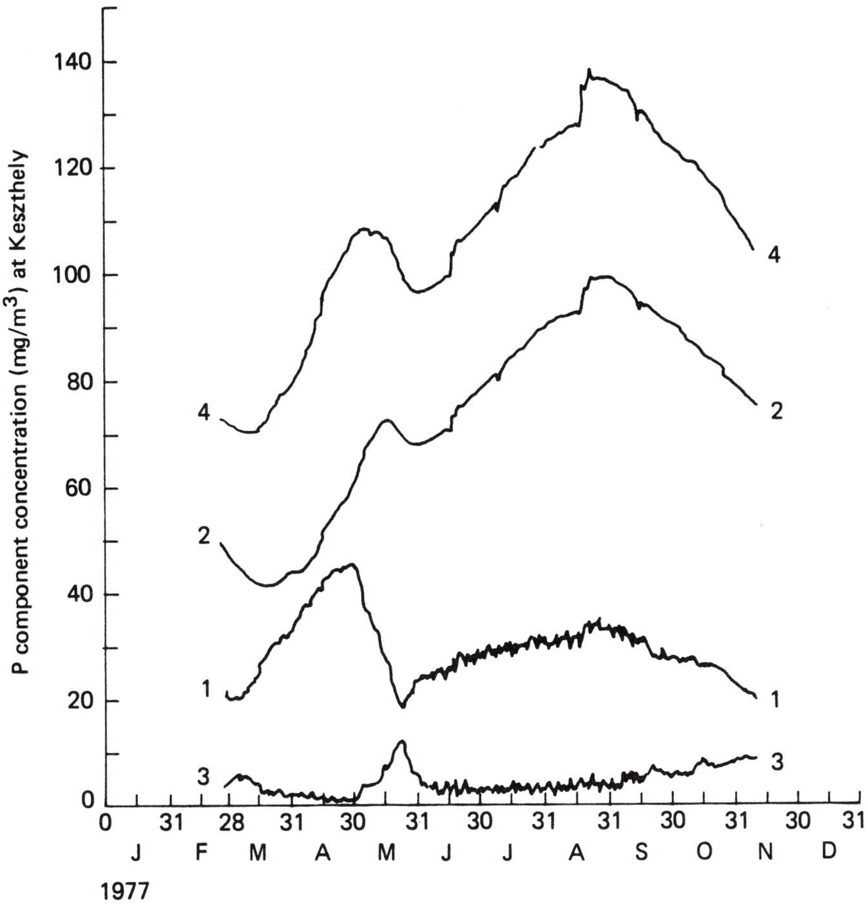

Figure 10.5. P dynamics in Keszthely Bay predicted by the coupled model base case. 1 = total algal P; 2 = detrital P; 3 = dissolved inorganic P; 4 = total P.

Figure 10.6 shows how nutrients accumulate and the concentration of total P grows rapidly between mixing events. However, the occurrence of a mixing event leads to a precipitous drop in concentration as P is transported by dispersion away from the Zala River mouth.

Comparative simulations

Additional simulations were performed to evaluate the importance of hydrodynamic influences on the model results and to study the relative effects of different factors upon the model predictions.

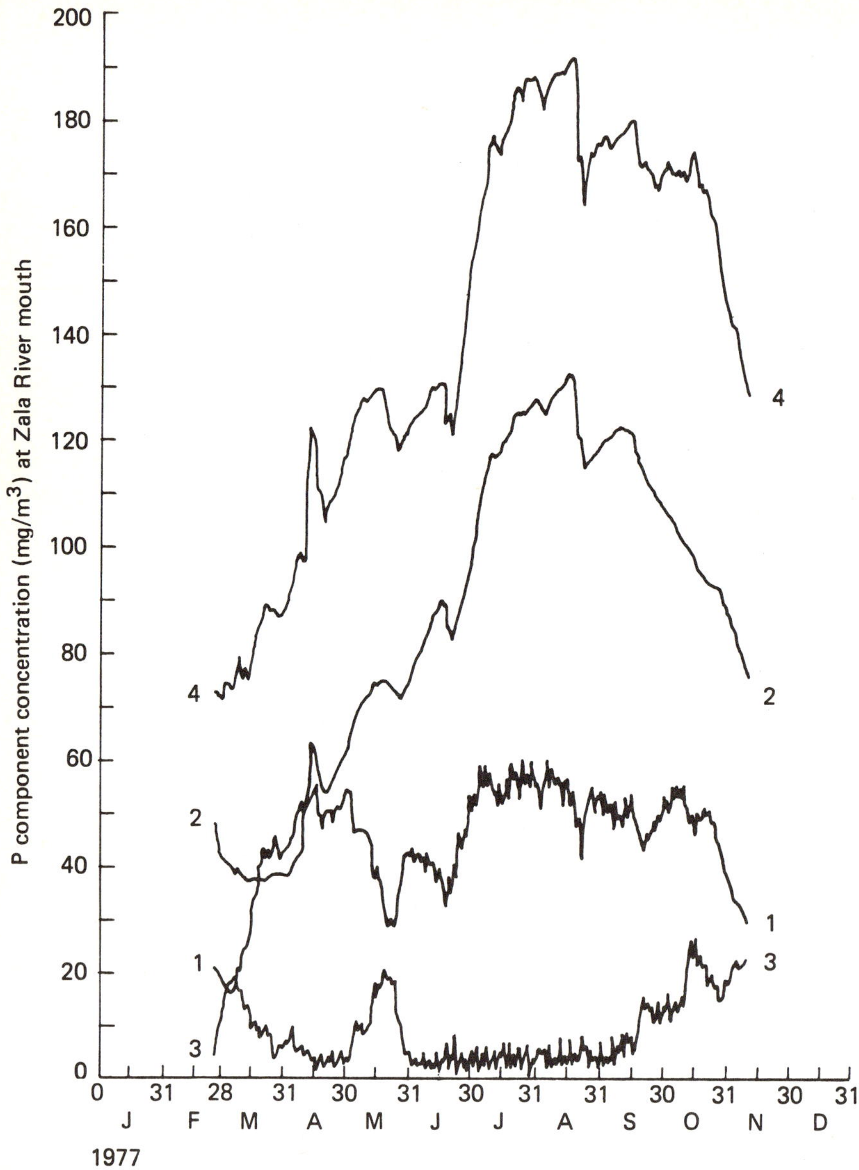

Figure 10.6. P dynamics at Zala River mouth predicted by the coupled model base case. 1 = total algal P; 2 = detrital P; 3 = dissolved inorganic P; 4 = total P.

Hydrodynamic Influences

The influence of the hydrodynamic component was evaluated in a series of simulations in which the representation of dispersion was varied. Simulation results from various runs are compared in Figure 10.7(*a*)–(*b*) on the basis

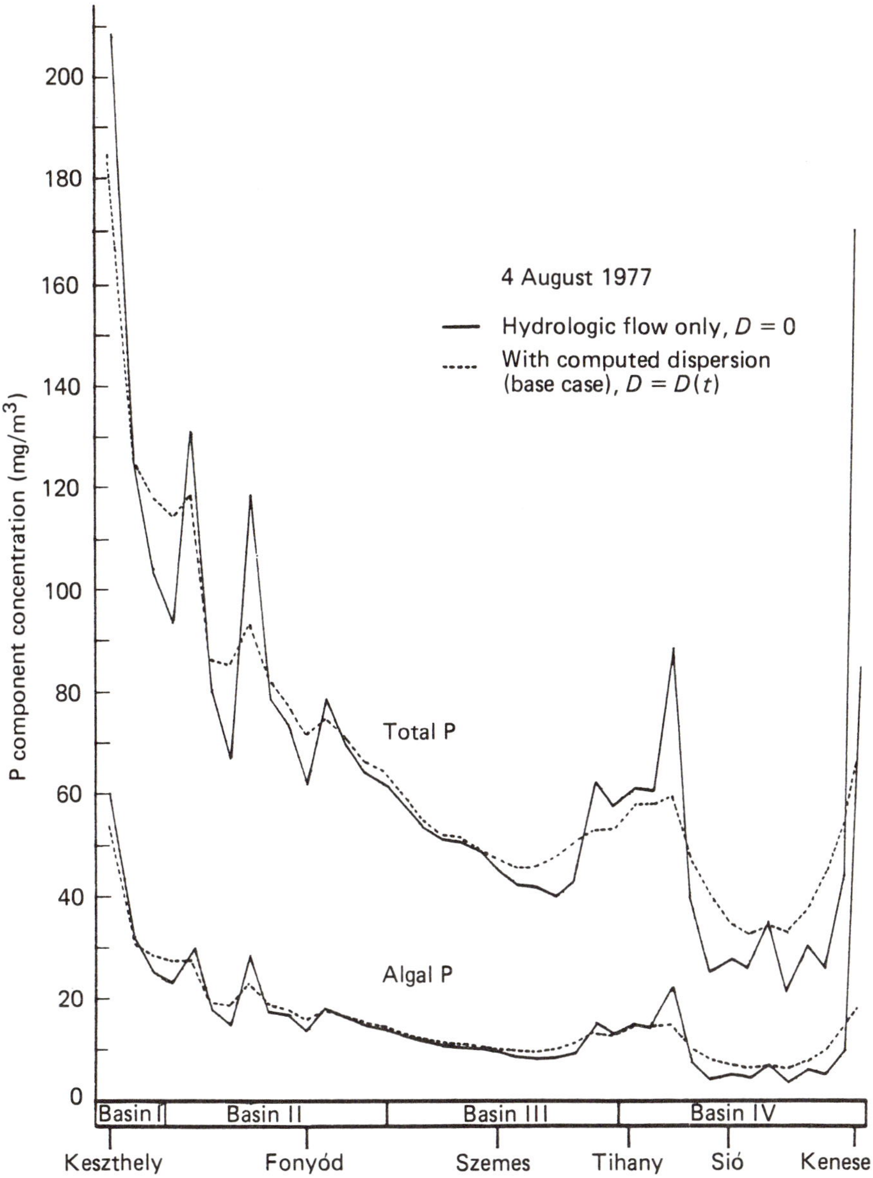

Figure 10.7. (*a*) Effect of model dispersion on predicted P distribution: comparison with zero dispersion.

of spatial concentration profiles of total and algal P for 4 August 1977, during the period when phytoplankton were near their peak summer concentration. The dynamic character of the simulations is compared in Figure 10.8(*a*)–(*d*), which shows histories of P component concentrations in Keszthely Bay (model grid number 2).

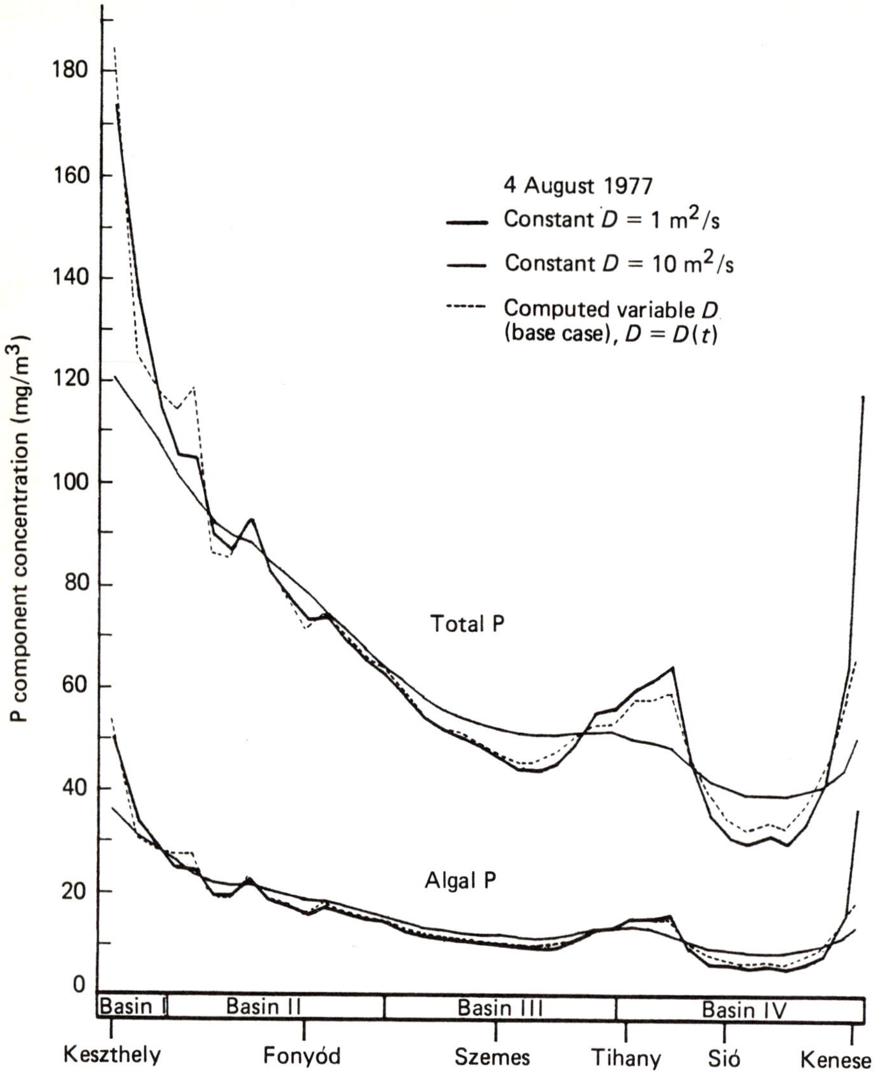

Figure 10.7. (*b*) Effect of model dispersion on predicted P distribution: comparison with constant dispersion.

Figures 10.7(*a*) and 10.8(*a*) show the results from a simulation in which the monthly average hydrologic flow was the only transport considered in the model (dispersive transport was not included in the simulation). The spatial profile (Figure 10.7) shows many peaks; each local peak corresponds to a tributary or sewage nutrient inflow. Without the influence of dispersive mixing, the inflowing P simply collects near the source. The effect is particularly striking at the lake's eastern end, which is outside the main flow path between the Zala River and Sió Canal: in the no-dispersion simulation, the

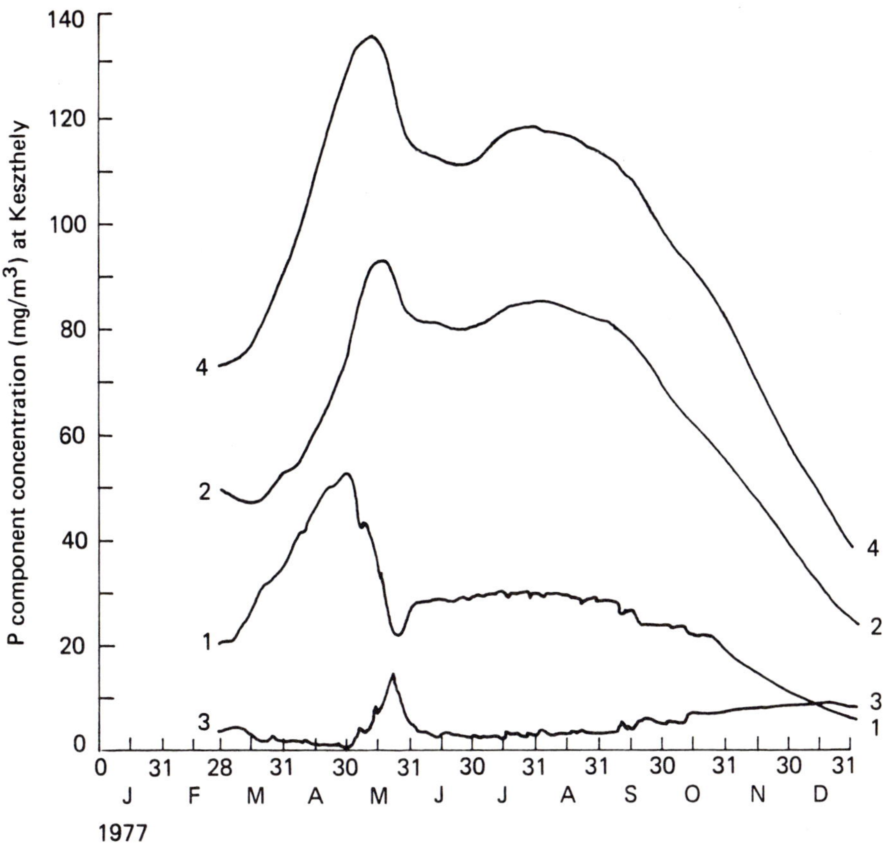

Figure 10.8. (*a*) Effect of model dispersion on predicted P dynamics in Keszthely Bay: with hydrologic flow only (no dispersion), $D = 0$. 1 = total algal P; 2 = detrital P; 3 = dissolved inorganic P; 4 = total P.

sewage discharge from Kenese is erroneously predicted to accumulate in the end grid of the model.

The changes in total P concentration through time differ considerably between the no-dispersion simulation [Figure 10.8(*a*)] and the base case dispersion run [Figure 10.8(*b*)]. Not only is the general character of the P history different, the no-dispersion simulation also lacks the detailed dynamics in the base case caused by transient mixing events. The influence of dispersive mixing is also seen in longitudinal profiles of P concentration: compare the profiles predicted by the base case simulation with dispersion and the no-dispersion simulation in Figure 10.7(*a*). Although local concentration peaks occur in the base case simulation – most prominently near the Zala River source at Keszthely – there is a distinct tendency for dispersion to smooth the profiles relative to the no-dispersion simulation.

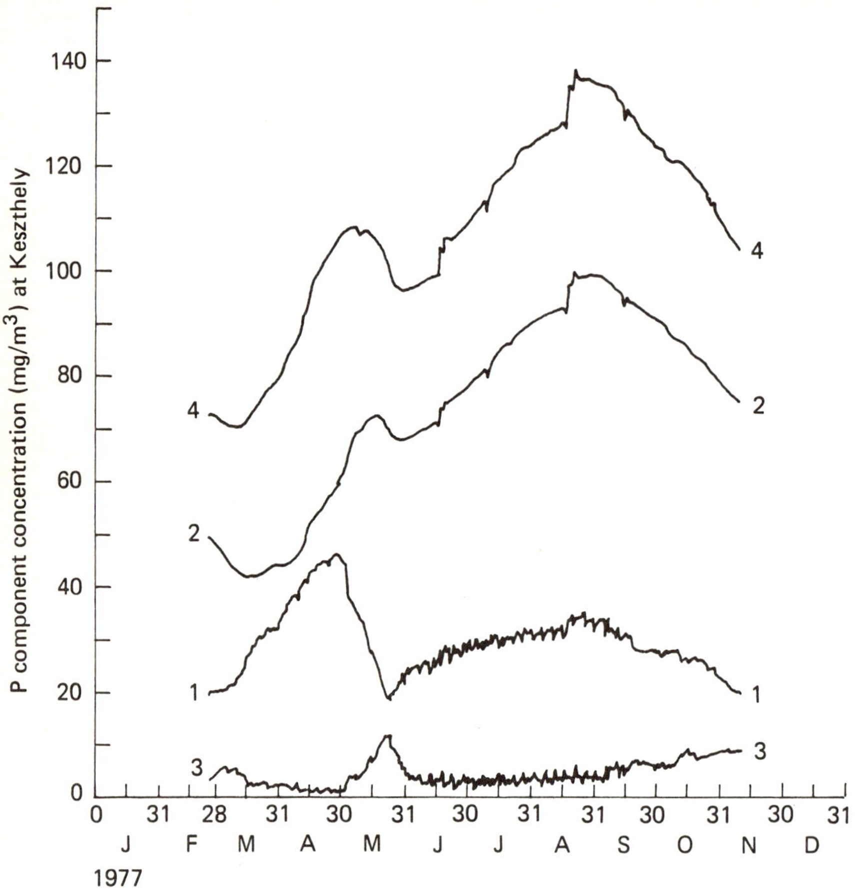

Figure 10.8. (*b*) Effect of model dispersion on predicted P dynamics in Keszthely Bay: with computed dispersion, $D = D(t)$ (base case). 1 = total algal P; 2 = detrital P; 3 = dissolved inorganic P; 4 = total P.

For comparison with the base case simulation, in which dispersion varies in both time and space, two simulations with fixed dispersion coefficients were run. The 4 August 1977 profiles for fixed dispersion coefficients of 1 m^2/s and 10 m^2/s are shown in Figures 10.7(*b*) and 10.8(*c*) and (*d*). The simulation with $D = 1$ m^2/s shows fair agreement with the base case simulation employing the computed dispersion coefficient, though the spatial profile [Figure 10.7(*b*)] is perhaps too high at the lake's western end and too low in the eastern part. P dynamics at Keszthely are also in good agreement, although the constant dispersion simulation [Figure 10.8(*c*)] lacks the transient dynamics seen in the base case [Figure 10.8(*b*)].

The simulation with D increased by an order of magnitude, to 10 m^2/s, shows the importance of the dispersion coefficient to the model results. The higher dispersion smooths the predicted profile substantially [Figure 10.7(*b*)],

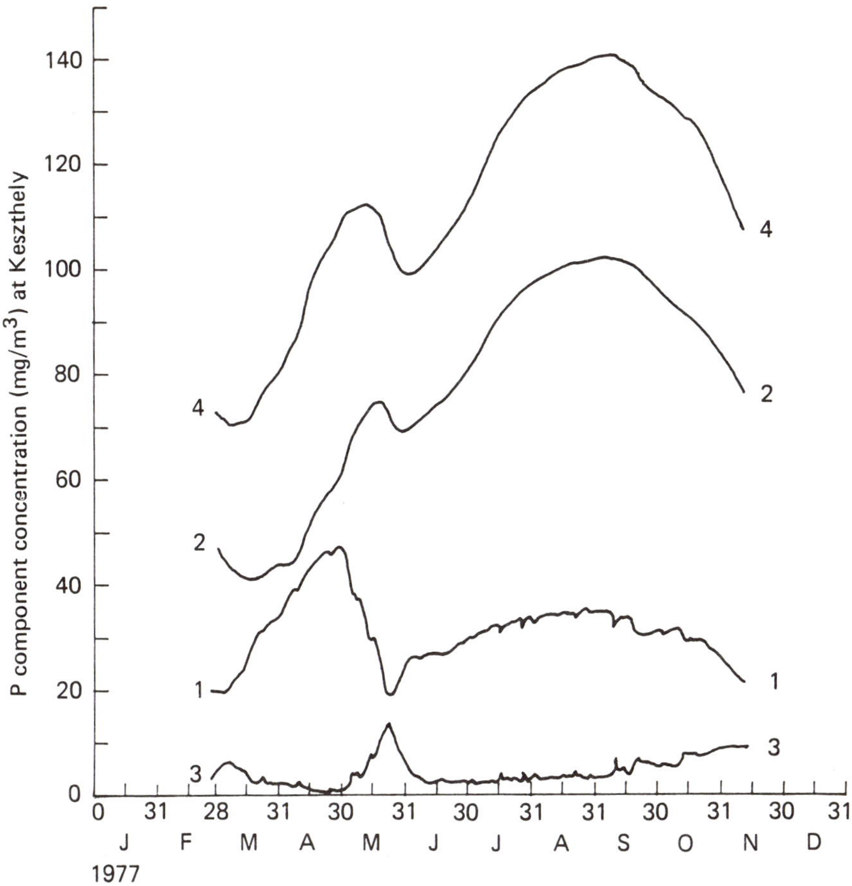

Figure 10.8. (*c*) Effect of model dispersion on predicted P dynamics in Keszthely Bay: with constant dispersion, $D = 1\ m^2/s$. 1 = total algal P; 2 = detrital P; 3 = dissolved inorganic P; 4 = total P.

removing all local concentration peaks. The portion of the lake east of Tihany (basin IV) is much more thoroughly mixed, virtually eliminating the concentration gradients of the base case simulation. The effects of increased mixing also occur in the Keszthely Bay dynamics [Figure 10.8(*d*)] where increased mixing produces a generally lower concentration of total P with time.

Influence of Loading

The distribution of P predicted by the coupled model is a consequence of the competing influences of hydrodynamic transport, biochemical reaction, and the distribution of loading in space and time. The importance of this last factor is explored in this section by means of a model simulation from which nutrient loading has been eliminated.

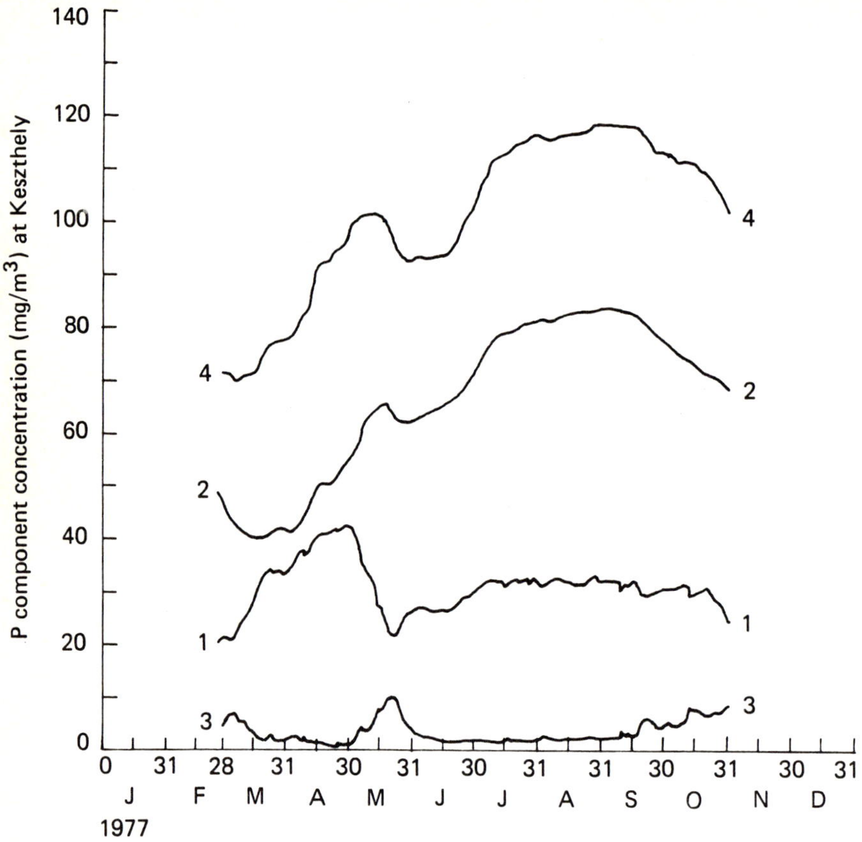

Figure 10.8. (*d*) Effect of model dispersion on predicted P dynamics in Keszthely Bay: with constant dispersion, $D = 10\ m^2/s$. 1 = total algal P; 2 = detrital P; 3 = dissolved inorganic P; 4 = total P.

Figure 10.9 compares the model results with and without nutrient loading. The cessation of loading has a great impact on the predicted spatial distribution for 4 August 1977. Almost all of the dominant west-to-east concentration gradient has disappeared and there remains only a residual indication of the initial conditions. The results without loading make it clear that longitudinal P gradients observed in the lake are sustained by the spatial distribution of the loading. Without continual loading inputs, advective and dispersive transport would tend to make the concentrations uniform throughout the lake. What is observed in the lake, and predicted in the base case simulation, is a lake-wide distribution dominated by the loading distribution. However, this distribution may be significantly altered locally by hydrodynamic influences and biochemical reactions.

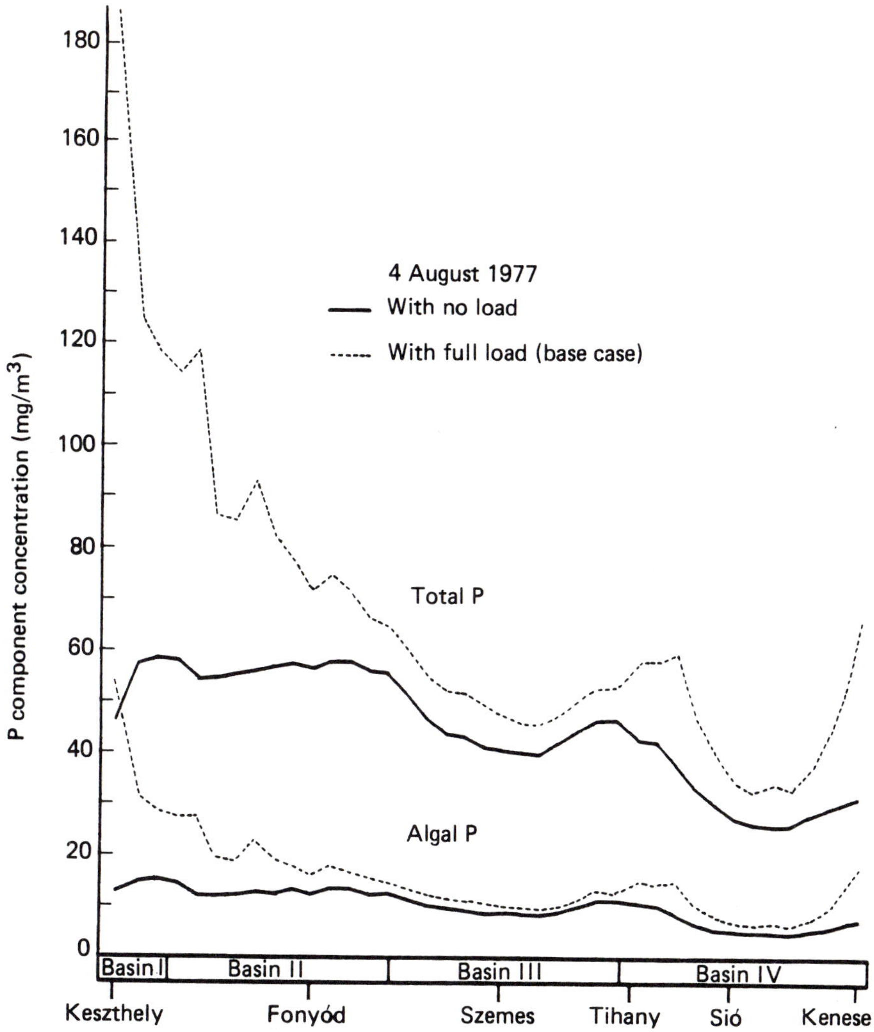

Figure 10.9. Effect of nutrient loading on predicted P dynamics in Keszthely Bay.

10.6 Comparison with Four-Box Model Results

The predictions of the four-box model are shown in Figures 10.10 and 10.11. The four-box model simulations employ the same hydrologic flow, nutrient loading, and other data as the coupled model; of course, spatial distributions are modified appropriately to fit the four-box discretization. These model inputs differ slightly from those originally used by van Straten (1980) for the SIMBAL model (see Chapter 11), so the predictions in this chapter differ somewhat from his.

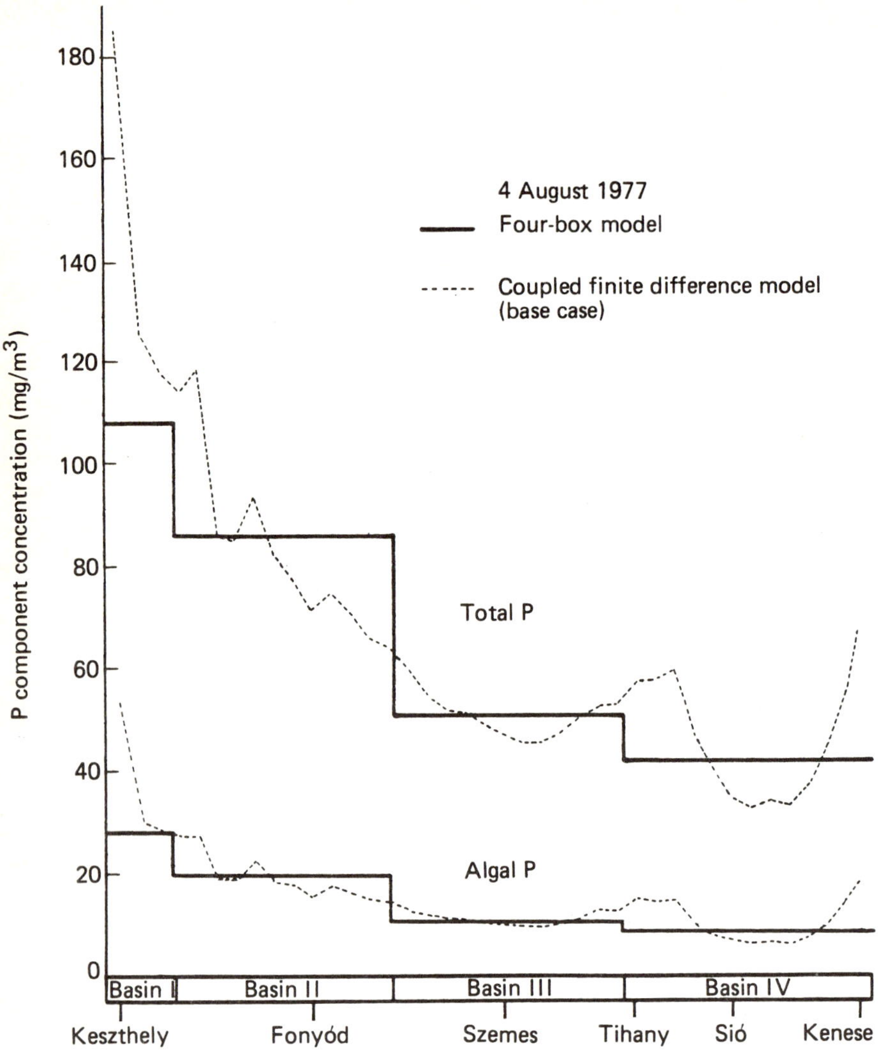

Figure 10.10. P distributions predicted by the four-box model and the coupled model base case.

In Figure 10.10, the spatial distribution of P predicted by the four-box model is compared with that predicted in the coupled model base case simulation. The loss of spatial detail in the four-box model is immediately obvious: local concentration peaks cannot be predicted by the four-box model. Nevertheless, it satisfactorily predicts the general character of the lake-wide longitudinal gradients. The only discrepancy of consequence is the lower concentration in Keszthely Bay predicted by the four-box model.

The concentration in Keszthely Bay is further detailed in Figure 10.11, which compares the four-box results for basin I with the coupled model

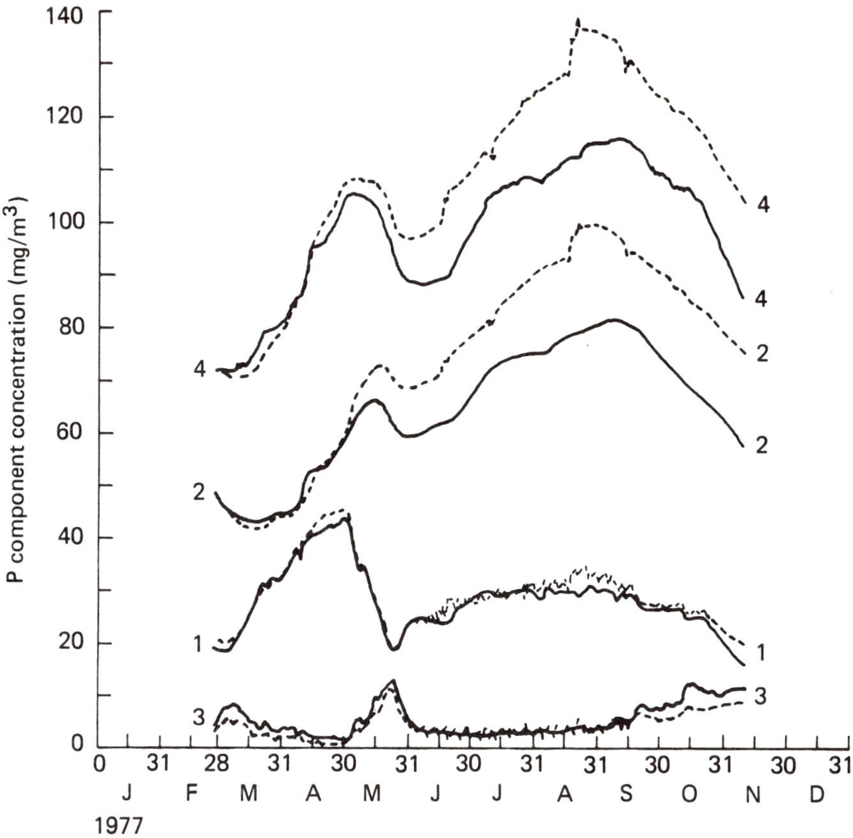

Figure 10.11. P dynamics predicted by the four-box model and the coupled model base case. 1 = total algal P; 2 = detrital P; 3 = dissolved inorganic P; 4 = total P. Full line = concentration in basin I predicted by the four-box model; dashed line = concentration in Keszthely Bay (model grid 2) predicted by the coupled definition model.

predictions for model grid 2, a roughly equivalent location. The P concentrations show approximately the same trends through time in either model; however, the four-box model predicts lower concentrations in all constituents during the summer. This is presumably caused by the greater mixing that occurs in the four-box model due to its implicit dispersion. The error is less for the lower concentrations of algal and dissolved inorganic P. Despite this error, we conclude that the four-box model is an adequate predictor of the lake's P dynamics. This is particularly true if one considers algal P to be the indicator of eutrophication and thus its prediction is the most critical. The predictions of algal P by the four-box and coupled models compare well (see Figures 10.10 and 10.11). Thus, the model comparisons generally confirm the findings reached earlier by analyzing the model formulations.

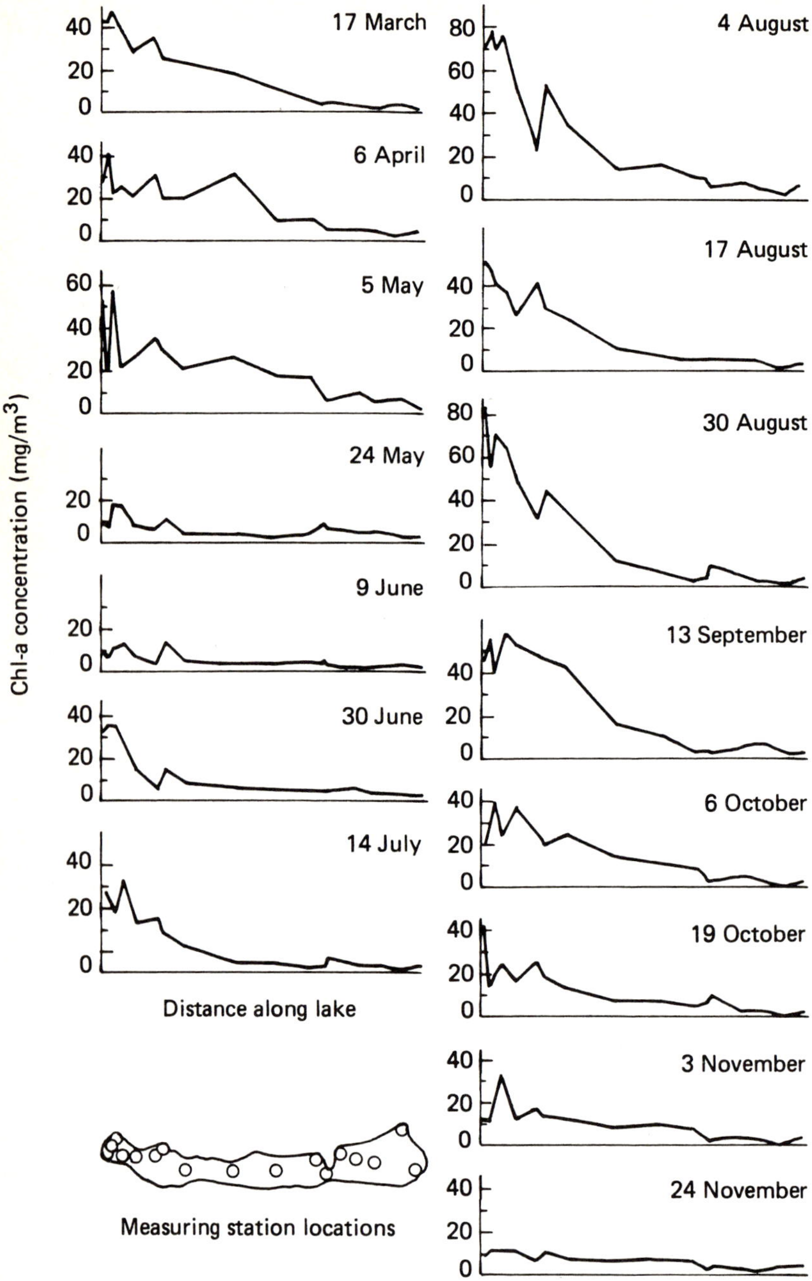

Figure 10.12. Measured Chl-a concentrations in 1977.

The greatest difference between the coupled model and the four-box model lies in their abilities to simulate spatial detail. Clearly, the coupled model produces a far more detailed description of the spatial distribution of P in the lake, but the water quality sampling network in the lake is far too sparse to assess these predictions. Nevertheless, the available data indicate the plausibility of the predictions. For example, Figure 10.12 shows the observed concentrations of Chl-a in Lake Balaton in 1977, the character of which is in agreement with the model predictions: local peaks in concentration are present and persist throughout the algal growth season.

10.7 Conclusions

The comparisons described in this chapter confirm the general ability of the Lake Balaton four-box model formulation to adequately model lake-wide trends in spatial distribution and dynamics of P. However, this is not a general finding: it applies only to the particular four-box model tested. The analysis shows that the adequacy of the four-box model depends upon the particular configuration used in the Lake Balaton models. Specifically, the definition of a small-volume box at Keszthely Bay is critical. In general, box models should be constructed with care, paying attention to the definition of box boundaries and the model's implicit dispersion. If this is done, the preceding discussion illustrates that box models can be useful and sufficient models of lake water quality.

The above notwithstanding, the results also indicate that fundamental conceptual differences distinguish the multiple-box model and the finite-difference approach. A mathematical analysis of the conceptual reactor models analogous to these two approaches illustrates their differences. The analysis reveals intrinsic difficulties in the application of multiple-box models owing to an artificial dispersion implicit in their structure. A step-by-step procedure is proposed to circumvent the shortcomings of box models and insure their sound construction.

The results from the coupled model lead to further conclusions concerning the interaction of hydrodynamics and water quality. In Lake Balaton, the observed P distribution is largely determined by the distribution of the nutrient loading sources; hydrodynamics has a weaker influence. While this is probably not an unusual situation, it is surely not universal. Thus, it is important to consider the role of hydrodynamic transport in lake eutrophication modeling.

The experimentation with different forms and values of dispersion shows that dispersion has an important influence upon the model predictions. Nevertheless, a constant, uniform dispersion coefficient is adequate – so long as the coefficient value is approximately correct. The short-term dynamics that occur with time-varying dispersion are minor and can be neglected, with the important implication that direct coupling of hydrodynamics and biochemistry is unnecessary. Rather, the value of the dispersion coefficient can be computed from hydrodynamic information and simply specified as a model parameter. Nevertheless, it remains worthwhile to separate the specification of this hydrodynamic parameter from the calibration of the biochemical model parameters. The processes are separable, and it strengthens the model formulation and calibration to treat them separately.

References

Baranyi, S. (1973) Method of determining the turnover time of the water resources of lakes with an outlet, in *Hydrology of Lakes Symposium, Helsinki*. IAHS–AISH Publication 109, pp 54–9.

Jolánkai, G. and Somlyódy, L. (1981) *Nutrient Loading Estimate for Lake Balaton*. Collaborative Paper CP-81-21 (Laxenburg, Austria: International Institute for Applied Systems Analysis).

Levenspiel, O. and Bischoff, K.B. (1963) Patterns of flow in chemical process vessels, in T.B. Drew, J.W. Hoopes, Jr., and T. Vermeulen (Eds.) *Advances in Chemical Engineering*, Vol. 4 (New York: Academic Press) pp 95–198.

Liu, H. (1977) Predicting dispersion coefficient in streams. *J. Env. Eng. Div., ASCE* 103:59–69.

Murthy, C.R. and Okubo, A. (1977) Interpretation of dispersion characteristics of oceans and lakes appropriate for numerical modeling, in *Proceedings of the Symposium on Modeling of Transport Mechanisms in Oceans and Lakes, October 1975*. Manuscript Report Series No. 43 (Ottawa, Canada: Marine Sciences Directorate Canada, Department of Fisheries and Environment).

Shanahan, P. and Harleman, D.R.F. (1982) *Linked Hydrodynamic and Biogeochemical Models of Water Quality in Shallow Lakes*. Report Number 268 (Cambridge, MA: Ralph M. Parsons Laboratory, Dept of Civil Engineering, MIT).

Shanahan, P. and Harleman, D.R.F. (1984) Transport in lake water quality modeling. *J. Env. Eng., ASCE* 110:42–57.

Thomas, H.A., Jr. and McKee, J.E. (1944) Longitudinal mixing in aeration tanks. *Sewage Works J.* 16:42–55.

Tuan, V.A., Thanh, N.C., and Lohani, B.N. (1980) Hydraulic model for biological reactors: application to activated sludge. *J. Water Pollution Control Fed.* 52:1931–6.

van Straten, G. (1980) Analysis of model and parameter uncertainty in simply phytoplankton models for Lake Balaton, in M. Dubois (Ed.) *Progress in Ecological Engineering and Management by Mathematical Modeling* (Liege, Belgium: Editions Cebedoc).

Zvirin, Y. and Shinnar, R. (1976a) A comparison of lumped-parameter and diffusional models describing the effects of outlet boundary conditions on the mixing in flow systems. *Water Res.* 10:765–79.

Zvirin, Y. and Shinnar, R. (1976b) Interpretation of internal tracer experiments and local sojourn time distributions. *Int. J. Multiphase Flow* 2:495–520.

PART THREE

Selected Topics

CHAPTER 11

Hypothesis Testing and Parameter Uncertainty Analysis in Simple Phytoplankton–P Models

G. van Straten

11.1. Introduction

System identification and parameter estimation are necessary steps preceding the application of any mathematical model designed for management and control. As outlined in Chapter 2 (Figure 2.2) in a typical modeling procedure first a model is postulated, usually based on some *a priori* knowledge of the system under study; then an attempt is made to estimate a unique set of parameters by matching model results with available data; and finally the error sequence is examined in order to detect structural deficiencies in the model. Although one should attempt to follow this procedure as closely as possible, its application is often hampered in a number of practical cases, particularly in the early stages of a lake eutrophication study, for three major reasons:

(1) Large uncertainty in observation data, mostly because of sampling errors, and partly because of chemical identification uncertainty.
(2) Uncertainty in forcing functions and input data due to stochastic variability combined with deficient recording.
(3) Incomplete knowledge of biological, chemical, and hydrophysical processes.

Several of these problems have been addressed in previous chapters as well as in the literature. The effect of observation errors on parameter reliability and prediction uncertainty is discussed in Di Toro and van Straten (1979) and Beck *et al.* (1979). Fedra *et al.* (1981) draw attention to the fact that very often there is no unique parameter set if data and input variability are properly accounted for. Consequently, rather than a unique prediction, a probability density function must result. The nonuniqueness of the parameters was earlier pointed out by Spear and Hornberger (1978), who attempted to separate the parameter space in a region that gives rise to predefined model behavior, and a region that does not. The target behavior was defined as a range of permissible values derived from scarce field data. The aim of this procedure was to test various assumptions on the P cycle in Peel Inlet, a shallow bay in Western Australia. This chapter describes the application of a similar procedure to Lake Balaton, performed in the initial phase of the eutrophication study. At that time much of the information described elsewhere in this book was not yet available. In those situations where uncertainty is still very large the approach discussed subsequently is particularly appropriate.

In most shallow lakes water–sediment interactions are believed to play a significant role in the eutrophication phenomenon, although very little is known about this process from direct experimentation. The principal aim of this chapter is to investigate the major modes of P transport to and from the sediment on the basis of available, but uncertain, data for in-lake P fractions. Hence, our objective is hypothesis testing, that is, to see whether or not certain assumptions must be rejected in the light of the data. For this purpose alternative simple models are formulated, using as much information about parameters, processes, and forcings as possible. The remaining unknown parameters, typically associated with water–sediment interactions, are considered as "tunable" parameters. These parameters can be varied manually, to see whether parameter sets exist for which the model shows a P behavior that matches the observed behavior. Rather then defining behavior as the time pattern of the actual data as such, we define it as a set of (simple) constraint conditions around the actual data, to allow for data uncertainty. Thus, a model solution is said to show the behavior if the concentration patterns fall within the boundaries specified in the behavior definition. Furthermore, rather then selecting the parameter vector by hand, we can do this by Monte Carlo simulation, i.e., by automatically running a large number of runs with parameters selected randomly from preset ranges. The result, finally, is two sets of parameter vectors – those that do and those that do not give the behavior – which can then be subjected to further analysis.

11.2. The Data

The data used for this study are the regular monitoring data on chlorophyll-a (Chl-a) and various P fractions, collected about ten times a year from nine locations (cf. Chapter 1, Figure 1.2). The uncertainty in Chl-a is best illustrated by comparing these data with an independent set collected by another agency. Figure 11.1 gives an impression of the variability involved. The P fractions reported are summarized in Figure 11.2. The uncertainty is not the same for these fractions: total P, total dissolved P, orthophosphate, and particulate inorganic P are observed directly, the other fractions have

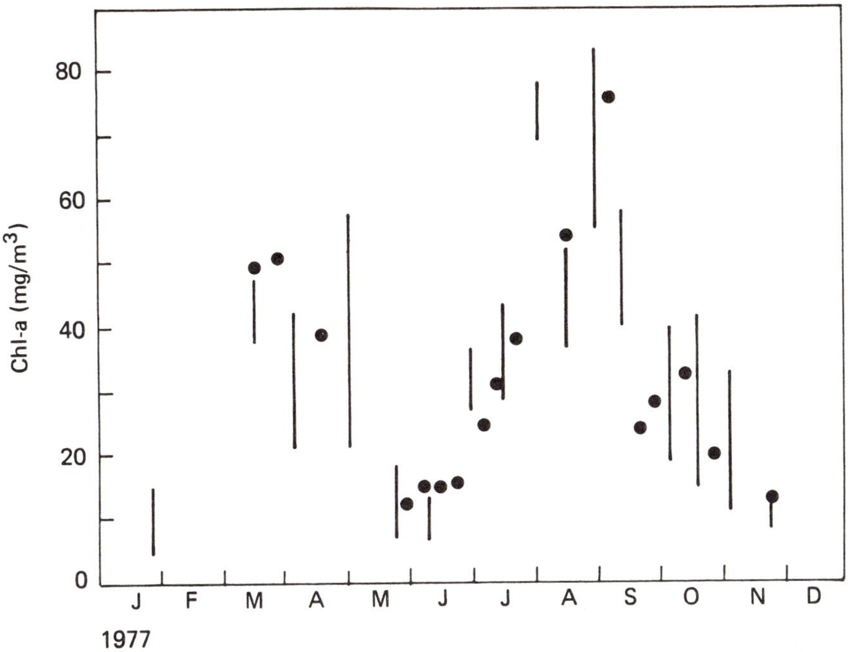

Figure 11.1. Example of data variability. Bars, range of Chl-a data for various locations within the Keszthely basin, collected by the Transdanubian Water Authority. Dots, mid-bay data collected by the Research Center for Water Resources Development.

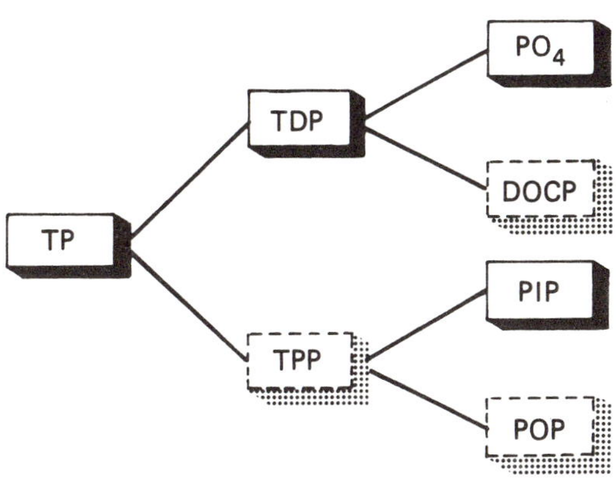

Figure 11.2. Reported P fractions for Lake Balaton. Direct measurements (solid boxes): TP, total P; TDP, total dissolved P; PO_4, orthophosphate; PIP, particulate inorganic P. Calculated (dotted boxes): TPP, total particulate P; DOCP, dissolved organic and condensed P; POP, particulate organic P.

Table 11.1. Geometric mean of 1977 P data, for the four basins and for the lake as a whole, stormy days excluded (mg P/m^3).

		I	II	III	IV	Lake
Volume (10^6m^3)		82	413	600	802	1897
Total P	mean	81.1	63.4	42.1	29.9	43.3
	s.d.	*8.0*	*8.7*	*8.2*	*4.2*	
Total dissolved P	mean	25.2	18.8	14.1	13.1	15.2
	s.d.	*8.1*	*6.1*	*4.3*	*3.9*	
Orthophosphate P	mean	4.7	4.7	4.4	3.4	4.0
	s.d.	*2.1*	*1.1*	*0.9*	*0.7*	
Dissolved organic P	mean	20.6	14.1	9.8	9.6	11.1
	s.d.	*7.5*	*6.5*	*4.4*	*4.1*	
Total particulate P	mean	55.9	44.6	27.9	16.8	28.0
	s.d.	*11.6*	*7.8*	*10.0*	*6.7*	
Particulate inorganic P	mean	10.3	8.3	6.7	4.4	6.2
	s.d.	*5.0*	*2.4*	*2.4*	*1.8*	
Particulate organic P	mean	45.6	36.4	21.2	12.2	21.8
	s.d.	*10.1*	*7.0*	*8.4*	*7.4*	

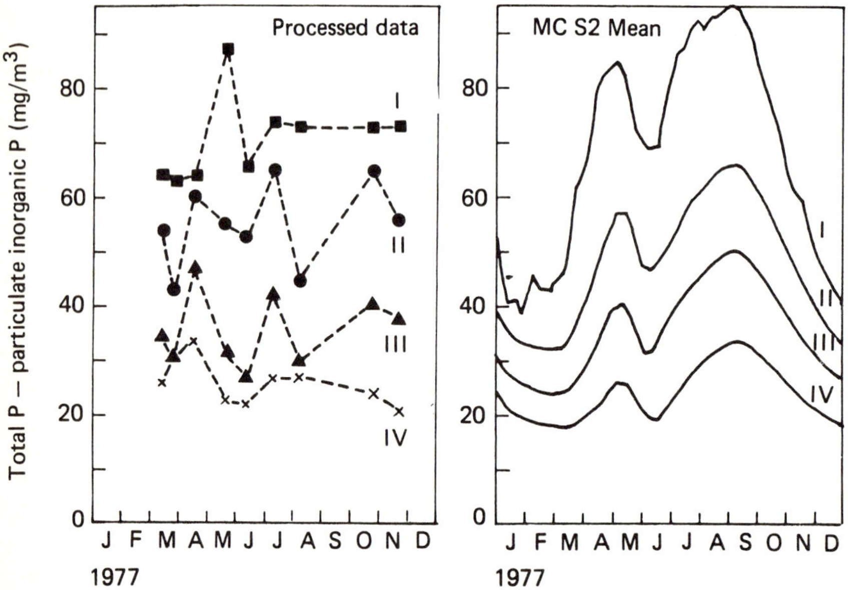

Figure 11.3. (*a*) Total P data (*left*, particulate inorganic P subtracted) and average of the seven behavioral runs (series 2) for the four basins.

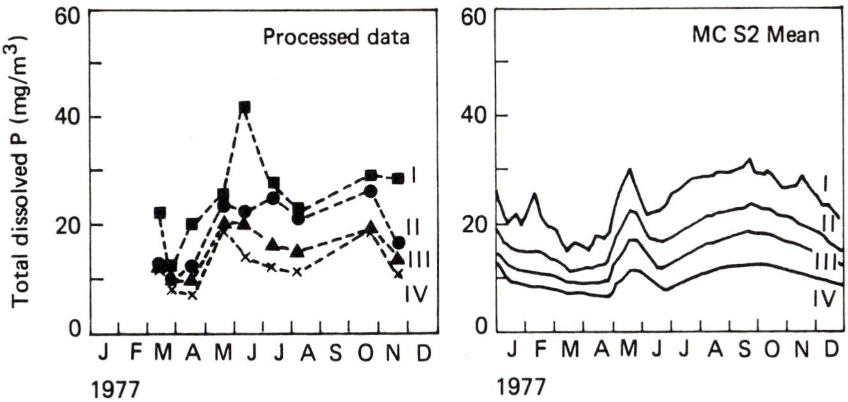

Figure 11.3. (*b*) Total dissolved P data (*left*) and average of the seven behavioral runs (series 2) for the four basins.

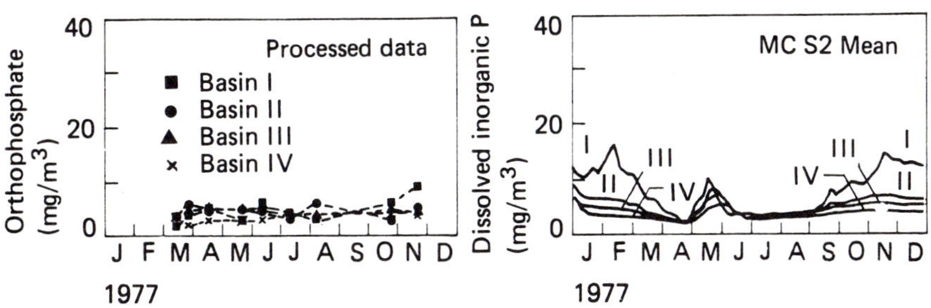

Figure 11.3. (*c*) Orthophosphate data (*left*) and average of the seven behavioral runs (series 2) for the four basins.

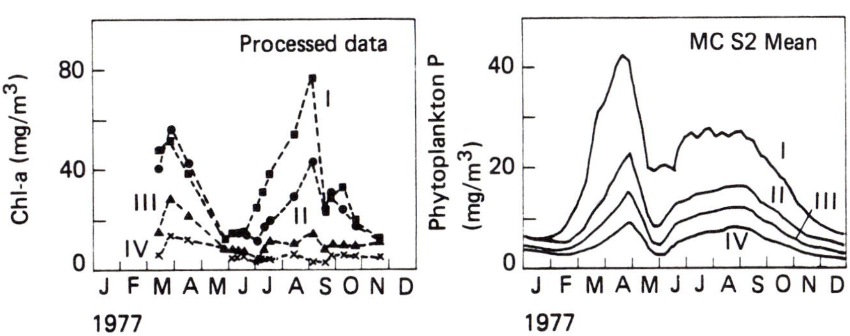

Figure 11.3. (*d*) Chl-a data (*left*) and average of phytoplankton P of the seven behavioral runs (series 2) for the four basins (note the difference in scale units).

been calculated, and are thus less accurate. For the purpose of the analysis the lake is segmented into four basins (cf. Figure 1.1) and basin-averaged values are computed from the nine measurement locations according to their position. Geometric averages over the year 1977 are presented in Table 11.1 for the four basins. The dynamic variations are shown in Figure 11.3 (*a*)–(*d*); in addition, the standard deviations of the individual points with respect to the mean are given in Table 11.1. The longitudinal gradient is immediately apparent from this table. The high level of total dissolved P is remarkable, indicating large dissolved organic and condensed polyphosphate concentrations, because orthophosphate levels are always very low. Of the particulate organic P, roughly half is phytoplankton. Thus, the data show that detritus P is a substantial fraction of total P in the lake. Roughly 10–15% of the total P is in the form of particulate inorganic P. This fraction is fairly constant throughout the year, except in stormy periods, when particulate inorganic P can reach up to 40 mg P/m^3. The ratio of particulate inorganic P to total suspended solids ranges between 0.5 and 1.5 mg/g. $CaCO_3$ is an important constituent of the suspended solids. During the year considerable Ca precipitation occurs (ca. 75%; Entz 1959), mostly as biogenic lime precipitation in the growing season. The pH is 8.3–8.7 throughout the year.

11.3. Modeling

Since the purpose of the modeling is to test assumptions on the major modes of P cycling, it was decided that the model(s) should be as simple as possible. This was also considered a necessity in view of the quantity and quality of the data. On the other hand, due regard had to be given to those aspects that are relatively well known for the lake, such as meteorologic and hydrologic data.

The load situation

An important input to the model, of course, is the P load. Basically, the known elements of the load are the total P and orthophosphate loads carried by the Zala River (obtained from weekly data), a tentative estimate of the total sewage load, and data on the total P and orthophosphate concentrations in atmospheric precipitation. For the model the contributions of the tributaries, as well as the distribution of the sewage loads over the four basins also had to be known. At the time of this research, the detailed estimates of Chapter 6 were not available, so that to obtain a reasonable estimate for the distribution, the particulate P load of the Zala River was extrapolated based on watershed surface areas and average slopes for the tributary watersheds. The sewage loads were distributed according to population density and sewage connection ratio. Consequently, the loads of the various basins could be estimated from the Zala River data (dynamically) and from the total sewage load (constant, but twice the normal value during the tourist season), as shown in Figure 11.4. The sewage load enters the lake either directly or through the tributaries. Other sources of dissolved P are included in the total sewage load estimate and are therefore not accounted for separately.

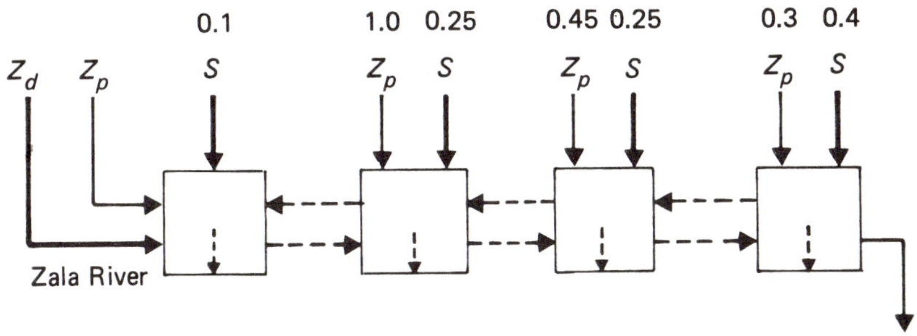

Figure 11.4. P load distributions. *S*, sewage; Z_p, Zala particulate P; Z_d, Zala dissolved P.

Hydrology and longitudinal mixing

Monthly data on precipitation, inflow, outflow, and evaporation were available, and these were used for the computation of basin-to-basin throughflow. In terms of the time scales relevant to ecological modeling the effect of throughflow is insignificant (flow velocity < 1 mm/s), although the throughflow does account for a considerable transfer of P from one basin to the next. In addition to throughflow significant currents exist in the lake due to wind action, and consequently some longitudinal interbasin exchange is to be expected. However, as pointed out in Chapter 10, the four-basin model itself already entails a certain amount of longitudinal mixing and the question is whether this implicit mixing is already sufficient, or whether the model should allow for additional mixing. The results of Chapter 10 suggest that in this case the first situation occurs. However, at the time of the research the work on hydrodynamics was still in progress, so that in the present study a provision was made for additional mixing. The construction adopted is shown in Figure 11.5. The additional dispersion effect is governed by an artificial forward and return circulation flow velocity v_{ret}. It should be emphasized that this parameter has no direct relation to the actual currents, although its value must certainly be less than the surface velocities (of the order of a few cm/s). In view of this parameter uncertainty, and given the fairly even distribution of wind over the seasons, no attempt was made to make v_{ret} time dependent, so that longitudinal mixing is assumed to be constant throughout the year (knowing the results of Chapter 10 now, this is quite a reasonable assumption after all). Experimentation with the model showed that v_{ret} had to be of the order of a few mm/s, because at higher values unrealistically low concentration gradients occurred. This also implies that the distribution of P and phytoplankton over the lake is rather sensitive to this parameter.

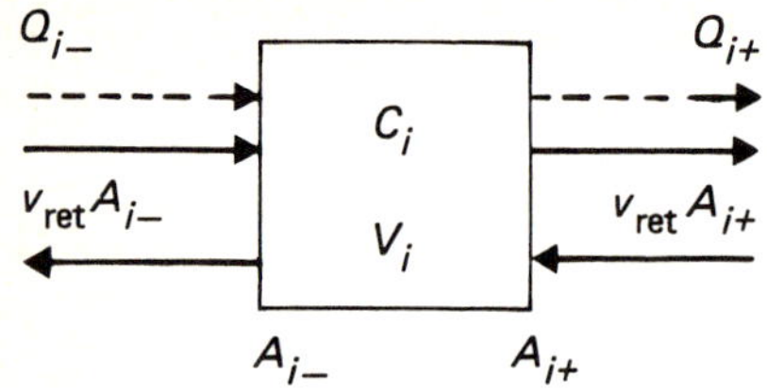

Figure 11.5. Exchange rate formulation. C, concentration; V, volume; Q, hydrologic flow rate, v_{ret}, exchange velocity; A, cross-sectional surface area.

$$dV_i C_i / dt = + (Q_{i-} + v_{ret}A_{i-}) C_{i-1} + v_{ret}A_{i+}C_{i+1} - (Q_{i+} + v_{ret}A_{i+}) C_i - v_{ret}A_{i-}C_i \; .$$

Algal dynamics

The lake is characterized by two algal blooms over the year (cf. Figure 11.3), although these are not always as distinct as in 1977. Algal counts indicate that the spring bloom is mainly associated with diatoms, in recent years mostly *Synedra acus* and *Nitzschia acicularis*. The water temperatures are below 12°C in this period. Later in the season a mixed phytoplankton prevails, dominated by *Ceratium hirundinella*, and in recent years blue–green species (mainly *Aphanizomenon flos-aquae*) also occurred, especially in the Keszthely and Szigliget basins. To account for the differences in environmental sensitivity over the year it was decided to introduce two groups of algae in the model, denoted "winter" and "summer" algae, respectively. The differences between the groups lie mainly in their temperature sensitivity and maximum growth rates. Both characteristics were derived from an analysis of the primary production data, from which it was also clear that light inhibition occurred at the surface, and consequently the light limitation was described with the depth- and day-averaged Steele formula. The equations and parameters used are presented in Table 11.2.

It should be noted that the maximum growth rate obtained from the primary production data analysis was extremely high compared with data from the literature. In the model, values of 2 and 6 d^{-1} are used for winter and summer algae, respectively. No explanation is known for these very high values. With such extreme growth rates the death process in the lake must be significant. Detailed zooplankton studies (Zankai and Ponyi 1976) have revealed a maximum filtering rate in summer of 1.4 cm^3/d/individual, while concentrations are at most 7 individuals/$10^{-3}\,m^3$. Consequently, zooplankton could not be the major cause of the algal death process. Since nothing is known about mortality it was assumed in the model that mortality is proportional to biomass. The rate coefficient was estimated as roughly 0.13 d^{-1} at 20°C by matching the autumn decline of phytoplankton in the model and in reality. However, the temperature dependence, described by an exponential function, was retained as a "tunable" parameter. The value range chosen was slightly higher than usual to give some account for a temperature-coupled zooplankton effect.

Table 11.2. SIMBAL Model Equations.

STATE EQUATIONS

$$\dot{A}_w = (I - O)_{A_w} + \mathrm{RGR}_w - \mathrm{RDTH}_w \quad (11.1)$$

$$\dot{A}_s = (I - O)_{A_s} + \mathrm{RGR}_s - \mathrm{RDTH}_s \quad (11.2)$$

$$\dot{D} = (I - O)_D + \mathrm{RDTH}_w + \mathrm{RDTH}_s - \mathrm{RMNRL} - \mathrm{RSETD} + L_D \quad (11.3)$$

$$\dot{P} = (I - O)_P - \mathrm{RGR}_w - \mathrm{RGR}_s + \mathrm{RMNRL} - \mathrm{RBIOP} \quad (11.4)$$

$$+ \mathrm{RREL} + L_P + \mathrm{REXP}$$

A_w, A_s, phytoplankton P, winter, summer algae, respectively
D, detritus P
P, dissolved inorganic P

INPUT FUNCTIONS

Inflow/outflow, $(I - O)$, see Figure 11.5
Temperature, T
Day total of global radiation, R
Fraction daylight, λ
P loads

Volumetric detritus load, L_D

$$L_D = d_i Z_P \varphi / V_i \quad (11.5)$$

d_i, distribution factor of particulate P among basins, see Figure 11.4
Z_P, Zala River particulate P load
φ, biologically available fraction
V_i, volume of basin i

Volumetric dissolved inorganic P load, L_P

$$L_P = L_{\mathrm{prec}} + e_i S / V_i + Z_d \quad (11.6)$$

L_{prec}, volumetric precipitation load
S, sewage load
e_i, distribution of sewage among basins, see Figure 11.4
Z_d, Zala River dissolved P load (basin I) ($Z_d = 0$; all other basins)

RATES

Growth rates, RGR_w, RGR_s

$$\mathrm{RGR}_w = k_{gw} f_P f_I f_{TW} A_w \quad (11.7)$$

$$\mathrm{RGR}_s = k_{gs} f_P f_I f_{TS} A_S \quad (11.8)$$

k_{gw}, k_{gs}, specific maximum growth rate coefficient, winter, summer, respectively
f_P, nutrient limitation factor

$$f_P = \frac{P}{P_k + P} \quad (11.9)$$

(cont.)

Table 11.2. (*cont.*)

P_k, half saturation constant

f_I, light limitation factor; Steele's equation, depth and day averaged with triangular light pattern

$$f_I = \frac{e\lambda}{\varepsilon H}\left\{\frac{1}{2L_H}[1-\exp(-2L_H)] - \frac{1}{2L_0}[1-\exp(-2L_0)]\right\} \quad (11.10)$$

ε, total extinction coefficient

$$\varepsilon = \varepsilon_0 + \alpha(A_w + A_s) \quad (11.11)$$

ε_0, extinction of water without algae

α, self-shading coefficient (on phytoplankton P base)

H, water depth

$$L_0 = R/\lambda I_s \quad (11.12)$$

$$L_H = L_0 \exp(-\varepsilon H) \quad (11.13)$$

I_s, optimal light intensity for algal growth

$$I_s = I_{sm} + I_{se} T \quad (11.14)$$

I_{sm}, base optimal light intensity

I_{se}, temperature correction coefficient to I_s

f_{TW}, f_{TS}, temperature reduction factors, winter, summer, respectively

$$f_{TW} = \frac{|T_{cw} - T|}{T_{cw} - T_{ow}} \exp\left[1 - \frac{|T_{cw} - T|}{T_{cw} - T_{ow}}\right] \quad (11.15)$$

$$f_{TS} = \begin{cases} \dfrac{T_{cs} - T}{T_{cs} - T_{os}} \exp\left[1 - \dfrac{T_{cs} - T}{T_{cs} - T_{os}}\right] & \text{if } T \le T_{cs} \\ 0 & \text{if } T > T_{cs} \end{cases} \quad (11.16)$$

T_{cw}, T_{cs}, critical temperature, winter, summer, respectively

T_{ow}, T_{os}, optimal temperature, winter, summer, respectively

Mortality rates, RDTH_w, RDTH_s

$$\mathrm{RDTH}_w = k_d \Theta_d^{T-20} A_w \quad (11.17)$$

$$\mathrm{RDTH}_s = k_d \Theta_d^{T-20} A_s \quad (11.18)$$

k_d, mortality rate coefficient at 20°C

Θ_d, mortality temperature coefficient

Mineralization rate, RMNRL

$$\mathrm{RMNRL} = k_m \Theta_m^{T-20} \mathrm{D} \quad (11.19)$$

k_m, mineralization rate coefficient at 20°C

Θ_m, mineralization temperature coefficient

(*cont.*)

Table 11.2. (*cont.*)

Settling rate, RSETD

$$\text{RSETD} = v_s\,(1-\gamma)D/H \tag{11.20}$$

v_s, settling velocity
γ, fraction dissolved of total detritus

Biogenic lime precipitation, RBIOP

$$\text{RBIOP} = xP\,(\text{RGR}_w + \text{RGR}_s) \tag{11.21}$$

x, biogenic lime coprecipitation coefficient

Release from the sediment, RREL

$$\text{RREL} = l_{\text{rel}}\Theta_{\text{rel}}^{T-20}/H \tag{11.22}$$

l_{rel}, rate of release of P per unit surface area
Θ_{rel}, release temperature coefficient

Sorption exchange, REXP (in model II only)

$$\text{REXP} = k_{\text{ex}}\,(P_{\text{eq}} - P) \tag{11.23}$$

k_{ex}, transport coefficient for sorption exchange
P_{eq}, effective equilibrium P concentration

P cycling

There is general consensus about the major processes in the in-lake cycling of P. Algae excrete organic P in dissolved form, and mortality leads to particulate detritus material. Part of this is lost from the cycle through settling, whereas the other part is hydrolyzed, thus contributing to the dissolved organic P pool. Finally, heterocyclic bacteria mineralize the dissolved organic P, perhaps through condensed polyphosphates to orthophosphate, which is then available for uptake by algae in the next cycle (cf. Leonov and Vasiliev 1981). In view of the purpose of the present work several simplifications were introduced in order to arrive at the simplest possible model that still represents the major features. First, the distinction between particulate and dissolved organic matter was dropped, and the sum of the two was simply called "detritus P". By this procedure one nonessential state variable was eliminated. Second, the bacteria were not modeled explicitly. Although there is little doubt of the significant role of bacteria, the inclusion of bacteria is not necessary for our present purposes. This statement is based on the argument that bacterial processes are comparatively fast, and are mostly governed by water temperature. Consequently, the effect of a time-varying bacteria population can be simulated in first approximation by introducing a strong temperature dependence of the mineralization rate (see Table 11.2). In addition, no systematic dynamic data exist on the bacteria population (although measurements suggest an increase since 1966), and thus it is not possible to check parameter assumptions against field data.

The sediment

As in many other shallow lake systems the sediment constitutes a considerable source of uncertainty. Initially, a simple sediment submodel was included in the model, such that about one third of the detritus was mineralized in the sediment (oxygen consumption measurements of lake water and sediment core suggest this ratio). Owing to the lack of systematic data, this submodel was abandoned and replaced by the simplifying assumptions that there is continuous sedimentation of (the particulate fraction of) detritus, and that dissolved inorganic P is released at a temperature-dependent but otherwise constant rate. Both sedimentation rate and release rate are essentially unknown, and are therefore treated as tunable parameters.

11.4. Transformation between Model Variables and Observation Variables

Figure 11.6 presents the structure of model I postulated on the basis of the previous considerations. The model has four state variables: winter algal P, summer algal P, detritus P (both dissolved and particulate), and dissolved inorganic P. Since this model-intrinsic representation does not match the data type of the measurements, a transformation is necessary both on the input and output side of the model. That is, the various P loads have to be allocated to each of the state variables, and the model results have to be expressed in terms of the measurements or vice versa.

For the distribution of the P load the following simplifying assumptions were made:

(1) All sewage is in dissolved inorganic form.

(2) The dissolved P load of the Zala River contributes solely to the dissolved inorganic pool (i.e., no detritus fraction).

(3) The particulate P load of the Zala River and the runoff in the other basins (see Figure 11.4) are divided into available and nonavailable fractions. The nonavailable fraction is believed to settle directly in the near-shore regions of the lake, so that this part does not show up in the mid-lake measurement data. A strong indication for such a phenomenon to occur is that about 95% of the total P entering the lake is retained. The available fraction is assumed to contribute to the detritus (because it is particulate, mineralizable matter).

The fraction of available P in the particulate load is, in fact, a model parameter as long as actual measurements (for instance, by more detailed chemical

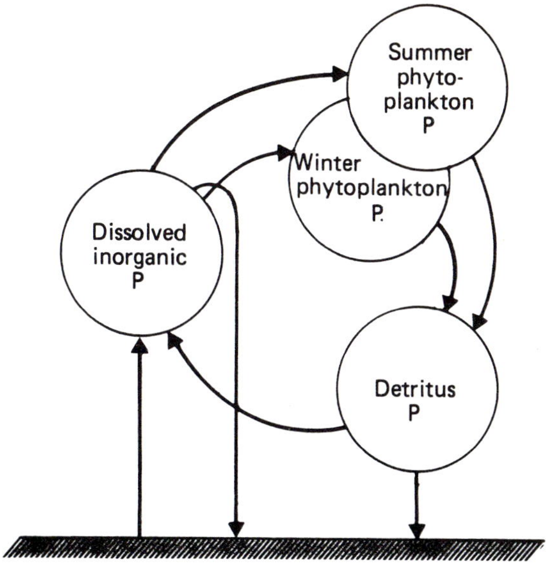

Figure 11.6. Model I structure.

fractionation of the Zala P input) are lacking. The value was set arbitrarily to 10% in the present application. At such low values the model is not very sensitive to this parameter because most of the available P load is in dissolved form (mainly sewage).

Next, on the output side a transformation is needed, in order to enable model results to be compared with actual data. There are, in principle, two ways of doing this: either by manipulation of the data to yield state variable values (e.g., detritus P equals particulate organic P minus phytoplankton P plus total dissolved P minus dissolved inorganic P) or by recombining the model results in terms of the measurements. The latter procedure is preferred because data manipulation results in large relative errors when subtraction of two uncertain numbers is involved, and would also cause problems if the measurements are not complete or asynchronous. Let **y** denote the vector of observations [i.e., Chl-a, total P (without particulate inorganic P), total dissolved P, and orthophosphate], and **x** the vector of state variables (winter algal P, summer algal P, detritus P, dissolved inorganic P), then the observation matrix is given by:

$$\mathbf{y} = \begin{bmatrix} C_1 & C_2 & 0 & 0 \\ 1 & 1 & 1 & 1 \\ 0 & 0 & \gamma & 1 \\ 0 & 0 & 0 & 1 \end{bmatrix} \mathbf{x}$$

where C_1 and C_2 are the Chl-a–P ratios of the winter and summer algae, respectively, and γ is the fraction of detritus in dissolved form. Both C_1, C_2, and γ are essentially unknown in the present case. The parameters C_1 and C_2 do not occur in the model itself, so it was decided to leave the results in the form of total phytoplankton P because a tentative comparison with Chl-a could always be made afterwards by assuming values for C_1 and C_2 without affecting the other model results. It should be noted that a very important implication of the lack of information on C_1 and C_2 is that one cannot hope to make accurate statements about the dynamics of the algal growth and mortality from a comparison of model results with measured data. In the present application this problem is not serious because the principal aim is to investigate the major P transport modes to and from the sediment, for which a rough estimate of the algal levels suffices. The situation is different with respect to parameter γ, which also occurs in the model itself [settling only operates on the particulate fraction $(1 - \gamma)$ of the detritus]. This is, in fact, the price that must be paid for reducing the number of state variables by combining particulate and dissolved inorganic P into one detritus term. An estimate for the value of γ is obtained from Table 11.1 as 0.4 by comparing the dissolved organic P (11 mg P/m^3) with the nonalgal part of the particulate organic P ($22 - 14/2 = 15$ mg P/m^3) assuming a Chl-a–P ratio of 2; see below).

Results for model I

Extensive experimentation with model I revealed an important deficiency. Figure 11.7 shows a typical pattern of dissolved inorganic P obtained with this model for the Keszthely basin. Data points are also shown. Most striking is the rapid rise of dissolved inorganic P in the model at the end of the year, which conflicts with actual observations. In this model the major mechanism to keep the orthophosphate levels low against continuous loads is uptake by algae. Since this uptake ends rather abruptly about mid-September, when temperature and light decline sharply, no removal mechanism is active in the model, and the dissolved inorganic P rises rapidly. Similar arguments apply to the period in May after the decline of the spring bloom. No improvement could be obtained by modification of the mineralization rate term, because the contribution of mineralization of decaying algal biomass was only a fraction of the P load. The results from model I strongly suggest that another mechanism regulates orthophosphate levels in autumn and winter.

The sorption hypothesis

The chemical composition of Lake Balaton certainly allows for an appreciable P coprecipitation with biogenically formed lime. This was accounted for in model I (see Table 11.2). However, biogenic lime formation is bound to the growing season, whereas the results suggest that P adsorption also occurs outside the season, thus producing the hypothesis that there is a continuous adsorption–desorption process. The sediment may play a direct role in this process, but it is equally possible that sorbents are continuously present and

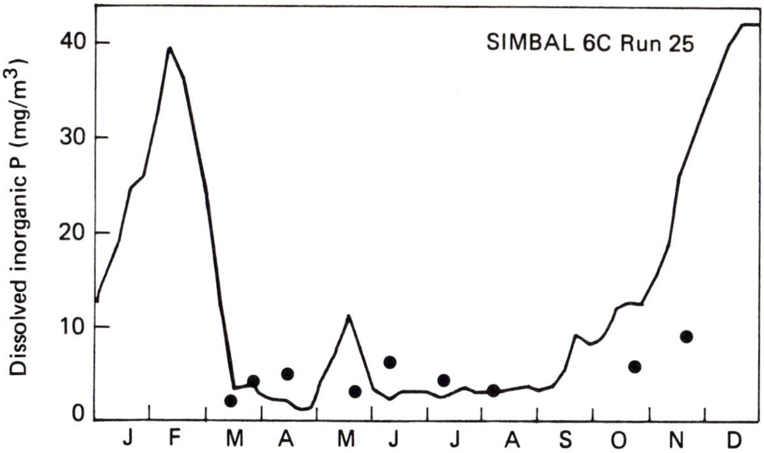

Figure 11.7. Typical result for model I. Comparison of simulation (solid line) with measurements (dots) for dissolved inorganic P in the Keszthely basin (basin I).

renewed by steady external loads and by resuspension and settling of sediment particles into and from the water.

The proper implementation of the sorption hypothesis would require the formulation of a sorption isotherm model, assuming that the adsorption–desorption is fast. Preliminary experimentation with such a model revealed that the most serious problem was the need to know the amount of (active) sorbents. Consequently, a dynamic model of suspended solids with a P adsorption capacity would have to be made. Although recent data are available from which such a model could probably be prepared (Chapter 7), another way was chosen here as a first approximation.

Table 11.3. Ratio of P_{eq} based on the particulate inorganic P (PIP) ratio.

Basin	PIP (mg/m^3)	Ratio
I	10.3	1
II	8.3	0.8
III	6.7	0.65
IV	4.4	0.4

In model II the sorption exchange reaction was postulated as a dynamic process: REXP = $k_{ex}(P_{eq} - P)$. From the data it was clear that the equilibrium concentration P_{eq} could not be far different from the actual dissolved inorganic P concentrations observed. The relatively heavy load in the Keszthely basin is likely to cause a relatively high P density on available sorption surfaces. Consequently, the equilibrium concentration would have to be different from basin to basin. Provisionally, the ratio of the average particulate inorganic P from basin to basin was used as an indication for the ratio in P_{eq}, as shown in Table 11.3. In view of the more recent results, reported in Chapter 7, that the adsorption is still in the almost linear part of the isotherm, this assumption is quite reasonable. The structure of model II is shown in Figure 11.8, and is referred to as SIMBAL (SIMple BALaton model).

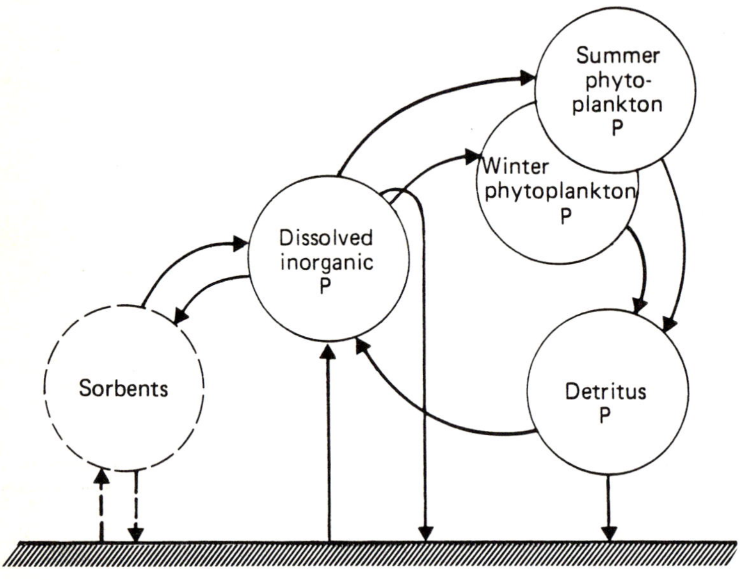

Figure 11.8. Model II structure with sorption hypothesis. The sorbent is not modeled explicitly.

11.5. Monte Carlo Simulation with Model II (SIMBAL)

A Monte Carlo simulation was performed to investigate whether parameter regions existed for which model II would give results that coincided with a predefined behavior derived from the data. The behavior definition used is given in Table 11.4.

It was decided not to vary all parameters in the simulation in order to reduce the number of runs required. At first, a series of 300 runs was done, varying nine parameter values according to Table 11.5. These were mostly parameters governing sediment–water interactions. The temperature factor for algal mortality was also included to allow for a reduction or increase in the mortality of cold-water algae, which might influence the speed of P regeneration after the algal blooms.

Out of the first series of 300 runs only one parameter combination was found that gave rise to the behavior, despite the fact that the constraint conditions were rather wide. One reason is that random and independent parameter selection ignores that some of the parameters of the model must be

Table 11.4. Behavior definition.

Period	Variable	Basin	Range (mg/m^3)
15 March–22 November	Total P	I	49.5–110.0
		II	31.5–82.5
		III	13.5–60.5
		IV	9.0–49.5
20 May–8 August	Total dissolved P	I	13.5–55.0
		II	13.5–38.5
		III	9.0–33.0
		IV	4.5–27.5
15 March–22 November	Dissolved inorganic P	I	0–16.5
		II	0–8.8
		III	0–8.8
		IV	0–8.8

Table 11.5. Parameter ranges.

Parameter	Range	
	Series 1	Series 2
v_{ret}, horizontal exchange flow (m/s)	0–0.003	0–0.003
P_k, Monod constant (mg P/m^3)	0–13	0–13
Θ_d, mortalitiy rate temperature effect	1.05–1.15	1.09–1.15
k_m, mineralization rate (d^{-1})	0.02–0.08	0.02–0.06
v_s, net settling velocity (m/d)	0.01–0.07	0.01–0.05
x, biogenic lime coprecipitation (m^3/mg)	0–0.03	0.015
l_{rel}, release from sediment (mg/m^{2}d)	0–1.0	0–1.0
k_{ex}, sorption rate coefficient (d^{-1})	0–0.5	0–0.2
P_{eq}, equilibrium concentration (mg P/m^3)	0–9	0–9

correlated in order to keep the state variables within a certain range. For instance, from the model equations (Table 11.2) it can be seen that k_m and v_s must be negatively correlated if the detritus is to be within certain limits. All parameter combinations generated outside this correlation region will not yield the required behavior. Several other correlations must be maintained, as discussed at some length in Chapter 3. Knowing this, it should be possible to set up a more "intelligent" Monte Carlo simulation, but in the present study this was not done.

An analysis of the position of k_m and v_s for runs for which the constraints were partially fulfilled (i.e., total dissolved P and dissolved inorganic P behavior, but not total P behavior) showed that k_m and v_s were confined to one corner of the parameter plane. Thus a reduction of the parameter region would be possible to increase the efficiency of the computation. Another interesting result of the first series was obtained from analysis of the position of k_{ex} and P_{eq} for partial behavioral runs. Figure 11.9(*a*) shows the runs resulting in total P (but generally not total dissolved P and dissolved inorganic

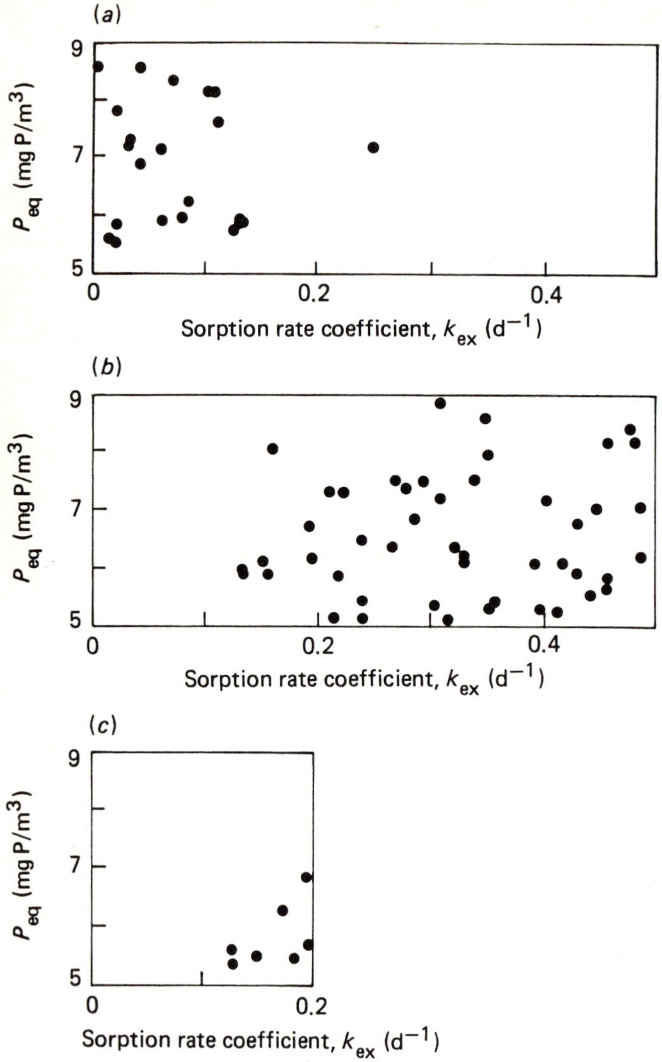

Figure 11.9. Monte Carlo simulation results for distribution of equilibrium concentration (P_{eq}) and sorption rate coefficient (k_{ex}). (*a*) Runs with total P behavior (series I); (*b*) runs with total dissolved and dissolved inorganic P behavior (series I); (*c*) full behavior runs (series 2).

P) behavior, and Figure 11.9(*b*) the runs resulting in total dissolved P and dissolved inorganic P (but not total P) behavior. It became apparent that in order to fulfill the total P conditions k_{ex} has to be low. This is because a rapid sorption rate would result in rapid P desorption in the summer, leading to higher algal growth and, consequently, to too much total P, whereas in autumn and winter P would be removed too rapidly. On the other hand, k_{ex} cannot be too low, because this would lead to insufficient adsorption in autumn and, thus, a violation of the total dissolved P and dissolved inorganic P conditions.

A second series of 300 runs was done after modification of the parameter ranges resulting from the first series analysis. The k_{ex} range was reduced, biogenic lime coprecipitation was taken to be constant because the model was insensitive to this parameter, and k_m and v_s were reduced as discussed previously. With these more limited parameter ranges the second series of 300 runs yielded seven parameter sets that fulfilled the behavior constraint conditions. In Figure 11.9(c) the position of these runs in the P_{eq}–k_{ex} plane are indicated. Table 11.6 summarizes the mean and ranges found for the parameters involved. The results confirm that k_{ex} must assume intermediate values, but, as seen from Figure 11.9(c), there is a correlation with the value of P_{eq}. As expected, k_m and v_s are also correlated.

Table 11.6. Parameter properties in behavioral runs (series 2).

Parameter	Minimum	Maximum	Mean
v_{ret} (m/s)	0.0002	0.0024	0.0016
P_k (mg P/m^3)	8.2	11.2	10.2
Θ_d	1.10	1.15	1.14
k_m (d^{-1})	0.030	0.047	0.035
v_s (m/d)	0.026	0.050	0.036
l_{rel} (mg/m^{2}d)	0.12	0.71	0.38
k_{ex} (d^{-1})	0.12	0.20	0.16
P_{eq} (mg P/m^3)	5.3	6.8	5.8

In order to obtain a visual impression of the quality of the model results obtained from the second series of runs, the mean over the seven behavioral runs was computed for every state variable on each time instant. The curves are shown in Figure 11.3, together with the actual data for each of the basins. Although not shown in the figure, the lines have to be seen as the approximate center of a probability density function rather than as deterministic curves. It is interesting that almost the same curves are found when the model is run with the average values of the parameters from the seven runs. In other words, the average parameter set is also a behavioral parameter set.

11.6. Discussion

Seven behavioral runs are too few for a proper statistical analysis, but for some parameters a particular behavior is possible for very specific regions in parameter space only, whereas others can take on various values (e.g., l_{rel}). Looking at the curves in more detail shows that total P in summer has a tendency to be too high. A run without the fixed temperature-dependent sediment release (l_{rel} = 0) yielded roughly 10 mg P/m^3 less in all basins, without much effect on phytoplankton or dissolved P. Thus, the fact that a parameter is not well determined indicates that the model is fairly insensitive to this parameter. Also it appears that once a sorption mechanism has been formulated there is no need to take into account release from the sediment as a separate term.

On the other hand, it seems that the sorption model does not yet give a complete answer, because the variability of the total P is much larger in the model than in the lake. In the model detritus decreases rapidly at the end of the season, even when mineralization slows due to the low temperatures, because of settling. The data suggest a build-up of particulate detritus material by the end of the season, which is not shown in the model. One possibility is that settling of particulate detritus material is a function of biogenic lime formation, so that settling is rapid in summer and slow in winter. Another factor that may contribute to more settling in summer is supersaturation of the lake water with Ca due to evaporation. These additional hypotheses need verification by further model and field experiments.

The role of the longitudinal exchange rate parameter v_{ret} is not clear from the results; both a high and a low longitudinal mixing could produce the behavior provided suitable values are used for the other parameters. Reversely, setting v_{ret} equal to zero, as suggested by the results of the hydrodynamic research discussed in Chapter 10, would reduce the allowable parameter space. This point underlines the importance of detailed subprocess research in reducing parameter uncertainty.

Finally, the phytoplankton P patterns need some discussion. At first sight the results do not look very good, but, as pointed out previously, no direct comparison with Chl-a data is possible. The results indicate that generally the Chl-a–P ratio is about 2 or somewhat higher. The model spring peak is slightly higher than the summer peak, which agrees with the data, except for the first basin. Here, most likely the dominant role of the blue–green algae, not contained explicitly in the model, disturbs the picture. It is also clear that the phytoplankton level between the two blooms is higher in the model than in reality. One should not forget that behavior parameters were sought for the four basins simultaneously. Thus, deviations from the model for one of the basins work through in the parameter estimates for the whole.

In the model, summer algal growth starts by the end of June, which is several weeks earlier than in the lake. It is not clear why this retardation occurs, because light, temperature, and P conditions are good in those weeks. It is possible to influence this behavior slightly by modifying the temperature dependences in the model. Another possibility for improvement is indicated in Chapter 13, where a comparison is made between the fixed cell quota models used here, and the so-called variable cell quota models, in which algae adapt their internal P concentration to external load conditions. These exercises suggest that perhaps the Monod P limitation term needs revision. It would be interesting to test this hypothesis using the approach of this chapter.

11.7. Conclusions

It is likely that a sorption mechanism, leading to P desorption during algal growth, when dissolved inorganic P is low, and to a strong adsorption outside the growing season occurs in the lake. This conclusion has stimulated further field and laboratory research into this phenomenon, and the results of Chapter 7 confirm the occurrence of such a process.

With respect to the model, one can conclude that within the crude behavior bounds – but not unreasonably crude given the level of crudeness of management goals –there are no reasons to reject the SIMBAL model, since it is possible to find parameter sets that give the defined behavior. At the same time this reveals a certain drawback of the behavior-type approach: there is no substitute for the error sequence analysis typical of classical deterministic parameter estimation. Only if one is prepared to attach larger credibility to the data does SIMBAL have deficiencies that call for model refinement.

Very often, systems analysis by model studies is the only way to extract the information hidden in existing data. The method used in this chapter provides a tool to give due account for uncertainty in comparing different alternative assumptions about the system's structure. The approach is particularly useful if a clear distinction between alternative hypotheses is possible, as in this case. In general, however, one must expect that uncertainty simply reduces our ability to discriminate. Still, even in such situations, the approach is valuable, because it gives indications as to where more detailed additional research is needed. And this is important, because, in the end, it is only proper field and laboratory experimentation that will allow us to achieve really decisive uncertainty reductions.

References

Beck, M.B., Halfon, E., and van Straten, G. (1979) *The Propagation of Errors and Uncertainty in Forecasting Water Quality, part I: Method.* Working Paper WP-79-100 (Laxenburg, Austria: International Institute for Applied Systems Analysis).

Di Toro, D.M. and van Straten, G. (1979) *Uncertainty in the Parameters and Predictions of Phytoplankton Models.* Working Paper WP-79-27 (Laxenburg, Austria: International Institute for Applied Systems Analysis).

Entz, B. (1959) Chemische Charakterisierung der Gewässer in der Umgebung der Balatonsees (Plattensees) und chemische Verhältnisse des Balatonwassers. *Ann. Biol. Tihany* 26:131–201.

Fedra, K., van Straten, G., and Beck, M.B. (1981) Uncertainty and arbitrariness in ecosystems modelling: a lake modelling example. *Ecological Modelling* 13(1,2):87–110.

Leonov, A.V. and Vasiliev, O. (1981) Simulation and analysis of phosphorus transformations and phytoplankton dynamics in relation to the eutrophication problem of Lake Balaton, in D.M. Dubois (Ed.) *Progress in Ecological Engineering and Management by Mathematical Modelling 1981* (Liege, Belgium: Editions CEBEDOC) pp 627–56.

Spear, R.C. and Hornberger, G.M. (1978) *An Analysis of Behavior and Sensitivity of a Poorly Defined System*. Report AS/R24, Centre for Resource and Environmental Studies, Australian National University.

Zankai, N. and Ponyi, J. (1976) Seasonal changes in the filtering rate of *Eudiaptonus gracilis* (G.O. Sars) in Lake Balaton. *Ann. Biol. Tihany* 43:105–16.

CHAPTER 12

A Complex Model for Simulating the Lake Balaton Ecosystem

T. Kutas and S. Herodek

12.1. Introduction

As discussed in Chapter 3, the response of water quality to changing nutrient loads is examined using simple empirical relations and complex dynamic models. The former are used to determine the P content of the water, chlorophyll-a (Chl-a) concentrations, and primary production as functions of nutrient load, water depth, and retention times, based on the statistical evaluation of measured data from many lakes. It is an advantage to work with few, easily measurable parameters, but they inherently assume an equilibrium state in the system (which is not true in most cases), and they cannot describe changes in water quality over time, i.e., how long it will take before the next steady state of the system is reached under new nutrient load conditions.

Dynamic models are thus used in the attempt to solve this problem. While earlier dynamic lake models sought for generality and could easily involve more than 30 compartments, it is now accepted that there can be no generalized lake model; for every lake a special model has to be built to reflect the characteristics of the lake and to answer particular questions, e.g., in the case of eutrophication, the causes of changes in algal biomass.

The modeling of such a complex ecosystem as Lake Balaton, undergoing eutrophication, has two goals: first, to improve our understanding of system behavior and to better simulate measured data; and second, to provide a basis for management and decision making. The present model is the latest in a series developed by the Balaton Eutrophication Modeling Group in Hungary (referred to as the BEM model), but it does not yet have its ultimate form, since model development evolves as ecological knowledge increases. The philosophy of modeling has not changed during this time: the model structure has to reflect the basic structure of the ecosystem, the main processes that take place within the lake must be incorporated, mathematical parameters have to carry biological meaning (as far as possible), and the model has to be suitable for management purposes.

Lake Balaton has been divided into four basins (Keszthely, Szigliget, Szemes, and Siófok) based on hydrodynamical considerations. These basins are modeled separately and the models are linked together by hydrologic throughflow. At present all the basins are described by the same model and parameter set, and differ only in the nutrient loads and initial values.

12.2. Model Description

The model describes the main mass transport processes in the open water of the four basins. Since the distribution of matter within each basin can be regarded as uniform, the open water is described by one point. The state variables are concentrations of P, N, and organic matter. A constant ratio of P and N to biomass is assumed, in both the primary production of organic matter and that produced during bacterial decomposition. Thus, the model is a constant stoichiometry model.

The forcing functions are global radiation, water temperature, and P and N loads. The model contains ten state variables for each of the basins: winter–spring (A_{b1}), summer (A_{b2}), and autumn (A_{b3}) phytoplankton; blue–green algae ($A_{b\,\mathrm{BG}}$); bacterioplankton (B_b); dead organic matter (D_b), dissolved inorganic P (P) and N (N); organic matter (D_{Sb}); and the P concentration in the sediment (P_S) (see Figure 12.1).

Nutrients circulate in two main cycles within each basin. In the first, inorganic P and N are incorporated into organic matter by algae during primary production; the algae become dead organic matter, which is then decomposed by bacterioplankton releasing the inorganic nutrients. In the second cycle dead organic matter, algae, and bacteria settle on the lake bottom, decompose in the sediment, and release P and N. The N returns directly to the water, while the P first enters the exchangeable P pool in the sediment. Nutrients can leave the system by outflow or by stabilization within the sediment, and some N can be removed by denitrification. A more detailed description of the Balaton ecosystem is given in Herodek *et al.* (1982b).

The BEM model rate equations are given in Table 12.1, and a description of the processes is given below. For all four phytoplankton compartments primary production RGR is formulated as follows:

$$\mathrm{RGR}_i = k_{gi} f_{Ti} f_{Li} A_{bi} \qquad i = 1, 2, 3, \mathrm{BG} \tag{12.1}$$

where k_{gi} is the maximal production rate, and A_{bi} is the phytoplankton biomass of species i. The factor f_{Ti} is the temperature limitation:

$$f_{Ti} = \begin{cases} \dfrac{T_{ci} - T}{T_{ci} - T_{oi}} \exp\left[1 - \dfrac{T_{ci} - T}{T_{ci} - T_{oi}}\right] & \text{if } T \le T_{ci} \\ 0 & \text{if } T > T_{ci} \end{cases} \tag{12.2}$$

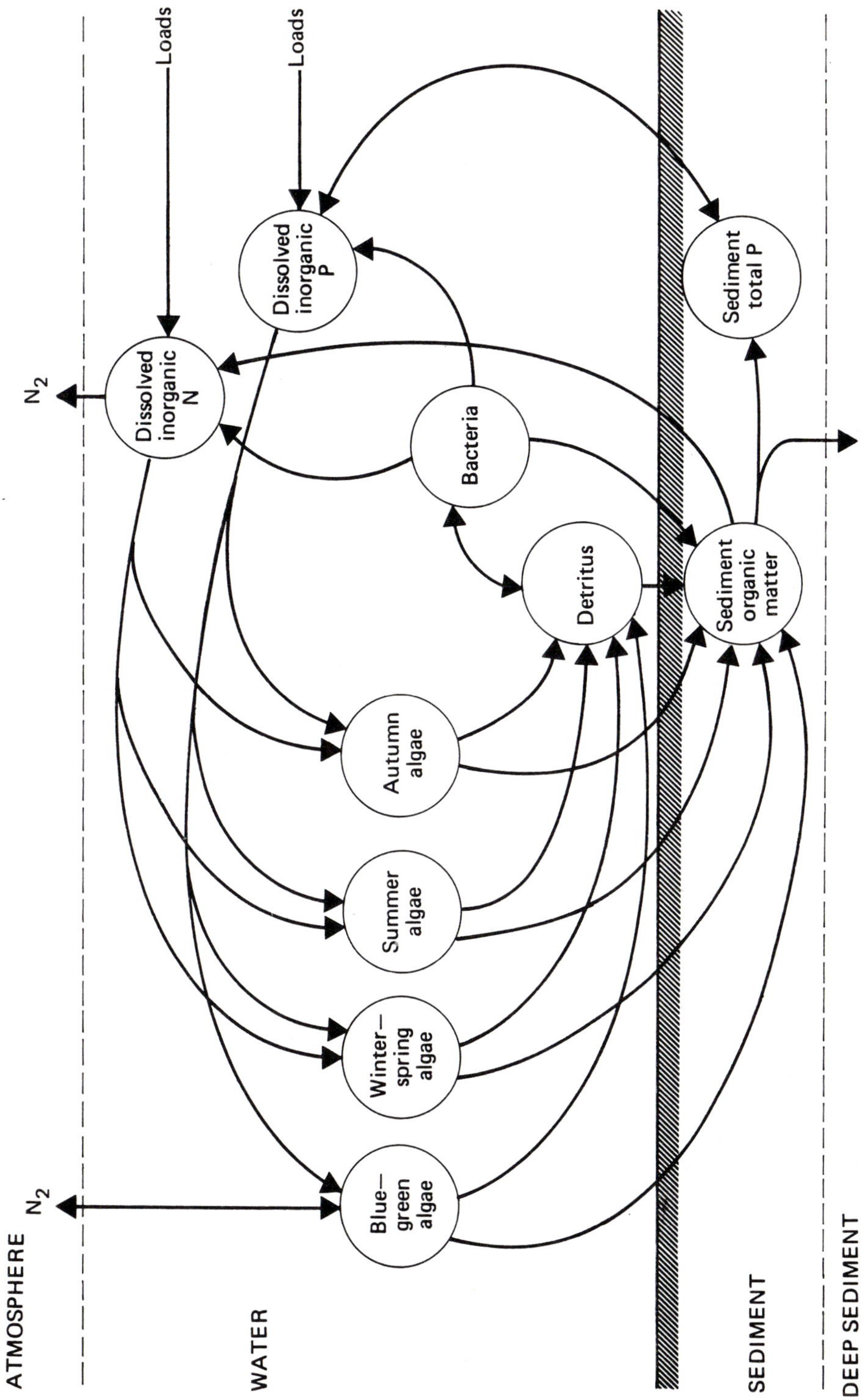

Figure 12.1. BEM model structure.

Table 12.1. BEM model equations.

$$\dot{A}_{bi} = (I - O)_{Ai} + \mathrm{RGR}_i - \mathrm{RDTH}_i - \mathrm{RSED}_i \qquad i = 1, 2, 3, \mathrm{BG}$$

$$\dot{B}_b = (I - O)_{Bi} + \mathrm{RGRB} - \mathrm{RDTHB} - \mathrm{RDEC} - \mathrm{RSEDB}$$

$$\dot{D}_b = (I - O)_D + \sum_i \mathrm{RDTH}_i + \mathrm{RDTHB} - \mathrm{RGRB} - \mathrm{RSEDD}$$

$$\dot{P} = (I - O)_P - \mathrm{RUPTP} + \mathrm{RDECP} + \mathrm{REXP} + L_P$$

$$\dot{N} = (I - O)_N - \mathrm{RUPTN} + \mathrm{RDECN} + \mathrm{REXN} - \mathrm{RDEN}$$

$$\dot{D}_{Sb} = \sum_i \mathrm{RSED}_i + \mathrm{RSEDB} + \mathrm{RSEDD} - \mathrm{RMNRLS}$$

$$\dot{P}_S = \mathrm{RPS} - \mathrm{REXP}$$

where

A_{bi},	phytoplankton biomass: $i = 1$, winter–spring; $i = 2$, summer; $i = 3$, autumn; $i = \mathrm{BG}$, blue–green algae
B_b,	bacteria biomass
D_b,	dead organic matter (detritus)
P,	dissolved inorganic P
N,	dissolved inorganic N
D_{Sb},	organic matter in the sediment
P_S,	the total P in the sediment
$(I - O)$	are inflow and outflow terms

All other rate terms are explained in the text.

where T_{ci} and T_{oi} are the critical and optimal temperatures, respectively, and T is the water temperature. The factor f_u represents the joint limitation of nutrients and light. For the first three phytoplankton compartments (A_{b1}, A_{b2}, and A_{b3}):

$$f_{ui} = (f_{NPi} f_{Ii})^{f_{oi}} f_{oi}^{1-f_{oi}} \tag{12.3}$$

where

$$f_{NPi} = \min\,(f_{Pi}, f_{Ni}) \tag{12.4}$$

$$f_{Pi} = \frac{P}{P_{ki} + P} \tag{12.5}$$

$$f_{Ni} = \frac{N}{N_{ki} + N} \tag{12.6}$$

where P and N are the inorganic P and N concentrations in the water, P_{ki} and N_{ki} are half saturation constants, and f_{Ii} is the light limitation:

$$f_{Ii} = \frac{1}{\varepsilon H}\left[\exp\left(1 - \frac{I_H}{I_{si}}\right) - \exp\left(1 - \frac{I_o}{I_{si}}\right)\right] \tag{12.7}$$

where

$$I_H = I_0 \exp(-\varepsilon H) \quad , \tag{12.8}$$

H is the depth of the water body, and ε is the extinction coefficient:

$$\varepsilon = \varepsilon_0 + \alpha \sum_i A_{bi} \qquad i = 1, 2, 3, \mathrm{BG} \tag{12.9}$$

where ε_0 is the extinction coefficient of the water without algae, α is a self-shading constant, I_{si} is the optimal light intensity, and I_0 is the light intensity just below the water surface. Finally, f_{oi} is defined as

$$f_{oi} = \min (f_{NPi}, f_{Ii}) \tag{12.10}$$

The factor f_{oi} represents the joint limitation according to the "minimum rule": the lowest factor (for P, N, or light) determines the growth reduction. In the model, however, a compromise is sought between this formulation and the "multiplicative rule" by calculating the weighted geometric mean of the minimum and the multiplicative forms, where the weight is the minimum itself [equation (12.3)]. The advantage of this is that when the limitations are near to 1 then the new formula is near to the multiplicative form, whereas when the minimum is low (i.e., when at least one of the limitations is near to zero) it approaches the minimum form to avoid the overlimitation effect. The primary production of blue-green algae has been formulated in a similar way; only the f_u term is different. Since blue-green algae can fix atmospheric N_2, N does not limit primary production. Also, blue-green algae can float in the optimally illuminated water layer, so it can be assumed that light does not limit primary production. In this case, f_u reduces to the P limitation, $f_{uBG} = f_{PBG}$.

The natural mortality (RDTH) of all living compartments is an exponential function of water temperature:

$$\mathrm{RDTH}_i = k_{di} \exp [\beta_{di} (T - T_{ci})] A_{bi} \qquad i = 1, 2, 3, \mathrm{BG} \tag{12.11}$$

$$\mathrm{RDTHB} = k_{dB} \exp [\beta_{dB} (T - T_{cB})] B_b \tag{12.12}$$

where k_{di} and k_{dB} are the specific mortality rates at the critical temperatures T_{ci} and T_{cB}, β_{di} and β_{dB} the temperature coefficients of mortality, and A_{bi} and B_b the biomasses of phytoplankton species and bacteria, respectively.

The growth of bacteria (RGRB) by uptake of dead organic material is

$$\mathrm{RGRB} = k_b \frac{D_b}{D_b + B_b} f_{TB} B_b \tag{12.13}$$

where k_b is the maximal uptake rate, D_b is the concentration of dead organic matter in the water, and B_b is the bacterioplankton biomass. The temperature dependence has the same form as for phytoplankton and is defined in equation (12.2).

The decomposition process (RDEC) is also dependent on temperature:

$$\mathrm{RDEC} = k_{rB} f_{TB} B_b \tag{12.14}$$

where k_{rB} is the decomposition rate coefficient. The rates of P and N excretion during decomposition are given as

$$\mathrm{RDECP} = y_{\mathrm{P}} \cdot \mathrm{RDEC} \tag{12.15}$$

$$\mathrm{RDECN} = y_{\mathrm{N}} \cdot \mathrm{RDEC} \tag{12.16}$$

where y_{P} and y_{N} are the P and N ratios in organic matter, which are assumed to be constant in all forms of organic matter. These values are therefore also used in determining the uptake of inorganic nutrients (RUPT) by phytoplankton:

$$\mathrm{RUPTP} = y_{\mathrm{P}} \sum_{i} \mathrm{RGR}_i \qquad i = 1, 2, 3, \mathrm{BG} \tag{12.17}$$

$$\mathrm{RUPTN} = y_{\mathrm{N}} \sum_{j} \mathrm{RGR}_j \qquad j = 1, 2, 3 \tag{12.18}$$

Blue–green algae do not consume N dissolved in the water but they use atmospheric N to produce new organic matter which constitutes an additional N input to the system.

The sedimentation of particulate materials (RSED) is proportional to the concentration

$$\mathrm{RSED}_i = k_{si} A_{bi} \qquad i = 1, 2, 3, \mathrm{BG} \tag{12.19}$$

$$\mathrm{RSEDB} = k_{sB} B_b \tag{12.20}$$

$$\mathrm{RSEDD} = k_{sD} D_b \tag{12.21}$$

where k_{si}, k_{sB}, and k_{sD} are the settling rate coefficients for phytoplankton species, bacteria, and detritus, respectively.

The mineralization of organic matter in the sediment (RMNRLS) is an exponential function of temperature

$$\mathrm{RMNRLS} = k_{mS} \exp\,(\beta_{mS} T)\, D_{Sb} \tag{12.22}$$

where k_{mS} is the mineralization coefficient rate, β_{mS} is the temperature coefficient of mineralization, and D_{Sb} is the concentration of organic matter in the sediment. Experiments with P have determined an adsorption–desorption process. In the model some P originating from decomposed organic matter enters the sediment P pool

$$\mathrm{RPS} = y_{\mathrm{P}} \cdot \mathrm{RMNRLS} \cdot (1 - r_s) \tag{12.23}$$

where r_s is the stabilized fraction; that is this part of the organic matter and its associated inorganic P and N is assumed to be permanently lost.

The rate of P exchange between the sediment and the water body (REXP) is

$$\mathrm{REXP} = k_{\mathrm{ex}} (y_{SP} P_S - P) \tag{12.24}$$

where k_{ex} is the P exchange coefficient, y_{SP} is a sediment P availability coefficient, and P_S is the total P content of the active sediment layer. This term appears with negative sign in the equation of the sediment P pool, and with positive sign in that of P in the water.

Measurements by Istvánovics *et al.* (1983) have shown a high temperature dependence of N release from the sediment. This is covered in the model because the release rate (REXN) is assumed to be directly proportional to the mineralization rate [equation (12.22)]:

$$\text{REXN} = y_{\text{N}} \cdot \text{RMNRLS} \cdot (1 - r_s) \tag{12.25}$$

Not all N released by mineralization in the sediment reaches the water. Some of it leaves the system during the process of denitrification (RDEN):

$$\text{RDEN} = y_{\text{N}} \cdot \text{RMNRLS} \cdot (1 - r_s) \cdot [1 - exp\ (\beta_{\text{den}} \cdot \text{RMNRLS})] \tag{12.26}$$

where β_{den} is the temperature coefficient of denitrification.

When the lake is covered with ice there is no wind-induced stirring, and this was simulated by a higher settling rate. As a consequence large amounts of organic matter accumulate at the sediment surface, but after ice break this enters the water. This process was simulated by returning a part of the organic matter of the sediment into the water as inorganic P and N. The process takes place during one month beginning one week after ice break, i.e., when the water temperature rises after a period of zero in simulation runs. The circumstances of this process are not clear, and result in spring algal biomass peaks in simulation runs earlier than they do in reality. For a more detailed description of the process equations of the BEM, see Kutas and Herodek (1982).

12.3. Calibration of the Model

As shown in Chapter 10, the separation of Lake Balaton into four basins eliminates the need to account explicitly for interbasin mixing. Since the hydrologic throughflow is quite small, this offers a special opportunity for calibration, because for a first try the basins can be treated as fairly independent. After calibration of each one-basin model the complex model is formed by simply multiplying the structure four times and coupling the four segments by the hydrologic throughflow. In this way first the one-basin models were calibrated, and later the complex model needed only some fine tuning of the initial values.

Simultaneous field measurements were carried out for primary production, algal biomass, and total extinction at Szemes in 1976–77 and at Tihany (Siófok basin) in 1977, and the calibration of the one-basin models was based on these data. The calibrations were made simultaneously for both basins; since the model was run with the same parameter set, only the forcing functions were different. After calibrating the one-basin version, the four-basin version run was begun.

The next step was the inclusion of blue–green algae into the model. These do not play an important role in the Szemes and Siófok basins, so their absence did not affect the calibration of the original one-basin model. The

Table 12.2. Model parameters.

Constant	A_{b1}	A_{b2}	A_{b3}	A_{bBG}	B_b	D_b	P	N	D_{Sb}	P_S
k_g (d^{-1})	7.0	3.2	1.1	2.0						
k_b (d^{-1})					1.6					
T_c (°C)	30.0	21.0	20.0	27.0	28.0					
T_o (°C)	23.0	11.5	6.0	24.0	24.0					
P_k (g/m^3)	0.004	0.006	0.009	0.012						
N_k (g/m^3)	0.04	0.09	0.11							
I_s ($kcal/m^2d$)	1200.0	800.0	300.0							
k_d (d^{-1})	3.3	0.85	0.25	0.25	0.2					
β_d (°C^{-1})	0.065	0.025	0.015	0.20	0.12					
k_s (d^{-1})	0.03	0.03	0.03	0.03	0.03	0.12				
k_{rB} (d^{-1})					0.88					
k_{mS} (d^{-1})									0.0004	
β_{mS} (°C^{-1})									0.155	
y_P (–)							0.0091			
y_N (–)								0.07		
r_s (–)									0.12	
k_{ex} (d^{-1})										0.35
y_{SP} (–)										0.07
β_{den} ($d/10^{-3} m^3 mg$)								1.7		

Table 12.3. Initial values (g/m^3).

Basin	A_{b1}	A_{b2}	A_{b3}	A_{bBG}	B_b	D_b	P	N	D_{Sb}	P_S
Keszthely	0.005	0.1	0.4	0.005	1.0	2.0	0.010	0.25	40.0	0.20
Szigliget	0.005	0.1	0.4	0.005	0.8	1.8	0.007	0.25	30.0	0.15
Szemes	0.005	0.05	0.3	0.005	0.7	1.5	0.005	0.25	12.0	0.10
Siófok	0.005	0.05	0.3	0.005	0.5	1.4	0.003	0.25	10.0	0.10

Table 12.4. Nutrient loads of basins (kg/d).

	Keszthely	Szigliget	Szemes	Siófok
P	98.5	93.8	79.6	135.1
N	1812.7	927.3	964.8	1104.0

blue–green algae are more important in the Keszthely and Szigliget basins in late summer. Besides an estimate for the parameters of blue–green algae the calibration of the new model version needed only a change in optimal temperature of summer phytoplankton. The final parameter and initial values are shown in Tables 12.2 and 12.3, respectively.

The P and N load calculations are based upon the proposals of van Straten and Somlyódy and corrections made by Jolánkai and Somlyódy (see Chapter 6). The model takes into account the available nutrient load, where "available" means those forms of P and N at particular points in the nutrient cycle that are available for uptake by algae. The annual average P and N loads (in kg/d) are given in Table 12.4.

12.4. Simulation Results

The results of the BEM model have been compared with actual Chl-a measurements (Tóth 1974). Chl-a concentration is the only indicator of water quality that has been measured simultaneously and frequently in all the lake basins (see Figures 12.2–12.5). Note that, in contrast with Chapter 11, the data used are not the calculated basin-averaged values. Rather, from the nine measurement points of the regular sampling program (see Chapter 1, Figure 1.2) those located approximately in the middle of each basin were selected (i.e., points 1, 2, 5, and 8 for Keszthely, Szigliget, Siófok, and Tihany, respectively). In order to obtain Chl-a concentrations, the phytoplankton biomass (a model state variable) was multiplied by a constant, 17.8, estimated from parallel biomass and Chl-a measurements (Herodek *et al.* 1982a); biomass is expressed in g dry weight/m^3, and Chl-a concentrations in mg/m^3. The model was run for the years 1976–77 using actual meteorological and load data.

According to the data, the Chl-a concentration has spring and summer peaks, and these were successfully simulated by the model. The heights of

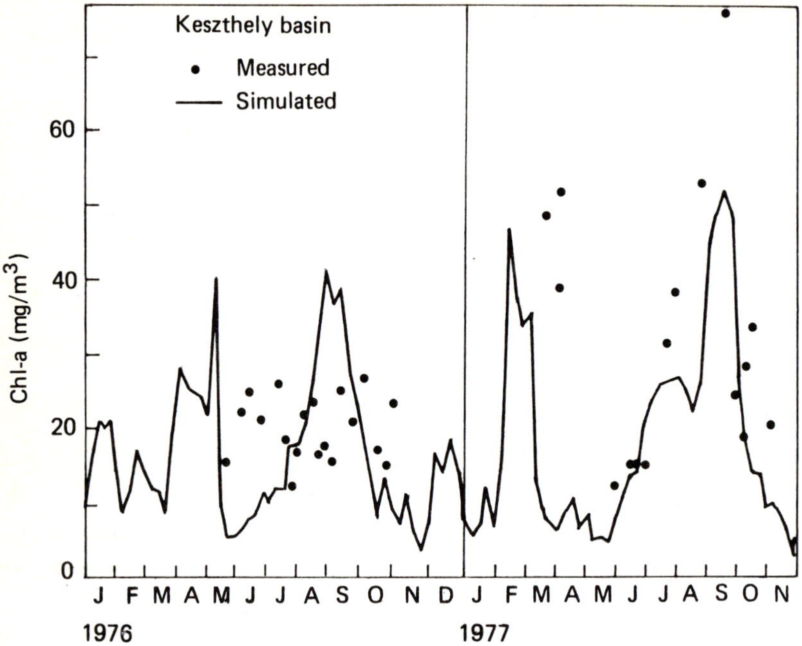

Figure 12.2. Measured and simulated Chl-a concentrations in Basin I (Keszthely basin), 1976–77.

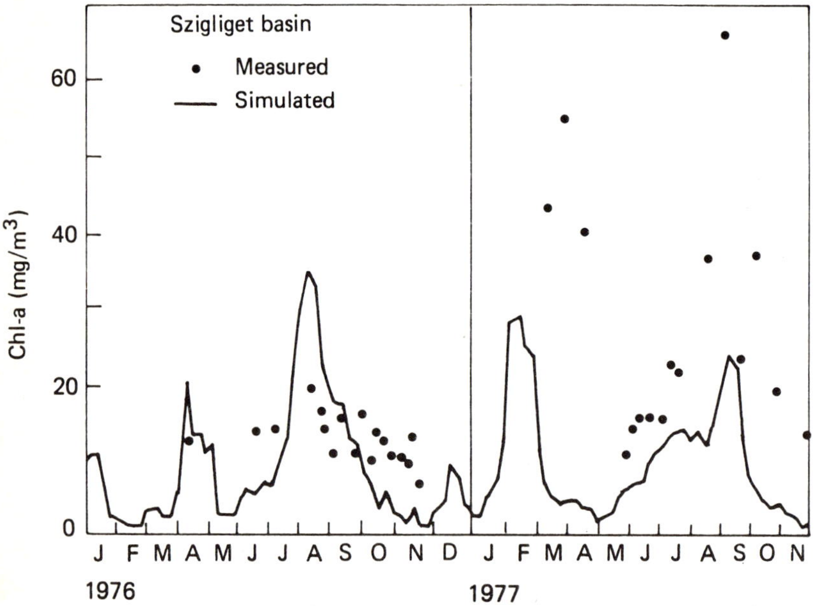

Figure 12.3. Measured and simulated Chl-a concentrations in Basin II (Szigliget basin), 1976–77.

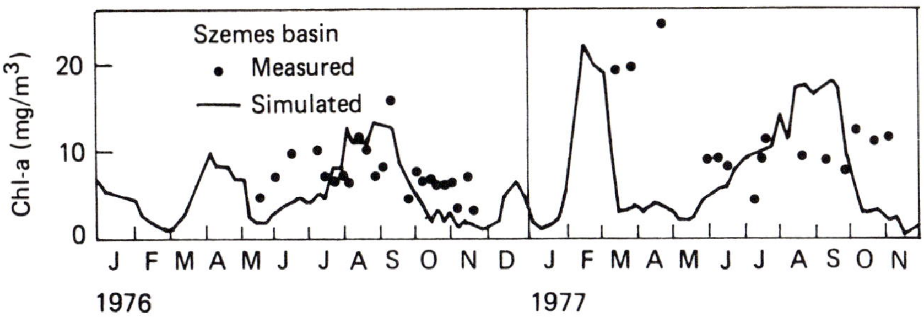

Figure 12.4. Measured and simulated Chl-a concentrations in Basin III (Szemes basin), 1976–77.

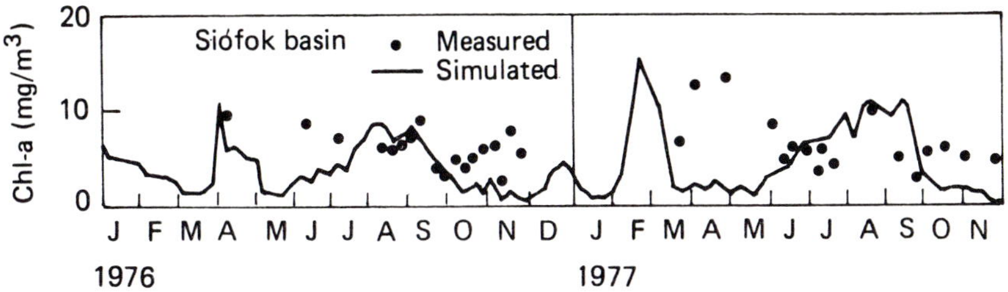

Figure 12.5. Measured and simulated Chl-a concentrations in Basin IV (Siófok basin), 1976–77.

the first peaks were simulated correctly, but in the model they appeared quicker than in the measurements, due to the unclear effects of ice break on the ecosystem. In the Keszthely and Szigliget basins the simulated summer peaks were almost the same in both years, but the measured values were lower in 1976 and higher in 1977. The model simulated the phytoplankton dynamics in Szemes and Siófok basins quite well. The differences between the simulations represent differences in the trophic states of the basins.

12.5. Nutrient Load Reduction Simulation

In Lake Balaton water quality is worst in the Keszthely basin, where algal blooms often render the water unsuitable for bathing. From the point of view of lake preservation it is most urgent to begin restoration measures in this area, and since the Zala River carries most of the nutrient load (92%) it is logical to concentrate efforts on a reduction from this source in the first instance. In fact, the largest proportion of the biologically available P originates from the sewage effluent of Zalaegerszeg and from liquid manure produced at large-scale animal farms, point sources which could be controlled at relatively low cost by tertiary treatment or similar techniques. As was pointed out in Chapter 4, tackling these sources proved to be most cost-effective, as compared to other methods, such as pre-reservoirs. P removal should be preferred over N removal, not only because N removal is far more difficult and expensive than that of P, but also because a successful N load reduction would be counteracted by increased N fixation by blue–green algae.

The effects of reduction in the P loads entering the lake via the Zala River have been simulated over six years using meteorological data from 1976. Since it is impossible to compare the output of six years' simulations, six indicators were chosen for all simulated years: annual average phytoplankton biomass and primary production, summer average phytoplankton biomass and primary production, and summer maximal phytoplankton biomass and primary production (where summer extends from the beginning of May until the end of September). Here we discuss only the summer average biomass, since this shows best the level of water quality, but all six indicators are given in Table 12.5.

If nutrient loads remain at the 1976 level, the summer average phytoplankton biomass increases quickly at first, and later slows down, without reaching a steady state. If the P load is reduced to 75% of this level, the phytoplankton biomass remains nearly constant. With half the nutrient load the algal biomass decreases by 20%, and the corresponding steady state is reached within three years. Reducing the P load to 25%, the summer average biomass decreases to half, and a new equilibrium is attained in five-to-six years. The greater the change in the nutrient load, the longer the lake takes to respond. If the load were reduced to zero, the algae would eventually disappear; the phytoplankton biomass would decrease to two thirds after one year, and then to a quarter after six years according to the model.

The slow response of the lake is due to the large amount of biologically available P that accumulates in the sediment. In Lake Balaton, as in all shallow lakes, far more nutrients are stored in the sediment than in the water, so the long-term response is determined by the state of the sediment. Some of the P content of the sediment is returned to the water, while another portion is stabilized in the sediment and thus leaves the system. Both losses are proportional to the sediment P concentration. If this concentration is below an equilibrium level, so less P leaves the sediment than enters it, the P content of the sediment increases, the internal load originating from the sediment increases, and thus the algal biomass increases. By reducing the nutrient load from the watershed below a certain level, less P would enter the sediment than would leave it, and this would continue until the system reached a steady state corresponding to the new load situation.

Using the results of sediment research in progress the model can be developed to be more exact, although some practical conclusions can be drawn at this stage. The unchanged nutrient load does not mean that the water quality will remain the same. Water quality responds slowly to changes in nutrient load and it may take several years to reach a new equilibrium. The change is greatest in the first years, but subsequently the rate of change decreases. The algal biomass is not directly proportional to the nutrient load; in order to reduce the algal biomass by half in the Keszthely basin, the biologically

Table 12.5. Indicators of water quality in load reduction simulations for the Keszthely basin.

Reduction	Year					
	1	2	3	4	5	6
(1) Annual average phytoplankton biomass (g dry weight/m^3)						
1.0	1.03	0.89	0.91	0.95	0.99	0.99
0.75	0.90	0.75	0.75	0.78	0.81	0.80
0.5	0.79	0.61	0.59	0.60	0.61	0.59
0.25	0.69	0.48	0.44	0.42	0.41	0.39
0.0	0.58	0.34	0.28	0.24	0.21	0.19
(2) Annual average primary production (g/m^{3}d)						
1.0	1.03	1.06	1.07	1.12	1.19	1.21
0.75	0.93	0.90	0.90	0.91	0.96	0.99
0.5	0.82	0.74	0.71	0.70	0.73	0.73
0.25	0.71	0.58	0.52	0.49	0.50	0.48
0.0	0.60	0.42	0.33	0.28	0.26	0.23
(3) Summer average phytoplankton biomass (g dry weight/m^3)						
1.0	1.17	1.25	1.24	1.29	1.34	1.37
0.75	1.05	1.06	1.03	1.05	1.07	1.09
0.5	0.93	0.87	0.81	0.81	0.81	0.81
0.25	0.81	0.69	0.60	0.57	0.55	0.54
0.0	0.69	0.50	0.39	0.33	0.29	0.27
(4) Summer average primary production (g/m^{3}d)						
1.0	2.06	2.17	2.27	2.31	2.40	2.43
0.75	1.84	1.84	1.87	1.88	1.92	1.94
0.5	1.63	1.52	1.49	1.45	1.46	1.46
0.25	1.43	1.21	1.10	1.03	0.99	0.97
0.0	1.22	0.89	0.71	0.60	0.53	0.48
(5) Summer maximal phytoplankton biomass (g dry weight/m^3)						
1.0	1.51	1.67	1.69	1.75	1.79	1.83
0.75	1.36	1.42	1.41	1.42	1.43	1.45
0.5	1.22	1.18	1.13	1.10	1.09	1.09
0.25	1.08	0.94	0.85	0.79	0.75	0.73
0.0	0.94	0.71	0.56	0.47	0.40	0.37
(6) Summer maximal primary production (g/m^{3}d)						
1.0	2.97	3.23	3.32	3.39	3.24	3.35
0.75	2.68	2.76	2.75	2.76	2.61	2.67
0.5	2.41	2.30	2.19	2.14	1.99	2.01
0.25	2.13	1.84	1.64	1.53	1.37	1.35
0.0	1.85	1.38	1.10	0.91	0.74	0.68

available P load in the Zala River would have to be reduced to well below half of its present level[1].

References

Herodek, S., Kutas, T., and Csáki, P. (1982a) Simulation of phytoplankton dynamics in Lake Balaton. *ISEM Journal* 4(3, 4):97–126.

Herodek, S., Vörös, L., and Tóth, F. (1982b) The mass and production of phytoplankton and the eutrophication of Lake Balaton III: The Balatonszemes basin in 1976–77 and the Siófok basin in 1977. *Hidrol. Közl.* 62:220–9 (in Hungarian with English summary).

Istvánovics, V., Herodek, S., and Entz, B. (1983) The effect of sediment dredging on the water quality of Lake Balaton II: The phosphorus and nitrogen cycles. *Hidrol. Közl.* 63(3):125–9 (in Hungarian with English summary).

Kutas, T. and Herodek, S. (1982) *A Short Description of Balaton Eutrophication Model* (BEM). SZTAKI Working Paper MS/9.

Tóth, L. (1974) Examination of the water quality of Lake Balaton. *VITUKI Report No.* 7782/4/8 (Budapest: VITUKI) (in Hungarian).

[1]As with other models, these results apply to a model description that does not take into account the long-term adaptation of the sediment adsorption characteristics resulting from load changes (see Chapter 7). In the light of the discussion in Chapter 3 we may expect that model coefficients, such as the "burial" ratio r_s and the "availability factor" y_{SP} will change with changing loads, and consequently the outcome of model computations will be different. So, large uncertainties still exist regarding our ability to accurately predict the long-term behavior of the lake. This does not mitigate the conclusion that reducing the Zala River load is an urgent task, but the effect may finally turn out to be more beneficial than presently expected (the editors).

CHAPTER 13

A Comparison of Water Quality Models and Load Reduction Predictions

R.A. Luettich, Jr. and D.R.F. Harleman

13.1. Introduction

Recent lake water quality management efforts have attempted to balance the dual need to protect and use lakes as water resources, based on the premise that both protection and use are compatible. To be successful, managers must quantify the effects of climatic forcing, intrinsic lake characteristics, and human activity within the watershed, and select control measures to maintain predetermined water quality standards. In response to this need, computer modeling is often attempted with the hope that models calibrated to one set of data will reproduce observations contained in alternative data sets and thus can be confidently used in a predictive capacity. To date, such modeling efforts have had only marginal success.

It is not difficult to rationalize the limited success of modeling efforts. Data describing external nutrient loads to a lake are often quite poor and better in-lake measurement techniques are clearly a necessity. For example, the inability to measure accurately biologically available P, the difficulty of separating and measuring detritus and living phytoplankton, and the poor temporal and spatial resolution in measurements all contribute to data uncertainty and, therefore, ultimately to modeling uncertainty. Eutrophication models are inevitably a considerable simplification of reality and contain a number of somewhat arbitrary assumptions as to which processes to include or exclude and with which specific mathematical form to express them (cf. Fedra *et al.* 1981).

Historically, most of the uncertainty has been lumped into the model parameters, which are calibrated so that the model response in some manner matches measured data. An extreme case of this is when chemical and/or biological data [e.g., chlorophyll-a (Chl-a) and/or P fractions] are used to simultaneously calibrate the chemical, biological, and hydrodynamic parameters in a complex eutrophication model (see Chapter 10, and Wang and Harleman 1982). Unfortunately, the result is a model that all too often cannot be verified using an independent data set, and which leaves the modeler with little indication of where the principal errors exist, i.e., in the loading data, measurement data, process expressions, parameters, or some combination of each. Therefore, little can be learned about the most critical issue confronting any modeler, the correctness of the model structure (Beck 1983).

Recently this problem has become more widely recognized (cf. Beck and van Straten 1983) and various attempts have been made to deal with uncertainty in field data and parameter estimation (Beck 1979, Di Toro and van Straten 1979, van Straten 1981) and in model structure (Fedra *et al.* 1981, Somlyódy 1982, Fedra 1983). This work usually maintains a probabilistic approach to water quality modeling and as such requires Monte Carlo simulations or recursive parameter estimation, which unfortunately become more costly and less straightforward as model complexity increases.

The work described here was initiated to examine model structure by comparing the behavior of several models which differ in the representation of basic processes. This work was carried out using load, water temperature, and climatic forcing data for the Keszthely basin of Lake Balaton during the summer of 1977 (see also Baker 1982). Chl-a and total P measurements made in the basin during the same time period, and the known fact that phosphate levels are usually less than 1 mg P/m^3 in the lake (Chapter 8), were used to define a desired model behavior. Studies by Herodek and associates (cf. Chapter 8, and van Straten and Herodek 1982) have shown the existence of several algal blooms throughout the year, characterized by distinct algal groups with markedly different optimal and critical growth, mineralization, and mortality parameters, as well as other parameters that vary on an approximately seasonal time scale. To avoid the need to specify time-varying parameters, only the summer bloom, which typically contains the highest biomass levels, was specifically considered in this modeling effort. Presumably this work could be extended by considering several algal groups having different parameter sets (or even model structures), which would reflect the predominantly temperature-influenced behavior (cf. Leonov 1980, and Chapters 11 and 12). However, any such extension must be made cautiously so that it does not simply make the model more adept at "curve fitting" due to an increased number of degrees of freedom.

13.2. Previous Modeling of Lake Balaton

Three models have been proposed for use in Lake Balaton: BALSECT (Leonov 1980), BEM (Chapter 12), and SIMBAL (Chapter 11); for a brief comparison see Chapter 3. SIMBAL (Figure 11.6) forms the basis of the models used in the present study. SIMBAL has 24 calibrating parameters, 15 of which were assigned values based on the literature, and nine of which were initially limited to a specific numerical range and then calibrated further through

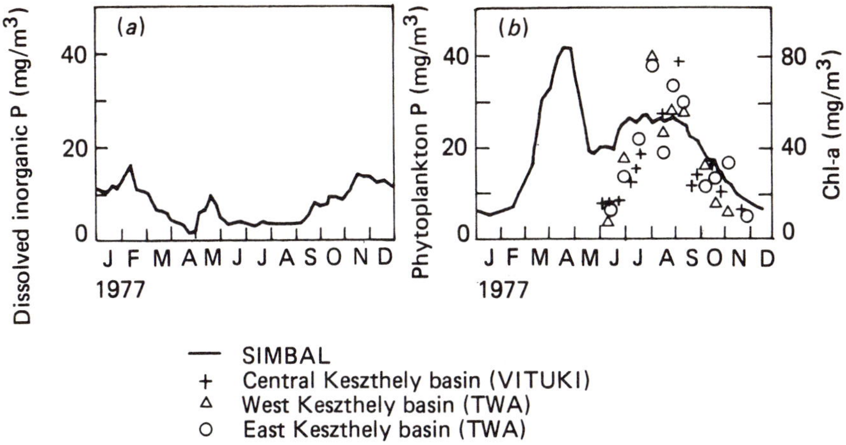

Figure 13.1. (*a*) Dissolved inorganic P and (*b*) phytoplankton P or Chl-a in the Keszthely basin. SIMBAL prediction (see Figure 11.3). VITUKI is the Research Center for Water Resources Development and TWA is the Transdanubian Water Authority.

Monte Carlo simulations. These simulations were run for the entire year of 1977 in each of the four lake basins, which were coupled via hydrologic throughflow and a calibrated basin-to-basin exchange flow. This effort led to seven parameter sets which gave model output in each basin for total P, total dissolved P, and dissolved inorganic P that fell into an acceptable range determined from data measurements made over the entire lake. Figure 13.1 shows van Straten's results for phytoplankton P and dissolved inorganic P in the Keszthely basin using the mean of the seven parameter sets described above. For the sake of compactness, this figure has been adapted taking a rather significant liberty, i.e., the superposition of several sets of Chl-a data onto the model results for phytoplankton P. The justification is that SIMBAL contains a constant cell quota growth formulation which implicitly assumes that phytoplankton P is directly proportional to the algal biomass. A basic crux of the modeling, as it has developed, is that Chl-a data can be used as a measure of algal biomass. Typically, a constant ratio is assumed, thus allowing the direct comparison in this case of Chl-a with phytoplankton P. Van Straten suggests an average ratio of Chl-a/phytoplankton P = 2, which was used in Figure 13.1 (cf. Di Toro and Matystik 1979, Chl-a/phytoplankton P = 2 in Saginaw Bay, Lake Huron; Wang and Harleman 1982, Chl-a/phytoplankton P = 2.5 in Canadian experimental lakes area).

Caution must be used in interpreting Figure 13.1 in the context of the current study. There is no reason to believe that the parameters that "best" fit an entire year of data in the whole lake would likewise yield the "best" representation of a summertime bloom if used in a similar model of the Keszthely basin only. This is particularly true since a one-box model of the Keszthely basin would allow no interbasin exchange flow – a quantity for which van Straten (1981) found that "the distribution of phosphorus and phytoplankton over the lake is rather sensitive [to]" (see also Shanahan and Harleman 1982). In spite of these limitations, Figure 13.1 suggests that SIMBAL did a rather poor job of representing the algal dynamics that occurred in Keszthely during summer 1977. Similarly, predicted dissolved inorganic P values (in the model assumed to be equivalent to phosphate) were in the range 2–10 mg P/m^3, which is representative of chemically determinable soluble reactive P, but probably much larger than actual phosphate values (Rigler 1966; and Chapter 8). Finally, van Straten found it necessary to include phosphate adsorption/desorption between the water and the sediments in SIMBAL to control unrealistically high phosphate values that were otherwise predicted in the autumn and winter months. However, it is interesting to investigate whether this conclusion is in part dependent on the high dissolved inorganic P values maintained by the model throughout the entire year and on the choice of a constant cell quota growth formulation.

13.3. Model Comparison

The shortcomings of SIMBAL, together with a running debate in the literature about the importance of modeling the effect of variable algal stoichiometry, which is known to occur in nature, and the remaining question of the importance of sediment interaction, particularly under external load reductions, prompted this investigation into the behavioral impact of structural changes in the uptake/growth and sediment interaction features of SIMBAL.

Model I

Figure 13.2 shows schematically and mathematically the basic model, hereafter called model I, used in this study. Comparison with Figure 11.6 illustrates that several changes were made to SIMBAL to obtain this model. Although it might have been tidier to use SIMBAL as model I, only the Keszthely basin was considered, so that there was no possibility of including a basin-to-basin exchange flow. Therefore, a part of SIMBAL was immediately lost. Several additional changes seemed appropriate to obtain a simpler yet realistic model in terms of the time period modeled (winter algae were dropped) and in the light of recent experimental evidence suggesting that coprecipitation of P with biogenic lime (Dobolyi and Herodek 1980) and temperature-dependent release of P from sediment (Istvánovics *et al.* 1983) are insignificant in Lake Balaton. In addition, phytoplankton were permitted to settle out at the same rate as detritus.

All other SIMBAL features were maintained. The nutrient limitation function is a single expression of the Michaelis–Menten form (equation 13.I.8, Table 13.1), indicating a constant cell quota (CCQ) model. The sediment–water interaction is contained in equation (13.I.17), which in SIMBAL represented the adsorption/desorption process. Actually, equation (13.I.17) has the form

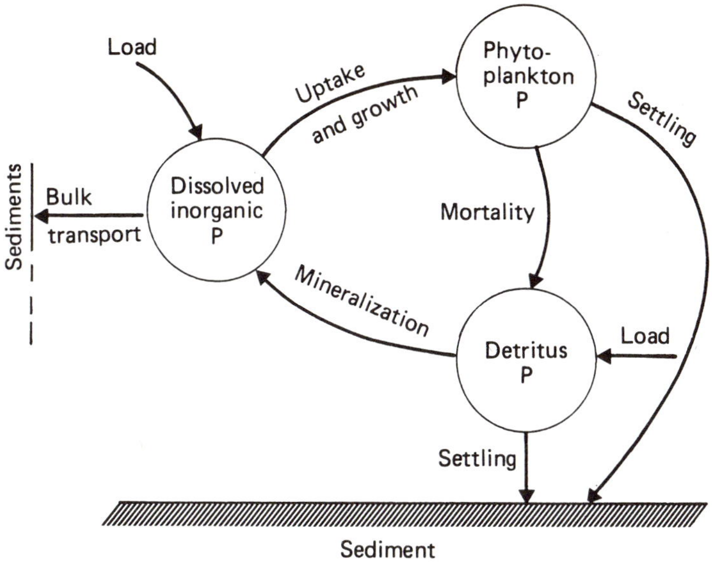

Figure 13.2. Model I structure.

of a bulk diffusion equation, in which P_{eq} represents an equilibrium concentration for diffusion. Actual phosphate adsorption/desorption with sediments is likely to be related to particle resuspension (Gelencsér *et al.* 1982), which has been shown to be correlated to wind speed, at least in the Szemes basin (Somlyódy 1982). SIMBAL does not explicitly include the effect of wind and therefore it is probably better to consider this as a two-way bulk transfer process controlled by the two calibrated parameters, k_{ex} and P_{eq}, and the predicted phosphate concentration in the water.

Initial conditions were determined from measured Chl-a and total P data. The starting date was chosen to correspond to the time of a distinct low point in the Chl-a level that occurred between obvious spring and summer Chl-a peaks. The initial phytoplankton P level was determined from the Chl-a data using the Chl-a/phytoplankton P ratio of 2; the initial phosphate level was arbitrarily set at 1 mg/m^3 (the model was reasonably insensitive to this value), while the initial detritus P was assumed to make up the remainder of the total P measured in the water at that time.

Model II

Model II is identical to model I except that no sediment–water exchange of phosphate is permitted.

Model III

Model III, shown in Figure 13.3, is identical to model I except that uptake and growth are modeled as separate processes (Table 13.2). Equation (13.III.6)

Table 13.1. Model I equations.

STATE EQUATIONS

$$\dot{A} = (I - O)_A + \text{RGRA} - \text{RDTHA} - \text{RSEDA} \quad (13.\text{I}.1)$$

$$\dot{D} = (I - O)_D + \text{RDTHA} - \text{RMNRL} - \text{RSETD} + L_D \quad (13.\text{I}.2)$$

$$\dot{P} = (I - O)_P - \text{RGRA} + \text{RMNRL} + \text{REXP} + L_P \quad (13.\text{I}.3)$$

A, phytoplankton P
D, detritus P (particulate and dissolved)
P, dissolved inorganic P

INPUT FUNCTIONS

Inflow–outflow

$$(I - O)_A = -AQ/V \quad (13.\text{I}.4)$$

$$(I - O)_D = -DQ/V \quad (13.\text{I}.5)$$

$$(I - O)_P = -PQ/V \quad (13.\text{I}.6)$$

Q, hydrologic outflow from Keszthely basin
V, volume of Keszthely basin
Inflows are incorporated in the load terms L_D and L_P

Temperature, T

Incident solar radiation, I

P loads

L_D, volumetric external detritus load
L_P, volumetric external orthophosphate load

RATES

Growth rate

$$\text{RGRA} = k_g f_P f_I f_T A \quad (13.\text{I}.7)$$

k_g, maximum specific growth and P uptake rate coefficient
f_P, nutrient limitation factor

$$f_P = \frac{P}{P_k + P} \quad (13.\text{I}.8)$$

P_k, half saturation constant

f_I, light limitation factor (depth averaged Steele equations)

$$f_I = \frac{e}{\varepsilon H}\left\{\exp\left[-\frac{I}{I_s}\exp(-\varepsilon H)\right] - \exp\left[-\frac{I}{I_s}\right]\right\} \quad (13.\text{I}.9)$$

H, water depth
ε, total extinction coefficient

$$\varepsilon = \varepsilon_0 + \alpha A \quad (13.\text{I}.10)$$

(cont.)

Table 13.1. (*cont.*)

ε, total extinction coefficient
ε_0, extinction coefficient without algae
α, phytoplankton P self-shading coefficient

I_s, optimal light intensity for algal growth

$$I_s = I_{sm} + I_{se} T \qquad (13.I.11)$$

I_{sm}, base optimal light intensity
I_{se}, temperature correction coefficient to I_s

f_T, temperature limitation factor

$$f_T = \begin{cases} \dfrac{T_c - T}{T_c - T_0} \exp\left[1 - \dfrac{T_c - T}{T_c - T_0}\right] & \text{if } T \le T_c \\ 0 & \text{if } T > T_c \end{cases} \qquad (13.I.12)$$

T_c, critical temperature
T_0, optimal temperature for algal growth

Mortality rate

$$\text{RDTHA} = k_d \Theta_d^{T-20} A \qquad (13.I.13)$$

k_d, phytoplankton mortality rate coefficient at 20°C
Θ_d, mortality temperature coefficient

Settling rates

$$\text{RSEDA} = v_s A / H \qquad (13.I.14)$$

$$\text{RSEDD} = v_s (1 - \gamma) D / H \qquad (13.I.15)$$

v_s, settling velocity
γ, fraction of detritus which is dissolved

Mineralization rate

$$\text{RMNRL} = k_m \Theta_m^{T-20} D \qquad (13.I.16)$$

k_m, mineralization rate coefficient at 20°C
Θ_m, mineralization temperature coefficient

Bulk sediment exchange

$$\text{REXP} = k_{\text{ex}} (P_{\text{eq}} - P) \qquad (13.I.17)$$

k_{ex}, transport coefficient for exchange between sediment and water
P_{eq}, sediment equilibrium P concentration

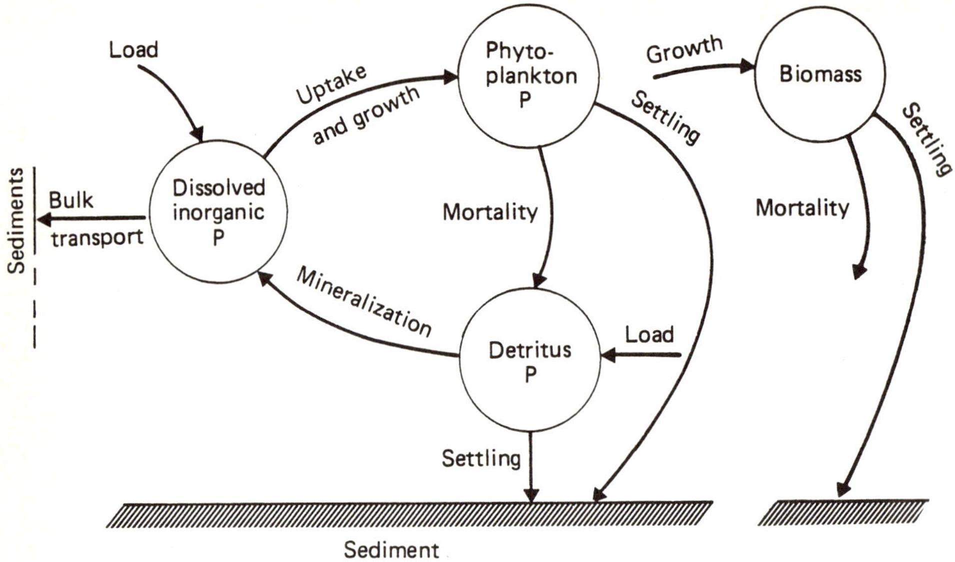

Figure 13.3. Model III structure.

expresses uptake with a Michaelis–Menten form which is modified by a "feedback" or "excretion" that linearly decreases the uptake as the internal nutrient store approaches an allowed maximum. The expression for nutrient limitation of growth, equation (13.III.8), is hyperbolic and of the form originally suggested by Droop (1968).

The separation of uptake and growth results in the inclusion of cell biomass as a state variable, and therefore requires the solution of a fourth differential equation. It also requires the specification of three additional parameters, the minimum and maximum cell quotas and a maximum phosphate uptake rate. Numerous investigators have used CCQ models, citing that variable cell quota (VCQ) models increased the numerical complexity (Kremer and Nixon 1978) and the number of degrees of freedom in the model. The argument of increased numerical complexity is extremely weak since it is no more difficult to program the numerical solution of four simultaneous differential equations than it is to solve three such equations. On the contrary, the results of this study suggest it may be easier to solve a VCQ model within the range of realistic phosphate concentrations (i.e., < 1 mg/m^3; see below). The latter argument is true since VCQ models require one additional state variable and three additional parameters (although in a model with 20+ parameters, it is not clear that an additional three will be significant). The inclusion of another state variable without additional data is open to criticism, however; at present the most-available measurement for model comparison is biomass (usually represented by Chl-a). To make this comparison using a CCQ model, phytoplankton P must be converted into biomass (Chl-a) and an additional state variable, albeit not a dynamic variable, is implicitly introduced.

Table 13.2. Model III equations.

STATE EQUATIONS

$$\dot{A} = (I - O)_A + \text{RPUPT} - \text{RDTHA} - \text{RSEDA} \quad (13.\text{III}.1)$$

$$\dot{A}_b = (I - O)_{Ab} + \text{RGR} - \text{RDTH} - \text{RSED} \quad (13.\text{III}.2)$$

$$\dot{D} = (I - O)_D + \text{RDTHA} - \text{RMNRL} - \text{RSETD} + L_D \quad (13.\text{III}.3)$$

$$\dot{P} = (I - O)_P - \text{RPUPT} + \text{RMNRL} + \text{REXP} + L_P \quad (13.\text{III}.4)$$

A, phytoplankton P
A_b, phytoplankton biomass
D, detritus P
P, dissolved inorganic P

INPUT FUNCTIONS

Inflow–outflow

$$(I - O)_{Ab} = -A_b Q / V \quad (13.\text{III}.5)$$

All others the same as for model I

RATES

P uptake rate

$$\text{RPUPT} = k_u \frac{P}{P_{k_u} + P} \frac{q_{max} - q}{q_{max} - q_{min}} A_b \quad (13.\text{III}.6)$$

q is the cell quota

$$q = A / A_b$$

k_u, maximum P uptake rate coefficient
P_{k_u}, half-saturation constant for uptake
q_{max}, maximum cell quota
q_{min}, minimum cell quota

Growth

$$\text{RGR} = k_g f_T f_I f_Q A_b \quad (13.\text{III}.7)$$

k_g, maximum specific growth rate coefficient
f_Q, cell quota growth reduction factor

$$f_Q = 1 - q_{min} / q \quad (13.\text{III}.8)$$

Mortality

$$\text{RDTH} = k_d \theta_d^{T-20} A_b \quad (13.\text{III}.9)$$

Biomass settling

$$\text{RSED} = v_s A_b / H \quad (13.\text{III}.10)$$

All other variables as defined for model I

Initial conditions for model III were determined in a manner similar to those for model I, with the ratio of biomass (mg dry weight/m^3)/Chl-a = 100, replacing the phytoplankton P/Chl-a ratio. This ratio was calibratable and the value used compares with those of Di Toro and Matystik (1979) (ratio = 100, Saginaw Bay, Lake Huron) and Herodek *et al.* (1982b) (ratio ~ 56) obtained from measurements in other basins in Lake Balaton. The measured Chl-a was converted into biomass and then into phytoplankton P assuming that the plankton were initially at their maximum cell quota.

Model IV

Model IV is identical to model III except that no sediment–water exchange of phosphate is permitted.

13.4. Results

The behavior of each model was examined using the 1977 data for the Keszthely basin. Initially, model III, which has the most degrees of freedom, was calibrated to the observed Chl-a data. Total P measurements were quite infrequent over the modeling period (less than one per month), so the total P levels predicted by the model were compared with the measured data qualitatively rather than concentrating on dynamic behavior. Parameters were initially either set to the values used in SIMBAL or chosen from the literature. Selected parameters were then adjusted within reasonable limits by hand until the "best" model calibration was found. The resultant parameter values are presented in Table 13.3 and compared with the original SIMBAL values in Figure 13.4(*a*). Initially there is rapid growth due to favorable light and temperature conditions and a high internal nutrient level. The detritus associated with the dying off of the spring bloom, together with the low living-biomass level, maintains the high (probably unrealistically high) phosphate concentration that prolongs this period of rapid growth. The model reasonably reproduces the bimodal Chl-a peaks and eventual population decline. Notice that except for periods at the very beginning and at the very end the predicted phosphate level is below 1 mg P/m^3, and even at its highest level remains below 8 mg P/m^3. It is worth pointing out that the maximum growth rate used in the model is 2.13 d^{-1}. This is quite reasonable in comparison with growth rates typically found in eutrophication models, although it is significantly lower than the value used in SIMBAL (6 d^{-1}) and suggested by field work in Lake Balaton (10–20 d^{-1}; van Straten and Herodek 1982). Numerous attempts were made to calibrate model III with higher maximum growth rates, but these invariably resulted in growth that occurred too quickly, no tendency for the summer peak to be bimodal, and an unrealistic fall algal bloom in November with Chl-a levels above 100 mg/m^3.

The structural importance of the sediment interaction on model behavior was examined by running model IV using the parameter values calibrated for model III [Figure 13.4(*b*)]. The results show enhanced growth during June and July and a large increase in the predicted maximum Chl-a level. This indicates

Table 13.3. Model parameters.

Parameter	Model							
	SIMBAL	I	Innl	II	III	IIIac	IV	IVac
k_g (d^{-1})	6	3	1.72	3	2.13	2.22	2.13	2.22
T_c (°C)	30	27.5	27.5	27.5	27.5	27.5	27.5	27.5
T_o (°C)	26	23.5	23.5	23.5	23.5	23.5	23.5	23.5
ε_0 (m^{-1})	3.2	3.2	3.2	3.2	3.2	3.2	3.2	3.2
α (m^{-1} mg P/m^3)	.015	.015	.015	.015	.015	.015	.015	.015
I_{sm} (cal/cm^2)	96	96	96	96	96	96	96	96
I_{se} (cal/cm^2°C)	9.6	9.6	9.6	9.6	9.6	9.6	9.6	9.6
P_k (mg P/m^3)	10.2	0.001	–	.001	–	–	–	–
k_d (d^{-1})	0.13	0.13	0.13	0.13	0.13	0.13	0.13	0.13
Θ_d (–)	1.14	1.2	1.2	1.2	1.2	1.2	1.2	1.2
v_s (m/d)	0.036	0.036	0.036	0.036	0.036	0.036	0.036	0.036
γ (–)	0.4	0.4	0.4	0.4	0.4	0.4	0.4	0.4
k_m (d^{-1})	0.035	0.05	0.05	0.05	0.05	0.05	0.05	0.05
Θ_m (–)	1.18	1.18	1.18	1.18	1.18	1.18	1.18	1.18
k_{ex} (d^{-1})	0.16	0.25	0.25	0	0.25	0.16	0	0
P_{eq} (mg P/m^3)	5.8	2	2	2	2	2	2	2
k_u (μg P/μg biomass/d)	–	–	–	–	0.01	0.01	0.01	0.01
P_{k_u} (mg P/m^3)	–	–	–	–	2.0	6.0	2.0	6.0
q_{max} (μg P/μg biomass)	–	–	–	–	0.02	0.02	0.02	0.02
q_{min} (μg P/μg biomass)	–	–	–	–	0.001	0.001	0.001	0.001
Chl-a/phytoplankton P	2	2	2	2	–	–	–	–
Biomass/Chl-a	–	–	–	–	100	100	100	100

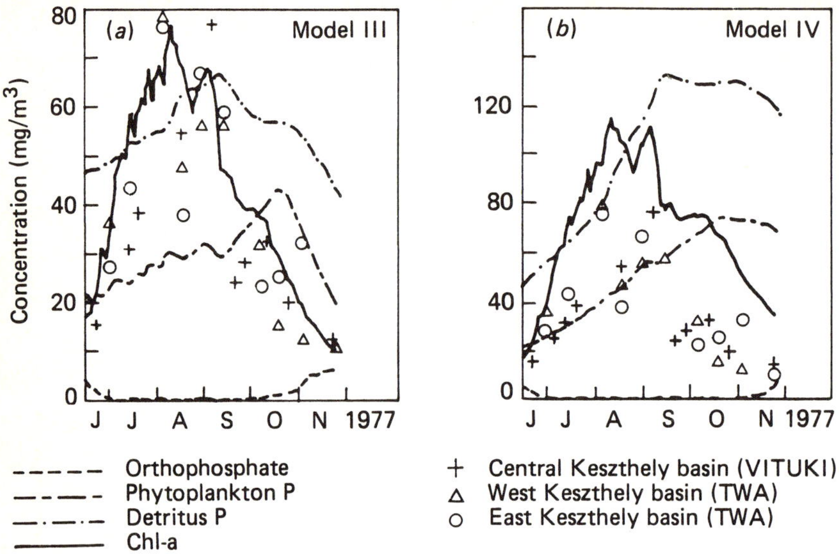

Figure 13.4. (*a*) Model III calibration; (*b*) model IV run using model III parameters (note the difference in scales).

that the sediments acted as a significant phosphate sink in model III during the critical early summer growth period. However, later in the summer the sediments seemed to have minimal importance since the absence of any sediment release of phosphate in model IV during this period did not seriously affect growth. Overall, this supports van Straten's hypothesis about the role of sediments in the phosphate cycle.

Unfortunately, further runs showed these results to be inconclusive. An alternative calibration of model III (called model IIIac) is shown in Figure 13.5(*a*) and Table 13.3. Comparison with Figure 13.4(*a*) clearly shows that either could be regarded as a "best" calibration. However, when model IV is run with the same parameters (called model IVac) it indicates a small decrease in Chl-a over the summer. A comparison of the parameter values of k_{ex} and P_{k_u} in Table 13.3 suggests the reason for the behavior. To calibrate model IIIac, k_{ex} was reduced by about 40% compared with model III, so that the overall role played by sediment interaction decreased. Also P_{k_u}, the uptake half-saturation constant, was increased, thereby decreasing the rate at which phosphate could be taken up and eventually used for growth. Together, these effects made growth less responsive to the additional phosphate available in the early summer in model IVac. The slight decrease in growth in model IVac in August shows that later in the summer the sediments are indeed acting as a slight source of phosphate in model IIIac.

The direct conflict between these two sets of results suggests that the sediment exchange part of model III is of superficial content and cannot be accurately calibrated with the given data set.

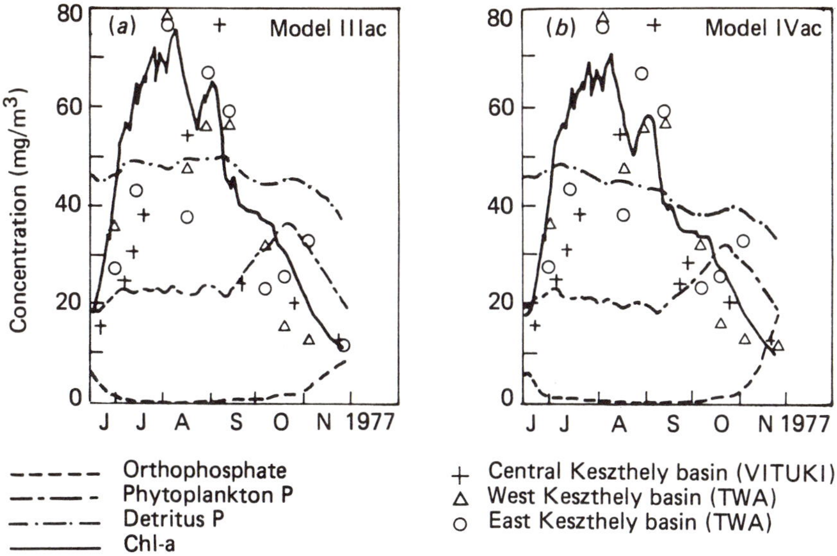

Figure 13.5. (*a*) Model III alternative calibration (model IIIac); (*b*) model IV run using model IIIac parameters (model IVac).

The importance of the uptake/growth relationship on model behavior was examined by trying to calibrate model I to 1977 data. To isolate the uptake/growth effects all recurring parameter values were kept equal to those in model III. This work produced several interesting results. A run was first made in which no nutrient limitation occurred, i.e., $f_P = 1$ (equation 13.I.8; Table 13.1). Results are shown in Figure 13.6(*a*) (model Innl). The Chl-a predictions show a remarkable ability to reproduce the observed data! (Note that in this case the predicted phosphate values have no meaning and therefore are not shown.) The success of the Chl-a predictions in this model suggest that nutrients may not limit growth during much of the summer. In order to represent this phenomenon using the classic Michaelis–Menten hyperbolic growth expression [equation (13.I.8)], phosphate levels must always remain much larger than the half-saturation constant so that the growth rate can remain close to its maximum. In this situation of minimal nutrient limitation, P $\gg P_k$ and equation (13.I.7) can be written

$$\text{RGRA} = k_g f_I f_T \frac{P}{P_k + P} \approx k_g f_I f_T \quad .$$

During periods that f_T and f_I are near one, the total phosphate uptake, $U_{\Delta t}$ in a given time step can be approximated by

$$U_{\Delta t} \sim \Delta t \cdot A \cdot k_g$$

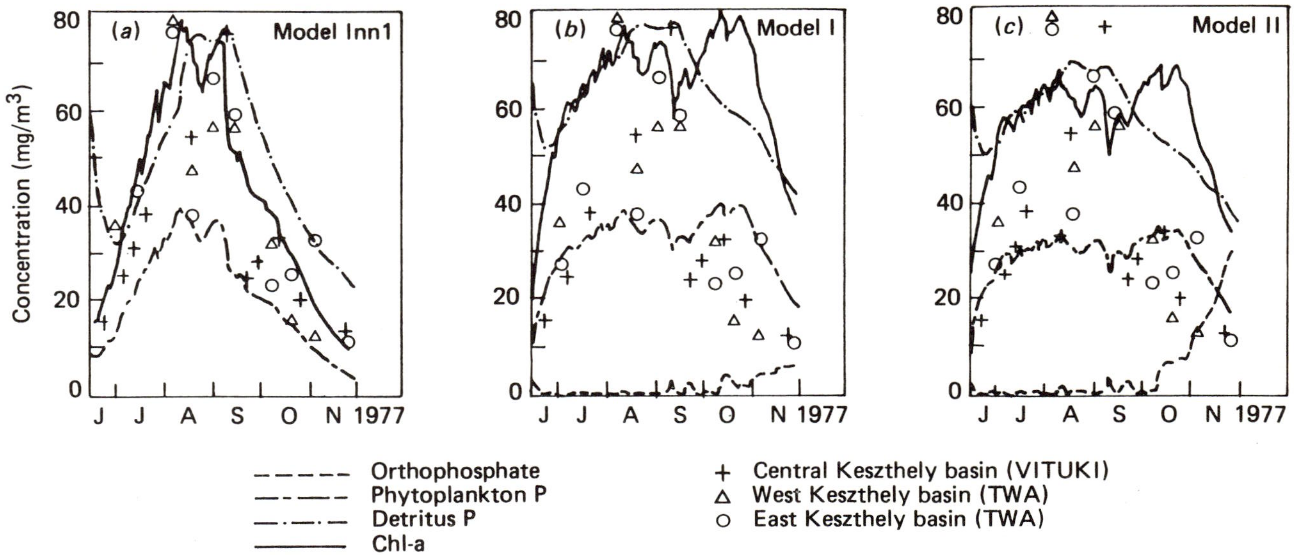

Figure 13.6. (*a*) Model I with no nutrient limitation calibration (model Inn1); (*b*) model I calibration; (*c*) model II run using model I parameters.

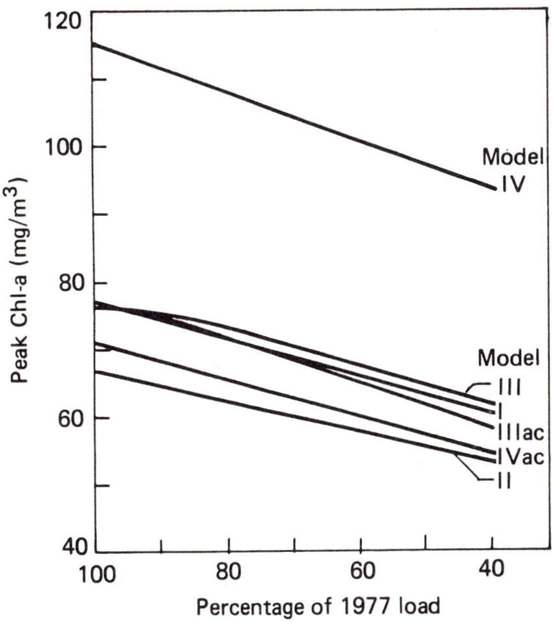

Figure 13.7. Predicted peak Chl-a concentrations for reduced loads.

where Δt is the numerical time step of the model. Using typical numbers for A and k_g, and a Δt of 3 h,

$$U_{\Delta t} \sim (3/24) \cdot 30 \cdot 2 = 7.5 \text{ mg P/m}^3$$

Often this is much more than the residual phosphate concentration in the water together with load and recycle. Therefore growth is nutrient limited, even though this was to be avoided. Various time steps were tried down to the ridiculous interval of 1 min, but nutrient limitation continued to be predicted. The consequence of this undesired nutrient limitation on the model behavior is clearly shown in Figure 13.6(*b*). In order to reach the peak Chl-a values a high maximum growth rate must be used. Therefore, at the beginning of the modeling period when the dying off of the spring bloom released a significant amount of phosphate into the water, growth was much more rapid than the data would indicate. After about two weeks, nutrient limitation occurred frequently, thereby reducing the growth. A comparison with Figure 13.6(*a*) shows that this resulted in predicted algal levels that increased too quickly and decreased too slowly. The higher maximum growth rate needed, 3 d^{-1}, is simply an artifact to overcome the periods of significant nutrient limitation.

A comparison of Figures 13.4(*a*) and 13.5(*a*) with Figure 13.6(*a*) shows that the VCQ formulation predicts some nutrient limitation after two to three weeks, although this is much less than the CCQ model because the growth is separated from the phosphate concentration in the water.

Figure 13.6(*c*) shows the results of running model II with model I parameters. The sediments obviously acted as a source of phosphate early in the modeling period, and as a sink late in the year. Van Straten (1981) found that an early version of SIMBAL which excluded the bulk transfer process with the sediments gave ridiculous phosphate levels late in the year, as shown in Figure 13.6(*c*). He used this as a justification for including this process in his final SIMBAL model. However, Figures 13.4(*b*) and 13.5(*b*) show that the formulation of the uptake and growth processes using VCQ expressions may soften or alter this conclusion.

The most common use of eutrophication models at present is to predict changes in future water quality under nutrient load reductions. Therefore, it is quite important to know how different model structures will effect the predicted response to load reductions. Figure 13.7 shows the response of the peak Chl-a level of each model to a reduction in the 1977 Zala River and tributary loads. Most striking is the similarity of the response of each model regardless of uptake/growth formulation, even though the VCQ models were calibrated to match the dynamic behavior of the data much more closely than the CCQ models. In fact, a greater difference occurs in the response of models III and IIIac than between either of the models and model I, suggesting that both CCQ and VCQ models may predict similar results for peak algal concentrations under load reductions as long as they are calibrated to the same peak values. The inclusion or exclusion of sediments seemed to have little effect, although this is mostly due to the initial conditions, which reflect only the history of high load levels, and to the short time period for which the model was run.

13.5. Conclusions and Suggestions

Several models have been developed specifically for application to Lake Balaton, but none has considered the importance of representing the uptake/growth expressions using a VCQ formulation. The work presented above suggests that a VCQ model may give more reasonable results when phosphate concentrations are kept below 1 mg P/m^3, as is typically observed in lakes. It is important to maintain a reasonable representation of the phosphate concentration in the water if there is to be any hope of properly modeling a sediment interaction that is dependent on this phosphate level. The use of a CCQ model under such circumstances requires an elevated maximum growth rate to produce significant growth at times of low phosphate concentrations, although this may cause too much growth at higher concentrations. The inclusion of VCQ expressions into a model formulation may mitigate the importance of the sediments and, in fact, make a sediment subroutine difficult or impossible to calibrate using data in which external loads are high.

It has been commented that a CCQ model may not completely capture the dynamics of algal blooms throughout a one- or two-year simulation, but that they accurately represent the trends in the data. However, as the number of algal compartments explicitly modeled increases (e.g., BEM, Herodek *et al.*

1982a), the time period over which each algal group is dominant decreases. For example, in a model containing three algal types – hypothetically cold-, mixed-, and warm-water algae – each type would be dominant for approximately 4–5 months of the year. As the dominant time scales are decreased the short-term dynamics become more important. Thus the work presented above suggests that it may be inconsistent to increase the number of algal groups without also increasing the detail with which each group is modeled, specifically without switching from a CCQ to a VCQ formulation.

Load reduction runs for each of the models showed very similar responses for peak Chl-a levels, regardless of the uptake/growth or sediment interaction formulation as long as the peak value was accurately reproduced by the calibration. This result is somewhat tenuous, however, due to the short time period used for the model response. It is not clear that a longer response period using the present models would have been more meaningful since the sediment interaction could not be calibrated confidently and the sediments contained no memory to reflect past loads.

The work presented is obviously a deterministic treatment of what is becoming recognized as a field in which probabilistic theory may be particularly applicable. Further information may be obtained about the importance of model structure from model comparisons employing parameter groups obtained from Monte Carlo simulations. The results contained herein are admittedly limited due to the use of data from only a part of one year. Additional comparisons using other years' data may add to the generality of these results and may also allow calibration of the sediment interaction terms.

References

Baker, R.M. (1982) *Comparison of Water Quality Models of Keszthely Bay, Lake Balaton, Hungary.* MSc Thesis (Cambridge, MA: MIT).

Beck, M.B. (1979) Model structure identification from experimental data, in E. Halfon (Ed.) *Theoretical Systems Ecology* (New York: Academic Press) pp 259–89.

Beck, M.B. (1983) Uncertainty, system identification, and the prediction of water quality, in M.B. Beck and G. van Straten (Eds.) *Uncertainty and Forecasting of Water Quality* (Berlin: Springer-Verlag) pp 3–48.

Beck, M.B. and van Straten, G. (1983) *Uncertainty and Forecasting of Water Quality* (Berlin: Springer-Verlag).

Di Toro, D.M. and Matystik, W.F., Jr. (1979) *Mathematical Models of Water Quality in Large Lakes. Part 1: Lake Huron and Saginaw Bay.* US Environmental Protection Agency Ecological Research Series (Washington, DC: EPA).

Di Toro, D.M. and van Straten, G. (1979) *Uncertainty in the Parameters and Predictions of Phytoplankton Models.* Working Paper WP-79-27 (Laxenburg, Austria: International Institute for Applied Systems Analysis).

Dobolyi, E. and Herodek, S. (1980) On the mechanism reducing the phosphate concentration in the water of Lake Balaton. *Int. Rev. ges. Hydrobiol.* 65(3):339–43.

Droop, M.R. (1968) Vitamin B12 and marine ecology IV. The kinetics of uptake, growth and inhibition in *Monochrysis lutheri. J. Marine Biol. Assn. U.K.* 48:689–733.

Fedra, K. (1983) A Monte Carlo approach to estimation and prediction, in M.B. Beck and G. van Straten (Eds.) *Uncertainty and Forecasting of Water Quality* (Berlin: Springer-Verlag) pp 259–92.

Fedra, K., van Straten, G., and Beck, M.B. (1981) Uncertainty and arbitrariness in ecosystems modelling: A lake modelling example. *Ecological Modelling* 13:87–110.

Gelencsér, P., Szilágyi, F., Somlyódy, L., and Lijklema, L. (1982) *A study on the influence of sediment in the phosphorus cycle in Lake Balaton.* Collaborative Paper CP-82-44 (Laxenburg, Austria: International Institute for Applied Systems Analysis).

Herodek, S., Kutas, T., and Csáki, P. (1982a) Simulation of phytoplankton dynamics in Lake Balaton. *ISEM Journal* 4(3, 4):97–126.

Herodek, S., Vörös, L., and Tóth, F. (1982b) The mass production of phytoplankton and the eutrophication in Lake Balaton III. The Balatonszemes basin in 1976–77 and the Siófok basin in 1977. *Hidrol. Kozl.* 62:220–9 (in Hungarian with English summary).

Istvánovics, V., Herodek, S., and Entz, B. (1983) The effect of sediment dredging on the water quality in Lake Balaton II: The phosphorus and nitrogen cycles. *Hidrol. Kozl.* 63(3):125–9 (in Hungarian with an English summary).

Kremer, J.N. and Nixon, S.W. (1978) *A Coastal Marine Ecosystem Simulation and Analysis*, Ecological Studies Series, Vol. 24 (Berlin: Springer-Verlag).

Leonov, A.V. (1980) *Mathematical Modeling of Phosphorus Transformation in the Lake Balaton Ecosystem.* Working Paper WP-80-149 (Laxenburg, Austria: International Institute for Applied Systems Analysis).

Rigler, F.H. (1966) Radiobiological analysis of inorganic phosphorus in lakewater. *Verh. Int. Verein Limnol.* 16:465–70.

Shanahan, P. and Harleman, D.R.F. (1982) *Linked Hydrodynamic and Biogeochemical Models of Water Quality in Shallow Lakes.* Technnical Report No. 268 (Cambridge, MA: Ralph M. Parsons Laboratory for Water Resources and Hydrodynamics, Dept of Civil Engineering, MIT).

Somlyódy, L. (1982) Water-quality modelling: A comparison of transport-oriented and ecology-oriented approaches. *Ecological Modelling* 17:183–207.

van Straten, G. (1981) Analysis of model and parameter uncertainty in simple phytoplankton models for Lake Balaton, in D.M. Dubois (Ed.) *Progress in Ecological Engineering and Management by Mathematical Modelling* (Liege, Belgium: CEBEDOC) pp 107–34.

van Straten, G. and Herodek, S. (1982) Estimation of algal growth parameters from vertical primary production profiles. *Ecological Modelling* 15:287–311.

Wang, M. and Harleman, D.R.F. (1982) *Hydrothermal–Biological Coupling of Lake Eutrophication Models.* Technical Report No. 270 (Cambridge, MA: Ralph M. Parsons Laboratory for Water Resources and Hydrodynamics, Dept of Civil Engineering, MIT).

CHAPTER 14

Role of the Zala River in the Eutrophication of Lake Balaton

O. Joó

14.1. Introduction

It is generally believed that the Zala River is the main source of artificial eutrophication of Lake Balaton. In the middle of the nineteenth century the navigation and regulation lock at the town of Siófok (Figure 14.1) was constructed with a discharge capacity of 15 m^3/s, which lowered the water level of the lake by 3–4 m. These conditions enabled the safe operation of railways and highways constructed along the lake's shoreline. The draining of wetlands along the southern shore also facilitated and enhanced animal husbandry. Recreational use of the lake was initiated and by the middle of the twentieth century the lakeside had become the most favored summer resort in Hungary.

Regulation of the Zala River was also initiated. Previously, inundations along the most downstream 20 km reach of the river had produced a thick peat layer. With regulation of the river the repeated and long-lasting inundation of this region has been eliminated. Since the 1920s the Zala River has been discharging directly into the shallowest and most western part of the lake (Keszthely bay) – at Fenékpuszta (Figure 14.1). Increasing eutrophication of the lake has been observed during recent decades (Chapter 1), with a marked propagation from west to east. This fact is explained primarily by the influence of the Zala River (the largest tributary, which drains about half of the total watershed of the lake), although the issue is a complex one since other direct and indirect sources, such as untreated effluents, erosion, large-scale farming operations, and the gradually increasing tourism, are also contributing considerably to the eutrophication problem of Lake Balaton (Chapter 1).

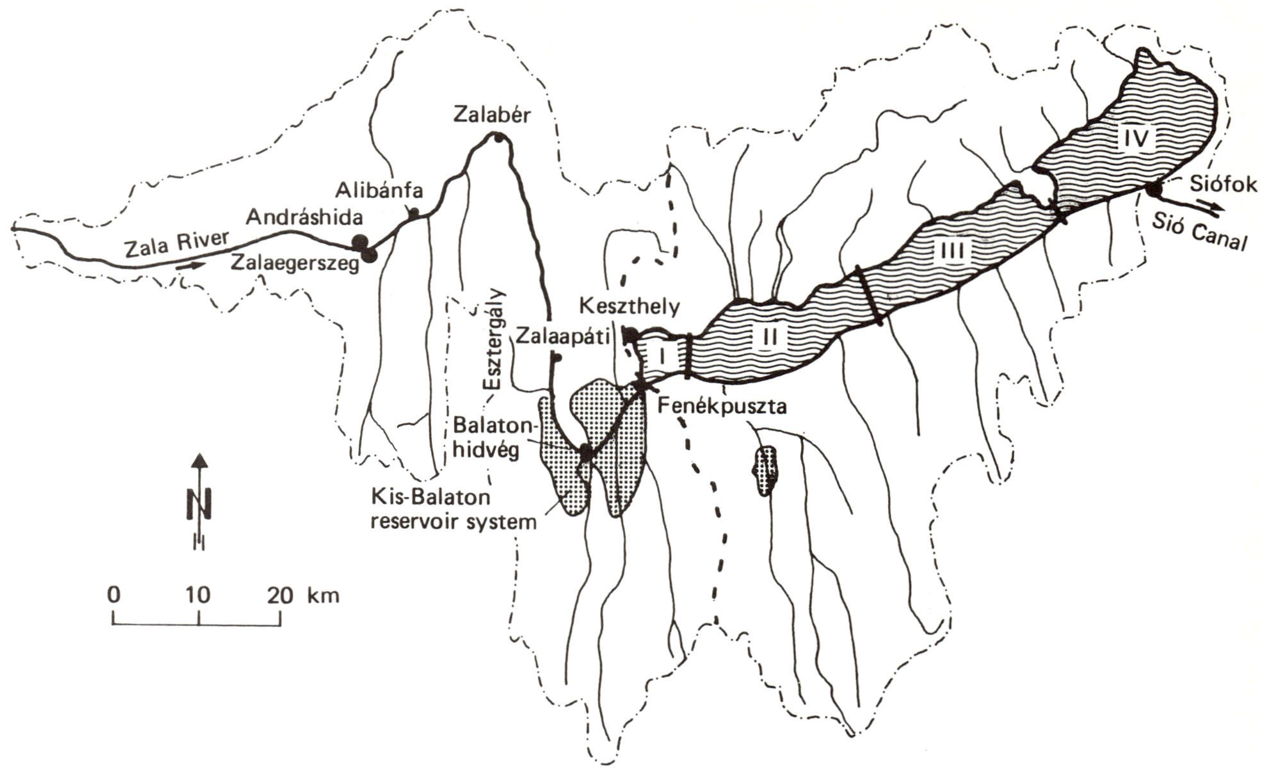

Figure 14.1. Lake Balaton and its watershed (with the Kis-Balaton reservoir system).

In this chapter water quality measurements of the Zala River are described. The objectives of the observations were:

(1) To quantify the pollutant loads carried by the river.
(2) To contribute to the elaboration of strategies for controlling the eutrophication of the lake.

As mentioned in Chapter 1 the Zala watershed area is 2622 km^2 (with an average elevation of 300–400 m above sea level). The river discharges into the smallest basin of the lake, 4.3% of the total volume.

The largest city on the Zala watershed is Zalaegerszeg (Figure 14.1) with a population exceeding 60 000. Some characteristics of the watershed, with relevance to water quality, are listed in Table 14.1 (which also shows variations during the past 50 years). The table shows that the agricultural area has decreased, large-scale farming with the intensive use of chemicals has replaced traditional cultivation methods, and livestock breeding has concentrated. Urbanization has been intensive and so public water supply and sewerage have gradually extended.

Table 14.1. Major factors related to water quality changes in the Zala watershed.

Factors	1935	1960	1975	1985 (estimated)
Agricultural land (km^2)	1917.0	1802.0	1670.0	1500.0
Large-scale farming land (%)	37.0	78.0	92.0	95.0
Number of animals in livestock breeding (%)	18.0	25.0	70.0	100.0
Fertilizer (effective material) (10^3 tons)	0.3	0.5	25.0	30.0
Pesticides (10^3 tons)	–	0.6	0.9	1.1
Urban population (%)	8.0	15.0	25.0	30.0
Piped water supply (10^{6}m^3)	0.1	1.5	4.5	15.0
Sewerage (10^{6}m^3)	–	0.5	4.0	12.0

14.2. Suspended Sediment and Nutrient Loads

To explore the role of the Zala River in the eutrophication of Lake Balaton, daily observations were initiated in 1975 at the mouth section, Fenékpuszta. Measured quantities were the rate of stream flow and the concentrations of suspended solids, total N, and total P obtained. The measurement program was extended in 1977 by monitoring the same parameters at Zalaapáti, 22 km upstream from the river mouth (Figure 14.1). This cross section accounts for 1523 km^2 of the catchment area. A 30-year hydrologic record is also available, thus enabling long-term interpretation of the water quality data.

Weekly measurements of some 20 other water quality components (including dissolved reactive P and nitrate N) at the two stations and similar biweekly observations at two sections further upstream (Zalaegerszeg and Alibánfa, Figure 14.1) completed the monitoring program.

The effluent from large treatment plants was sampled 3–4 times annually in earlier years and a program of monthly sampling has been recently launched. Occasional longitudinal water quality profile studies were carried out for low and mean water levels (3 times in 1982–83). A program to enable the quantification of agricultural nonpoint sources was launched in 1982, for which a pilot watershed, 20 km^2, on the Esztergály Creek has been assigned (Figure 14.1).

The measurements were designed to give reliable quantitative answers to the following questions:

(1) What is the total water discharge into the lake?
(2) How much of the annual suspended sediment load reaches Lake Balaton (the bedload is negligible)?
(3) What is the total P input to the lake and what fraction is biologically available?
(4) What is the distribution of load components between low and high flows?
(5) How much does the point source sewage contribute to the loads at the river mouth?
(6) What portion of the sewage load originating from the city of Zalaegerszeg enters Lake Balaton?
(7) How much do nonpoint source nutrients contribute?
(8) What are the sources of this diffuse contribution?
(9) What fraction is the contribution of the Zala River load to that of the entire lake?
(10) Based on answers to the above questions, what should be done to control eutrophication?
(11) What is the input load of the Kis-Balaton reservoir system, designed to control eutrophication of the lake (see later)?

Details of measurements carried out at the mouth section of the Zala River between 1975–82 are given in Figure 14.2(*a*)–(*h*). The figures include discharge and the concentrations of suspended solids, total P, and total N. Apparent from the observations is that, above the background values, the suspended solids and nutrient concentrations increase with increasing discharge, and during floods extremely high values (e.g., 500 g/m^3 for suspended solids, 10 g/m^3 for total N, and 2–3 g/m^3 for total P) can occur. July 1975, December 1976–February 1977, and the end of 1982 were especially rich in such events.

In Figure 14.3 a more detailed load record is given for the section of Fenékpuszta (see also Figure 6.3). Comparison of Figures 14.2 and 14.3 indicates that qualitatively there is a close relationship between flow and P load. With respect to precipitation, 1980 was an average year, while 1981 was drier, and 1979 wetter than the average. The frequency of the P peaks also reflects this sequence.

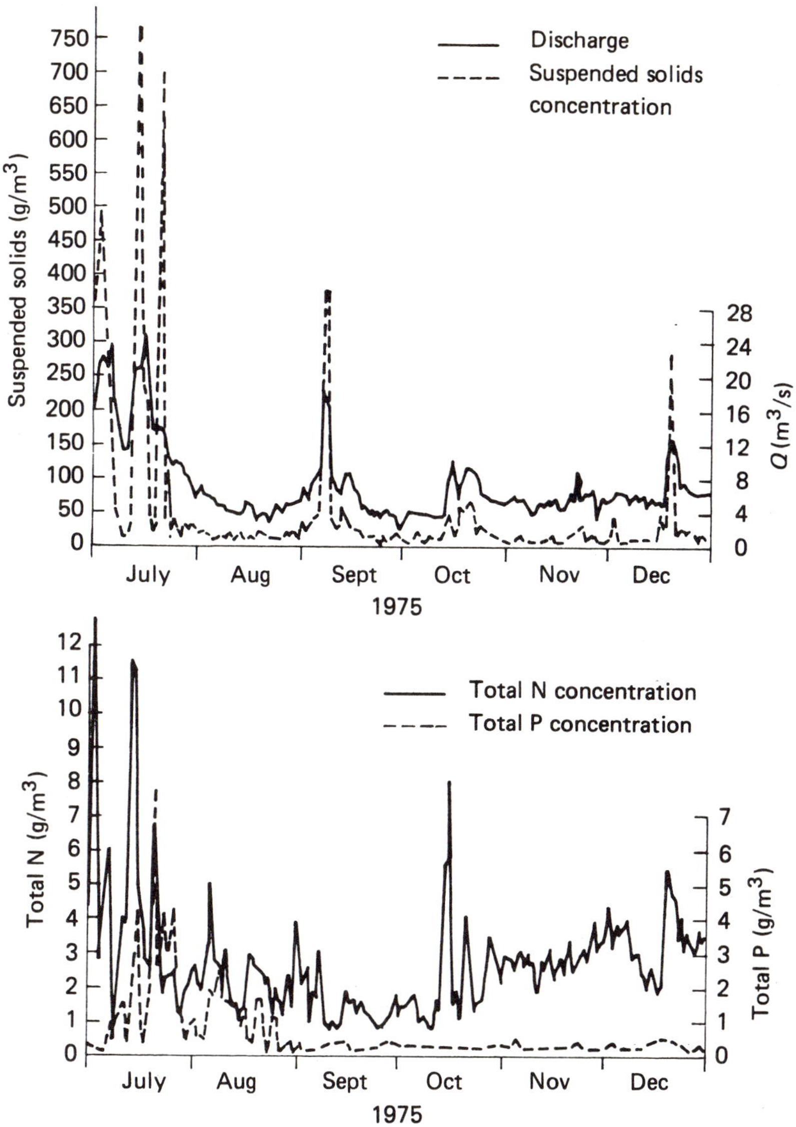

Figure 14.2. (*a*) Discharge and concentrations of suspended solids, total P, and total N for the Zala River, Fenékpuszta, 1975.

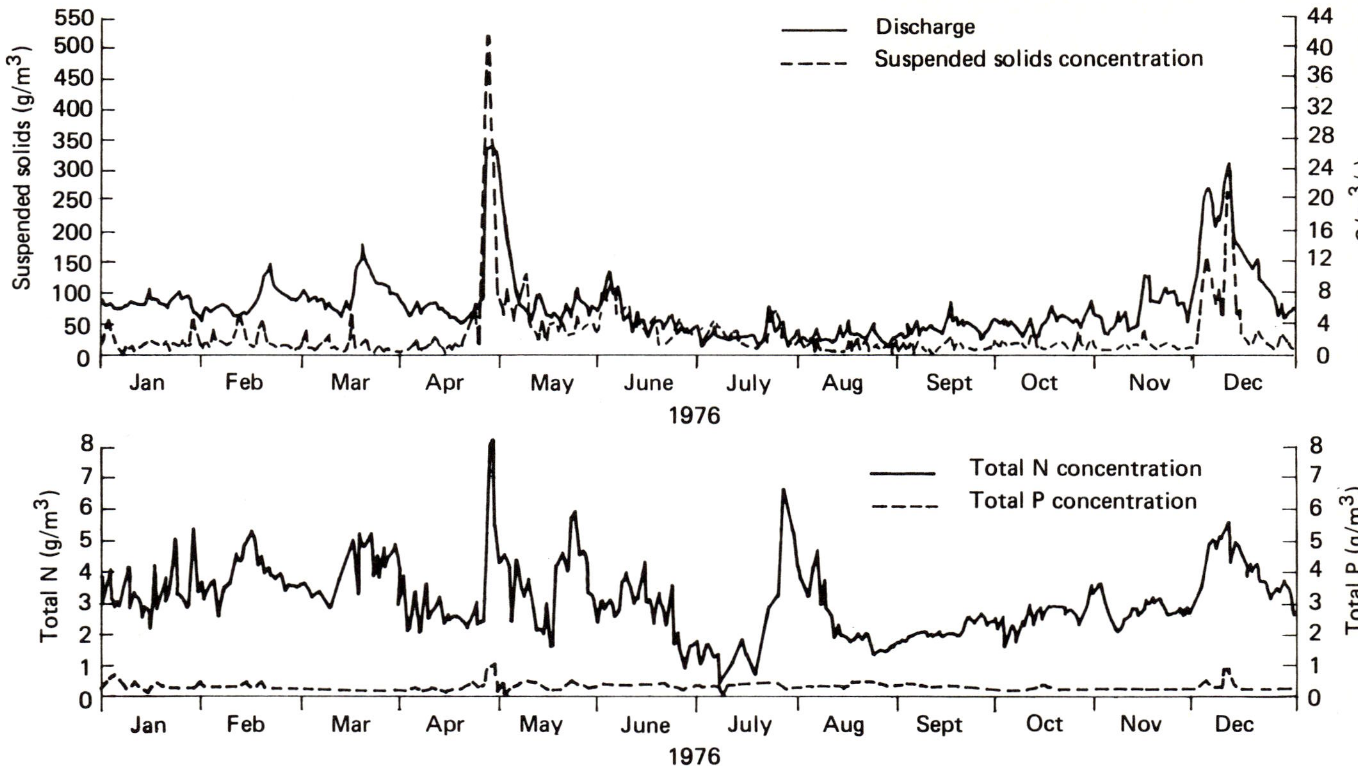

Figure 14.2. (*b*) Discharge and concentrations of suspended solids, total P, and total N for the Zala River, Fenékpuszta, 1976.

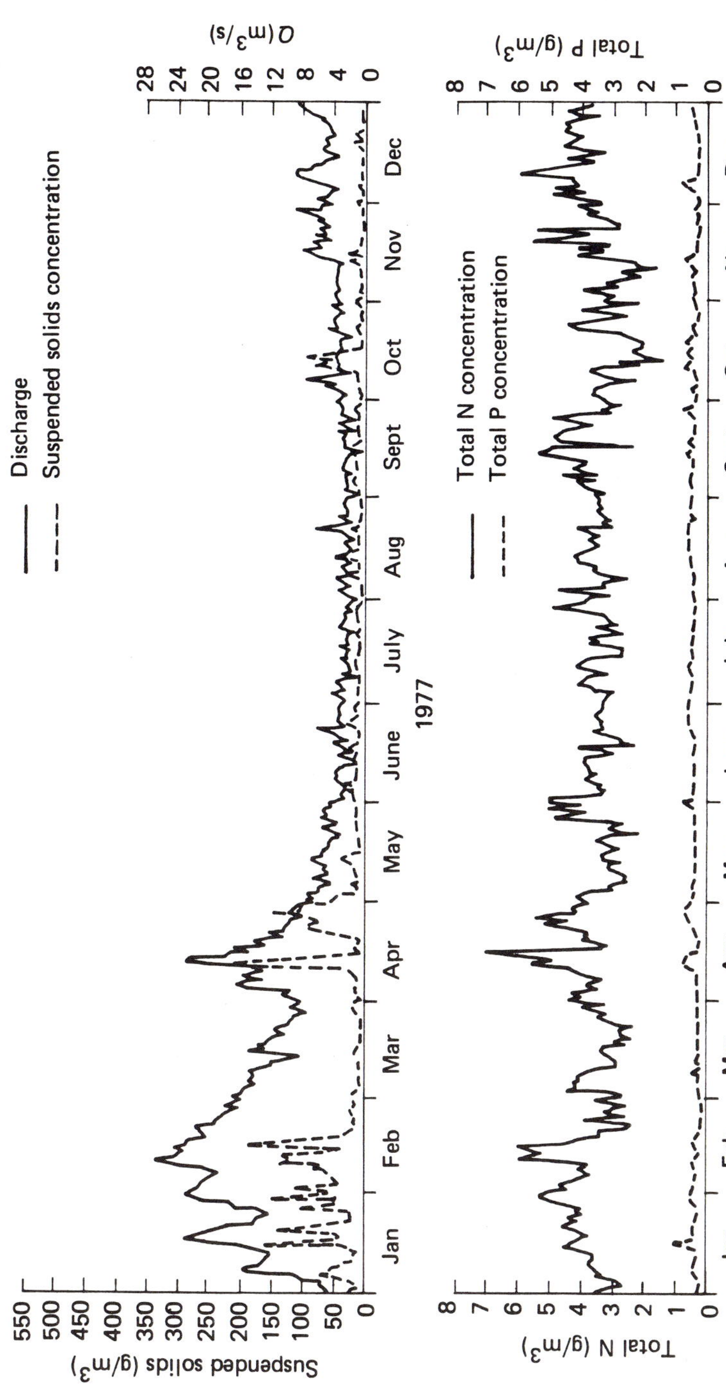

Figure 14.2. (*c*) Discharge and concentrations of suspended solids, total P, and total N for the Zala River, Fenékpuszta, 1977.

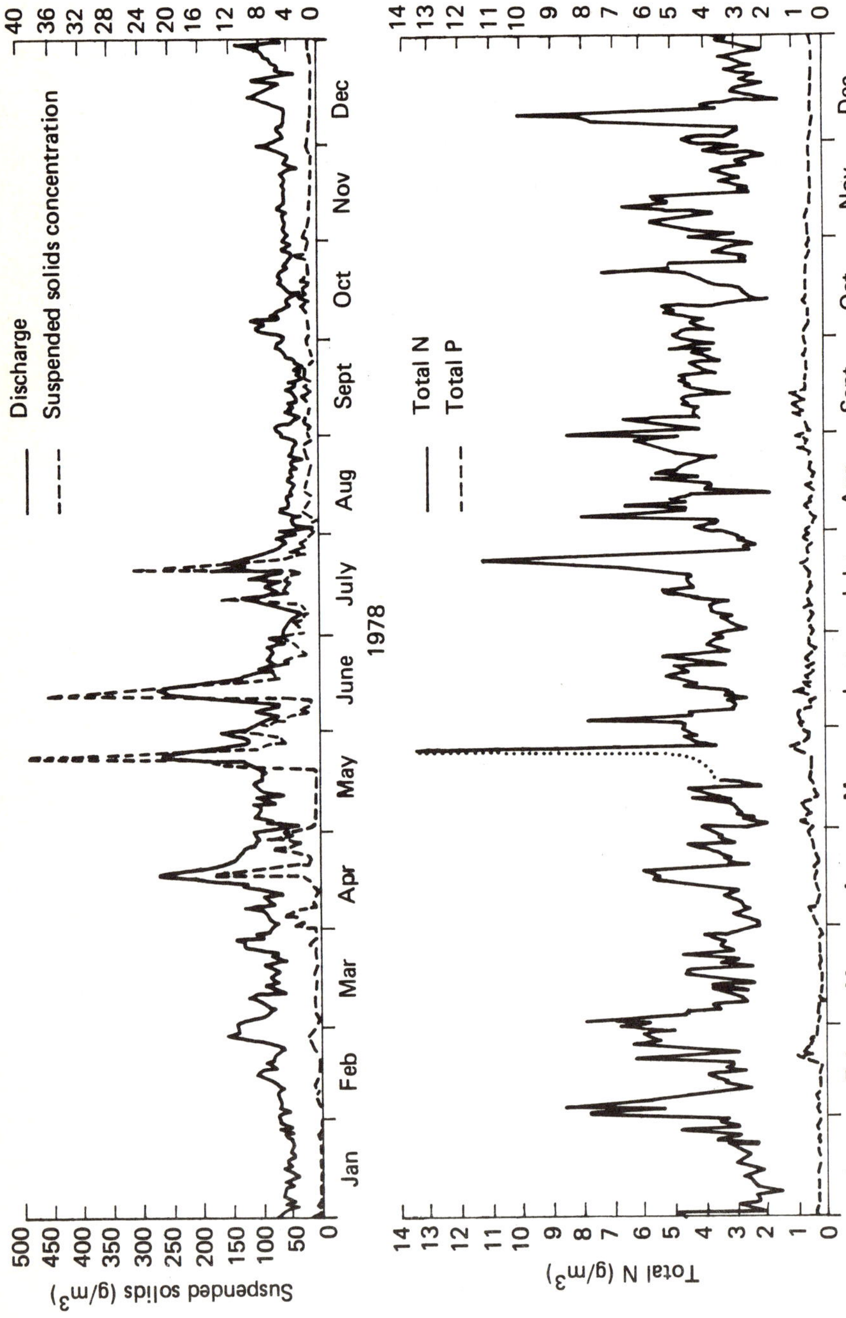

Figure 14.2. (*d*) Discharge and concentrations of suspended solids, total P, and total N for the Zala River, Fenékpuszta, 1978.

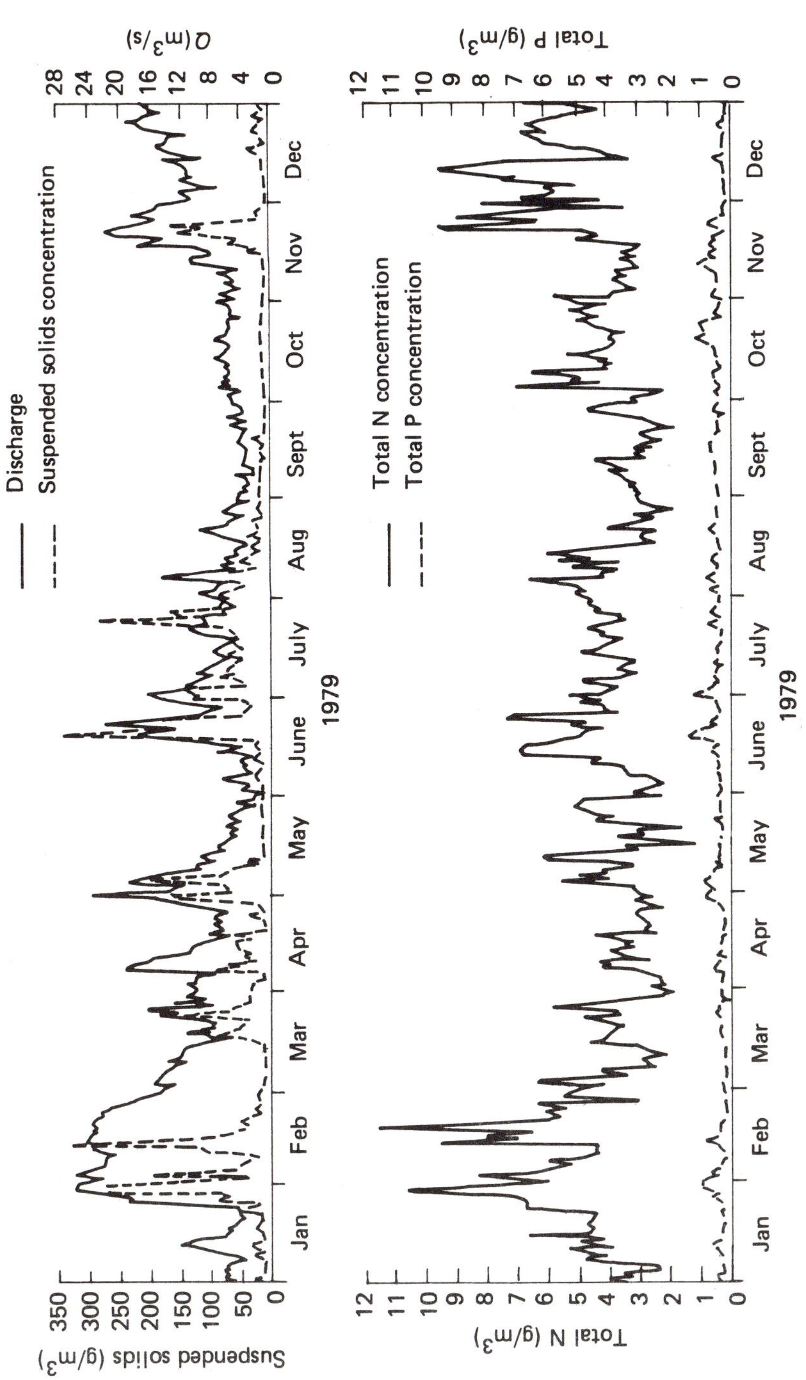

Figure 14.2. (*e*) Discharge and concentrations of suspended solids, total P, and total N for the Zala River, Fenékpuszta, 1979.

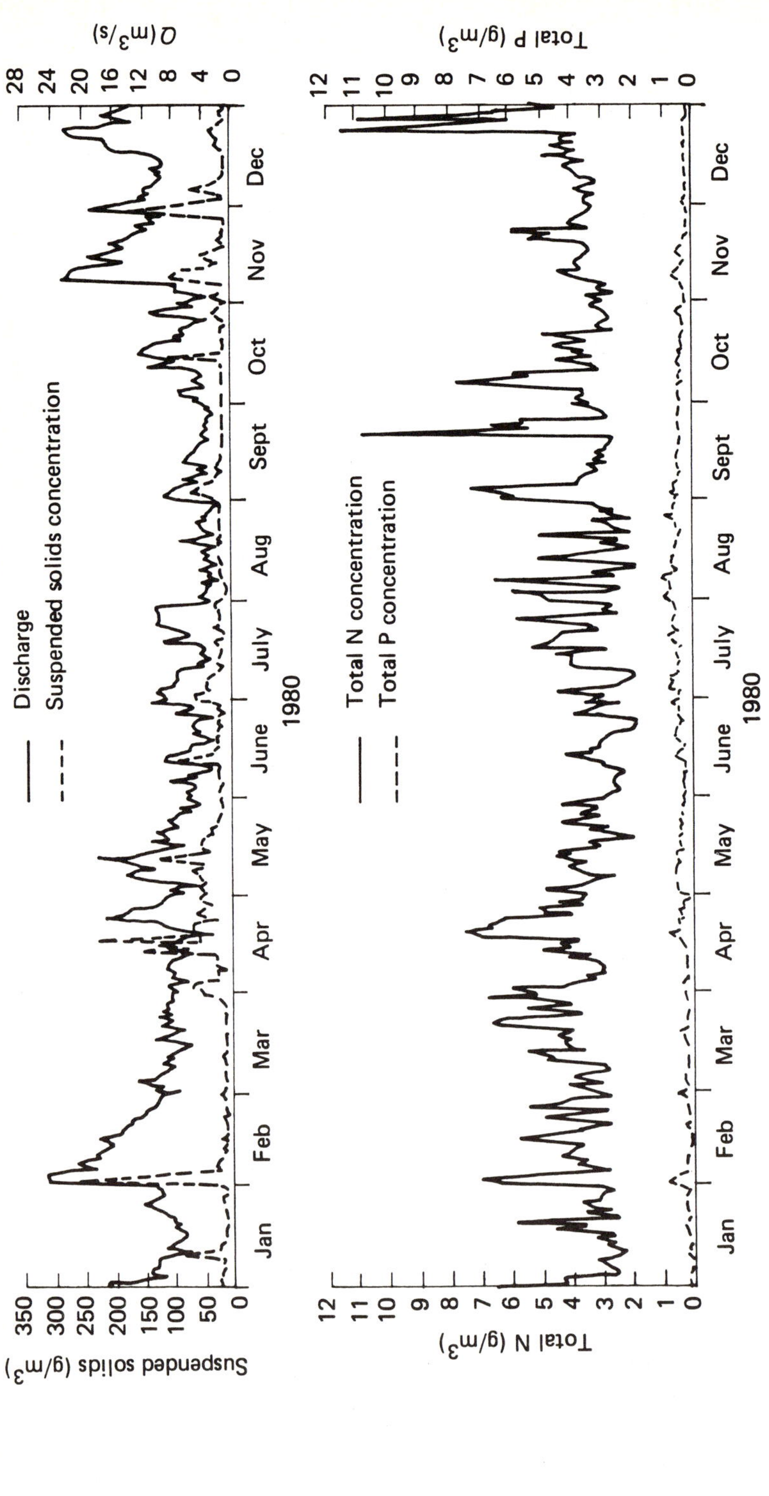

Figure 14.2. *(f)* Discharge and concentrations of suspended solids, total P, and total N for the Zala River, Fenékpuszta, 1980.

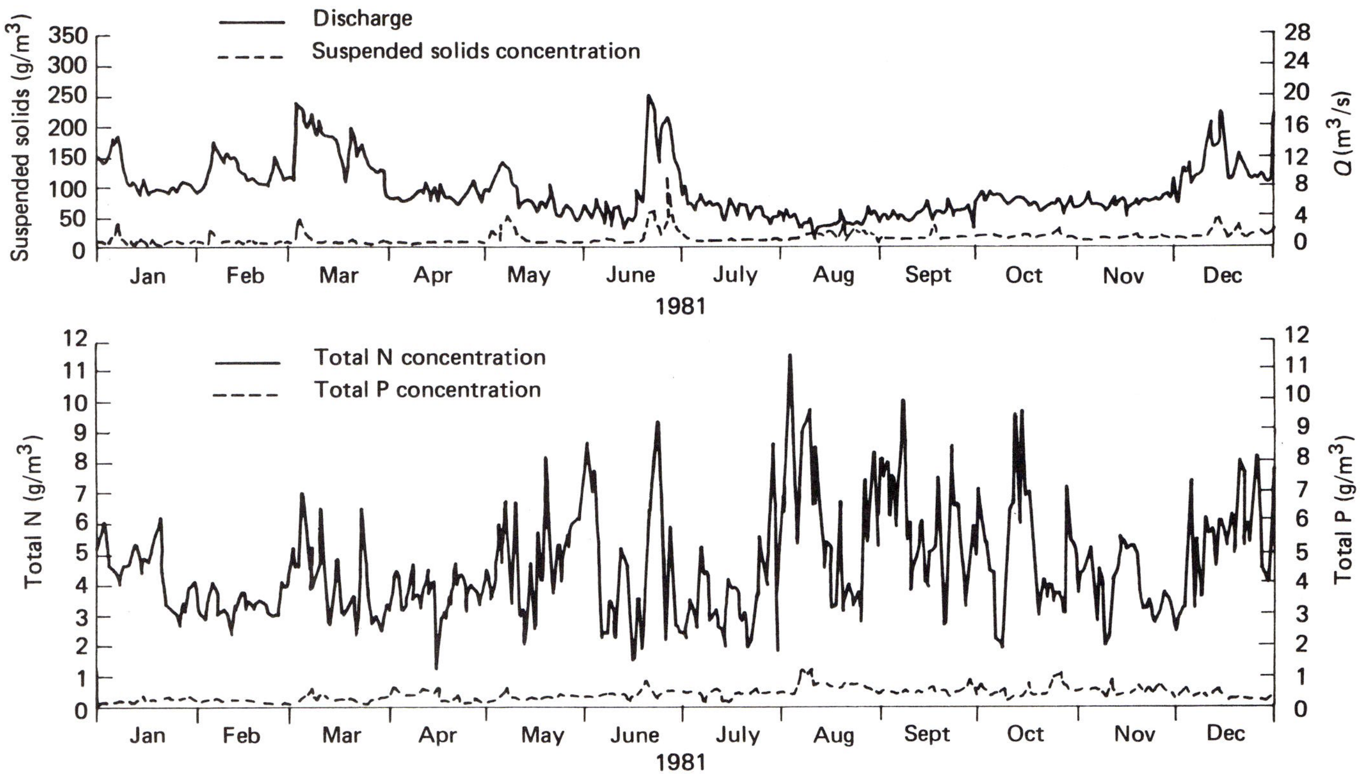

Figure 14.2. (*g*) Discharge and concentrations of suspended solids, total P, and total N for the Zala River, Fenékpuszta, 1981.

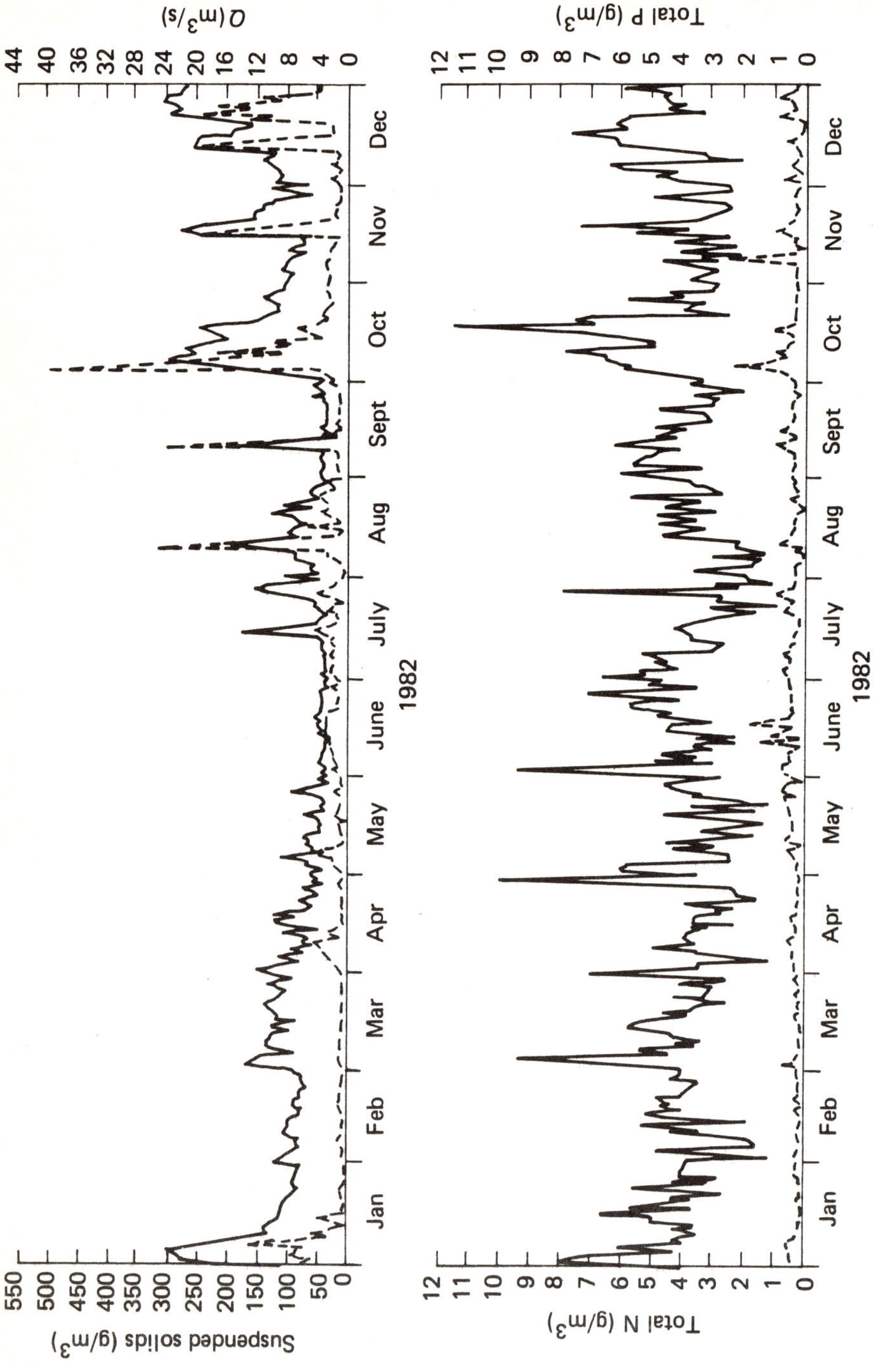

Figure 14.2. *(h)* Discharge and concentrations of suspended solids, total P, and total N for the Zala River, Fenékpuszta, 1982.

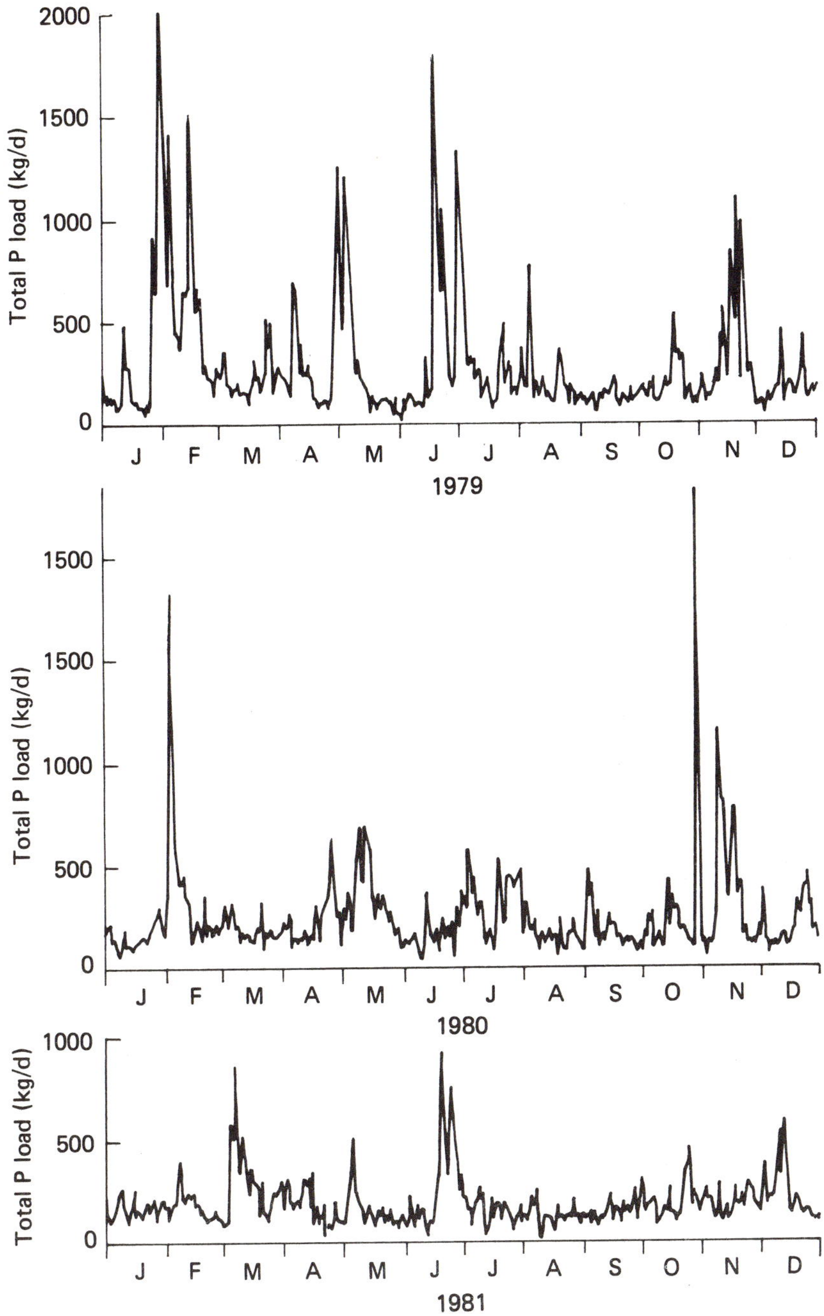

Figure 14.3. Total P load for the Zala River, Fenékpuszta, 1979–81.

A summary of the main data for the period 1977–82 obtained at Fenékpuszta and Zalaapáti is given in Table 14.2. Dividing the records into low water (LW) and high water (HW) periods[1] precipitation and discharge values are shown together with the suspended solids, total P, dissolved reactive P, total N, and nitrate N loads. The following main conclusions can be drawn from the analyses.

The period 1977–82 was drier than average. The most part of the total discharge into the lake was drained during the LW periods. High waters at Fenékpuszta contributed about three quarters of the total suspended solids load, less than half of the total P load, one third of the dissolved reactive P load, and half of the N load of the lake.

Through the measured section at Zalaapáti (despite its watershed being 40% less than that of Fenékpuszta) 1.5 times more suspended solids and only 10% less P passed during the period investigated. High waters at Zalaapáti contributed twice as much suspended solids and 20% less phosphate P than high waters at the mouth section, while total P loads remained the same. During all the years investigated the low water dissolved reactive P contribution was higher than that of the high water periods (suggesting a point source effect).

From analysis of the data a significant nonpoint source contribution is apparent. Another important conclusion is that the total suspended solids load entering Lake Balaton is only slightly more than 10000 tons annually, contradicting former estimates of the order of several hundred thousand tons.

Measurements have indicated that there is a significant settling of suspended solids between Zalaapáti and Fenékpuszta; they also suggest that retention of suspended solids by the protective impoundment system Kis-Balaton will be high.

In Figure 14.4(*a*)–(*b*) monthly average values of discharge and nutrient loads are given for the section of Fenékpuszta. It can be seen from the figure that, akin to the rainfall–runoff pattern, there is a relationship between monthly flow and N load and a somewhat less evident relationship between monthly flow and P load. However, no such explanation for the observed dissolved reactive P load can be given, suggesting again that this load component is primarily of sewage origin.

In Table 14.3 rainfall and flow records for the period of 1976–82 are summarized, as are the seasonal variations of the nutrient load components. The period between 1976 and 1982 was, on average, 33 mm drier than normal. These deficits have accumulated since 1976, resulting in a total deficit of 233 mm in 1982, which is almost one third of the annual total precipitation value. In spite of the nearly doubled discharge values during winter and spring, as compared to those of summer and fall, the P loads are about the same. It is

[1]Note that LW and HW periods were not separated by defining a "critical" discharge value, but rather a somewhat subjective procedure was employed, in which the entire record was subdivided into periods differing in character. A period was classified as belonging to the HW category if the peak discharge was at least doubled in comparison to that of the preceding period.

Table 14.2. Summary of precipitation, discharge, and loads for 1977–82.*

	Precipitation (mm/yr)		Discharge ($10^6 m^3$/yr)		Suspended solids load (tons/yr)		Total P load (tons/yr)		Dissolved reactive P load (tons/yr)		Total N load (tons/yr)		Nitrate N load (tons/yr)		Notes
	A	B	A	B	A	B	A	B	A	B	A	B	A	B	
Total	714	708	147	229	12800	8750	81	89	34	42	713	967	248	363	Long-term average precipitations are 743 and 723 mm/yr, respectively
High water (25% of the total)	–	–	49	45	82	75	52	45	35	35	53	50	56	55	% of the total
Low water (75%)	–	–	51	55	18	25	48	55	65	65	47	50	44	45	% of the total

*A – Zalaapáti, B – Fenékpuszta

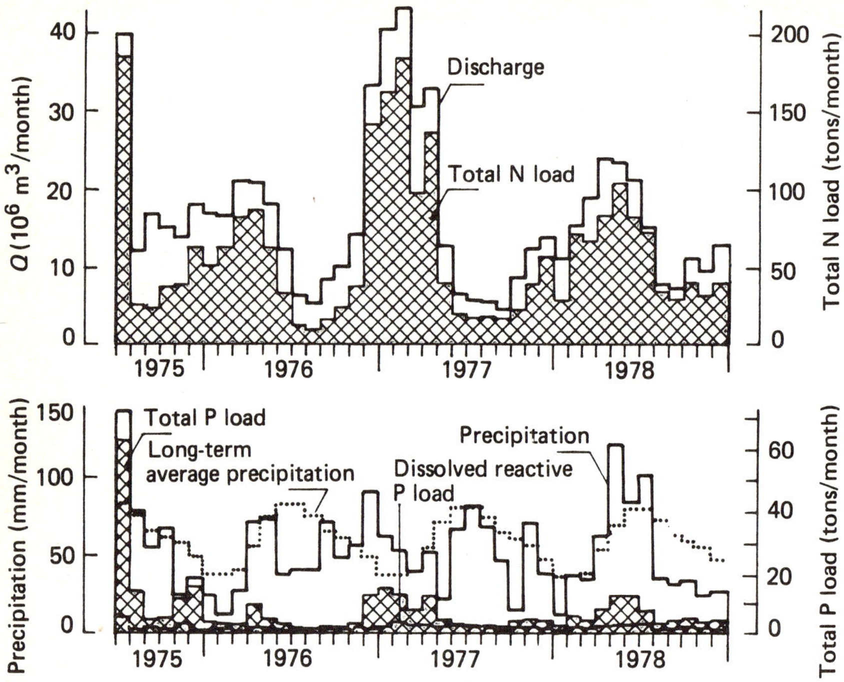

Figure 14.4. (*a*) Monthly average precipitation, discharge, and nutrient loads for the Zala River, Fenékpuszta, 1975–78.

interesting to note that equal loads of P and equal flows correspond to the spring and fall periods (fertilization periods).

Based on annual values it can be concluded that in years with abundant flow the export of P and N from the watershed is higher than in drier years. Data also suggest that the export of nutrients is increasing on a long-term scale.

On the basis of data already available the total nutrient and suspended solid loads carried into Lake Balaton can be accurately calculated for the Zala River. However, the contribution of waste waters (point sources) and diffuse or nonpoint sources to this total load is also important. Total N and P loads originating from waste water effluents on the Zala River basin are estimated to be 310 tons/yr and 54 tons/yr, respectively (average for 1975–82). According to our inventory there are 120 concentrated pollution sources in the Zala River basin. In 1982 these sources discharged 12×10^5 m^3 sewage into the recipient water bodies, with varying degrees of treatment. Fifteen plants have been investigated in detail, the total discharge of which amounted to about 80% of the total sewage load in the river basin. The highest discharge is that of the sewage treatment plant for the city of Zalaegerszeg, amounting to almost 20% of the annual total of 20×10^6 m^3 sewage load in the Zala River, and 40% of the annual 100 tons of total P load of sewage origin.

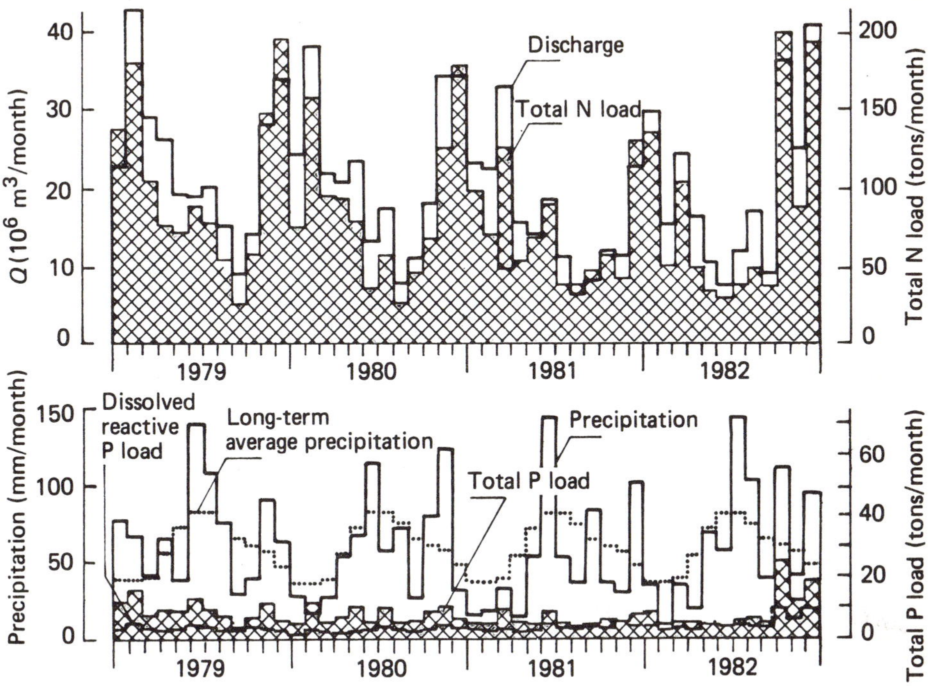

Figure 14.4. (*b*) Monthly average precipitation, discharge, and nutrient loads for the Zala River, Fenékpuszta, 1979–82.

Table 14.3. Averages of observations at Fenékpuszta for various periods (1976–82).

Period (months)	Precipitation (mm/yr)		Q ($10^6 m^3$)	Total P load (tons/yr)	Dissolved reactive P load (tons/yr)	Total N load (tons/yr)
	Long-term average	1976–82				
Dec–Feb	125	134	82	25	9	370
Mar–May	170	150	67	22	8	258
June–Aug	242	242	38	18	9	143
Sept–Nov	186	164	45	21	11	184
Dec–May	295	284	149	47	17	629
June–Nov	428	406	83	39	20	326
Jan–Dec	723	690	232	86	37	956

This figure clearly emphasizes the need to introduce P precipitation in the treatment plant.

In summary, on average the Zala River annually discharges into Lake Balaton 230 × 10^6 m^3 of water, about 10000 tons of suspended solids, 1000 tons of N, and 90 tons of P (less than half of which is dissolved reactive P, surely available for algal growth). The Zala River total P load contributes to about one third of the lake's load. The ratio of the loads suspended solids:total N:total P is about 100:10:1. Annual export rates corresponding to unit watershed area are about 4, 0.4, and 0.04 tons/yr km^2 for suspended solids, total N, and total P, respectively. These values are in good agreement with those in the literature.

Analysis of the observations for the Zala River and sewage P loads show that about half of the total P load and more than two thirds of the total N load are of diffuse origin. This fact indicates the need for engineering solutions (i.e., the impoundment system of Kis-Balaton) that can offer the retention of nutrients from all sources. The higher ratio of diffuse sources at Zalaapáti as compared to the mouth section and the high sediment yields from the watershed stress the need for increased erosion control.

14.3. Diffuse Sources

Nonpoint source or diffuse pollution is widely discussed in the relevant literature. For the watershed of Balaton, however, only a few measurements are available (see Chapter 6). Based on measurements (twice monthly) during 1981–82 the network of Plant Protection and Agrochemical Stations estimated the agricultural contribution to the load of nutrients as one quarter to one third of the total: an obvious underestimate, the reason for which is infrequent sampling (see later and Chapter 6).

According to data from the Plant Protection and Agrochemical Station of Zala County, 3000 tons of P (6900 tons of P_2O_5) and 10000 tons of N fertilizers were applied to the Zala watershed in 1981. Using diffuse P and N loads estimated previously the following calculations can be made. If all of these loads were of fertilizer origin, then 2% of the P and 8% of the N fertilizer (in effective material) applied in 1982 would have been lost to recipient waters: quite realistic values!

Unfortunately, no measurements on the farm scale are available to prove or disprove such estimates. If the agricultural diffuse portion of nutrient loads could be more accurately quantified then remedial action could probably be much better focused on the appropriate alteration of fertilization techniques (for example, on liquid or slow-release fertilizer applications).

An attempt was made to account for the P contribution of smaller subwatersheds of the Zala basin, by using weekly and biweekly measurements at various locations. First of all, discharges and P loads observed at Zalaapáti and Fenékpuszta were compared for 1979–81 on the basis of daily, weekly, and biweekly measurements. In the biweekly case data of odd and even weeks were also compared. The study shows that loads determined from daily measurements are consistently higher than those determined on the basis of weekly data (see also Chapter 6). For the wet year 1979 the loads calculated from biweekly observations were one quarter to one third less than those derived from daily sampling.

Table 14.4. P load estimates for subwatersheds of the Zala River based on biweekly measurements (1979–81).

Stations	Watershed area (km^2)	Q ($10^6 m^3/yr$)	Total P load (tons/yr)	P load of sewage origin (tons/yr)	P load of nonpoint source origin (tons/yr)	Unit areal total P load ($kg/km^2 yr$)	Unit areal P load of nonpoint source origin* ($kg/km^2 yr$)
Zalaegerszeg	436	42	13	–	13	30	30
Zalabér	1238	105	58	37	21	56	10
Zalaapáti	1533	147	67	39	28	30	24
Fenékpuszta	2622	243	81	51	30	13	2

*Unit areal loads refer to subwatersheds belonging to neighbor stations.

Bearing in mind the above inherent inaccuracies, a P load balance for the period 1979–81 has been calculated for the Zalaegerszeg–Zalabér and Zalaapáti–Fenékpuszta sections, as shown in Table 14.4. The distances of Zalaapáti, Zalabér, and Zalaegerszeg from the mouth section at Fenékpuszta are 23 km, 55 km, and 82 km, respectively. Load components originating from point-source effluent outfalls are also shown for all sections. The total P load attributable to diffuse sources upstream of the sections was taken as the average over three years. The data indicate that the diffuse P load is highest along the Zalaegerszeg–Zalabér reach. Similar conditions to those at Zalaapáti prevail. The diffuse load is very low on the reach between Zalaapáti and Fenékpuszta. It is interesting to note that the highest specific P load (30 kg/km^2yr) corresponds to the most upstream subwatershed. Between Zalaapáti and Zalabér the annual export rate is similarly high (24 kg/km^2yr), while a much smaller value of 10 kg/km^2yr is found for the section of Zalaegerszeg–Zalabér. Practically no diffuse source contribution can be determined for the lowest reach downstream of Zalaapáti, which has a watershed area of 1089 km^2.

14.4. Pollution Discharged by the City of Zalaegerszeg

The largest point source in the Balaton region is the city of Zalaegerszeg (Figure 14.1), with a daily sewage discharge of about 16000 m^3. The existing two-stage treatment plant is heavily overloaded.

Major problems to be studied were, among others, the magnitude and dynamics of the total P load from the treatment plant, the contribution of this load to that of the mouth section of the Zala River, and the effectiveness of P precipitation on the load of Lake Balaton.

To do this, detailed, longitudinal water quality profile studies were carried out by the West-Transdanubian Water Authority, jointly with the Research Center for Water Resources Development (VITUKI), three times in 1982 and 1983. During the three periods (each five days long) drought to average flow conditions were dominant. There was no precipitation during the measurements.

Sampling of hourly or two-hourly frequency was established at seven locations of the Zala River and at 32 tributaries of the reach studied. In addition to discharge measurements, water quality components such as total P, dissolved reactive P, suspended solids, chloride, etc., were determined. A summary of the data obtained during the three periods for the treatment plant of Zalaegerszeg is given in Table 14.5. Apparent from the measurements is the presence of strong diurnal variations: in general morning and evening peaks exist. There are about three fluctuations of discharge and dissolved solids within a day, five of sulphate, and the rest of the components vary within an order of magnitude. The unsteady nutrient release of the treatment plant of Zalaegerszeg can be traced along the downstream reach of the river as far as Zalaapáti (Figure 14.1).

Table 14.5. Summary of measurements taken in the effluent water of the Zalaegerszeg sewage treatment plant.

Parameters	Average	Minimum	Maximum
Sewage discharge (m^3/s)	0.18	0.08	0.25
Total P load (g/s)	1.6	0.40	3.20
Dissolved reactive P load (g/s)	1.2	0.20	2.40
Hour of occurence	–	3–7 a.m.	2 p.m.

In Figure 14.5 longitudinal profiles of P concentrations and discharges are shown, together with the corresponding catchment areas. From the above measurements the following conclusions can be drawn (see also Figure 14.1):

(1) Travel time for the river reaches Andráshida–Zalabér, Zalabér–Zalaapáti, and Zalaapáti–Fenékpuszta are 1.1, 1.7, and 0.5 km/h, respectively.
(2) P concentrations rapidly decrease to Alibánfa, with a reduced rate of decrease downstream of this section.
(3) There is an increase of dissolved and suspended solids concentrations until Alibánfa. Between Alibánfa and Zalaapáti the increase is less marked. Downstream of Zalaapáti a significant loss of suspended solids is observed, in good agreement with losses calculated on the basis of long-term average mass balances. Mass fluxes of total dissolved solids are proportional to the increase in the drainage area.
(4) Total P and dissolved P loads significantly decrease between Alibánfa and Zalabér, while downstream of Zalaapáti there is an increase which is smaller than those of the corresponding drainage area.
(5) Taking into consideration only the two measurements of 1983, there was no increase in the P mass fluxes downstream of the village of Balatonhidvég (this may be due to the effects of the large marshland in this region).
(6) The P load from the tributaries equals that from the city of Zalaegerszeg. During the observation periods (excluding weekends) the load from Zalaegerszeg was larger than the value derived from the regular (but few) measurements.
(7) A full P balance for the individual reaches of the river (taking tributary inputs into consideration also) shows a 25% P loss for the Alibánfa–Zalabér reach and a 42% loss for the Zalaapáti–Fenékpuszta stretch. For the mouth section at Fenékpuszta the gross retention of P is equal to the load from the effluent outfall of Zalaegerszeg. The most important practical conclusion is perhaps that at low and mean flow conditions a significant portion (about 40%) of the total P load from the treatment plant of Zalaegerszeg does not reach Lake Balaton.

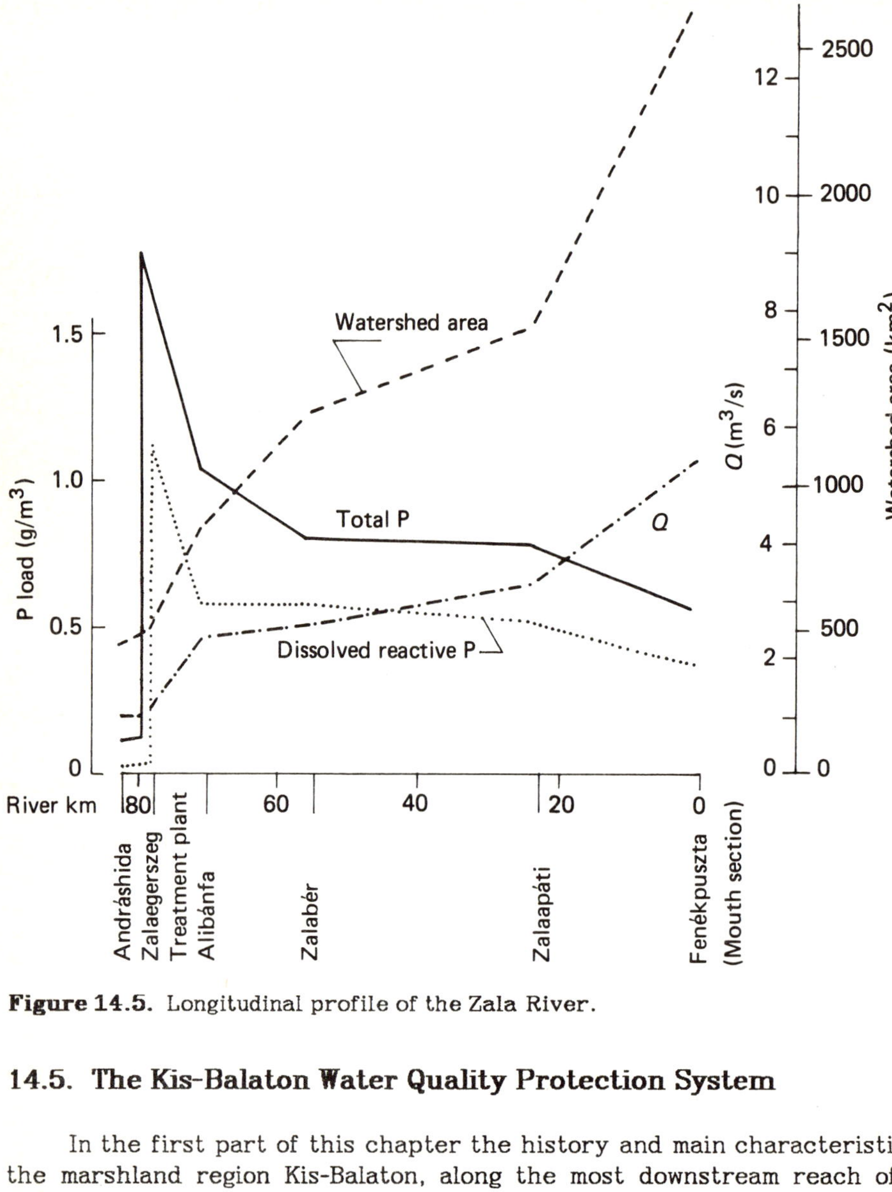

Figure 14.5. Longitudinal profile of the Zala River.

14.5. The Kis-Balaton Water Quality Protection System

In the first part of this chapter the history and main characteristics of the marshland region Kis-Balaton, along the most downstream reach of the Zala River (Figure 14.1), were briefly described, together with the results of water quality measurements in this region. It is known that P and N are the most important nutrients in terms of eutrophication. It was shown that total P and total N originate to a large extent from diffuse sources and the loads change strongly with time. The importance of the Zala River system in the overall nutrient balance of the lake has also been stated (see also Chapter 6).

Owing to the high proportion of diffuse sources in the Zala watershed, perhaps the best chances of reducing nutrient loads and reversing the deterioration of the lake's water quality are offered by a retention pond system constructed on the downstream portion of the Zala River. It should be expected that a considerable portion of suspended solids and associated nutrients will be retained in this system (according to measurements carried out in 1983, half of the total suspended solids content of the river is deposited in this region, under low flow conditions). Biodegradable organic matter carried by the river will be decomposed in this system, and not in Lake Balaton. The biogenic lime precipitation will probably also occur in the proposed reservoir. At present, this precipitation in the Keszthely Bay amounts to $10–20 \times 10^3$ tons annually. The resultant lime flocs of large surface area may have high P-removing capacities. From the literature and our own experiences there are only rough – and sometimes contradicting – estimates for nutrient removal capacities of marshland ecosystems (in reservoirs with an average depth of 1.5 m). Efficient removal of nutrients can be expected to be mostly effected by the (partly existing) reed zones, due to the activities of the microecosystem attached to the reed stems. Favorable purifying capabilities may also be offered by the peat layers of this region, which have depths varying between 2 and 8 m. Nutrients are expected to be retained in the soil as the result of ion exchange processes. Additional benefits may occur due to the extension of the habitat of water birds in the natural preserve, which has an area of 1600 ha.

Difficulties may also be encountered when converting the terrestrial ecosystem into an aquatic one of such a large area. Note that no reported results on such a vast and artificially created "water purifying" marshland are known to us. These circumstances add to the uncertainties and risks of this project, but in spite of the risks involved the project should be implemented (see Chapters 4 and 5).

The planned protective system is shown schematically in Figure 14.1. The inundation of the first phase reservoir began in April 1984 and the reservoir started to operate as a throughflow system in mid June 1985. The plans for the second phase are under elaboration.

The first phase reservoir lies along the right bank of the Zala River at a distance of 10 to 20 km upstream of the river mouth. The operational water level of the reservoir is 105.5 m above the level of the Baltic Sea. Water levels corresponding to 100 years of high flow can be expected at about 107.4 m above sea level. When operational the surface area will be 18 km^2 with a water volume of 21×10^6 m^3. For the mean flow of the Zala River the reservoir will provide a retention time of about one month. The cost of implementation is about 700×10^6 Ft. The inside structures have been designed on the basis of scale models with the purpose of providing a "uniform" water distribution over the area and maintaining an appropriate retention time. At the deepest part of the reservoir a system of boxes separated by coffer dams will provide an area of 400 ha, fed and drained by independent gates. Its function is to control the retention time if the nutrient loads are very large (e.g. rising part of flood waves). This structure will also provide for the "short circuiting" of the river flow. Namely, if there is an algal bloom at certain parts of the reservoir the river could be conveyed through this system of boxes, without necessitating release from the reservoir. Inundation and drainage will be facilitated also by a system of channels. Seepage water will be collected by a drain system along the left bank of the river and excess water (if any) will be drained by pumps. At the construction sites archeological investigations are under way. A forest area of about 300 ha, within the reservoir, will be clear cut, while protective forest strips along the reservoir will be planted. In the closing dam at the eastern end of the reservoir a navigation lock with a discharge capacity of 50 m^3/s will be constructed. The highest Zala River floods, of about 90 m^3/s, will be attenuated by the reservoir to 50 m^3/s.

The second phase will be constructed downstream of the straights of Balatonhidvég (largest surface area and volumes are 60 km^2 and 80×10^6 m^3, respectively). The deadline for this work, as determined by the Government, is 1987 (see Chapter 5). The second stage reservoir will also be surrounded by forest strips and drainage channels.

Areas inundated by an excess of water and the railway will be drained. The combined evapotranspiration losses of the two stages will be approximately 20×10^6 m^3 per year, causing no significant losses in the water balance of Lake Balaton. Further reduction of the presently prescribed 30 cm water level fluctuation range for Lake Balaton will be made possible by the equalizing effect of the reservoirs. If minimum water levels in the lake can be raised by 10 cm, this could provide for an additional 60×10^6 m^3 lake volume. The full cost of implementation of the Kis-Balaton project will most probably exceed 2×10^9 Ft.

14.6. Summary and Conclusions

The Zala River transports about 10000 tons of suspended solids, 1000 tons of N, and 100 tons of P annually into the western, shallowest basin of the lake, which also has the smallest volume and highest trophic level. Higher flows result in higher loads of solids and nutrients. About half of the total P load and about two thirds of the N load are estimated to be of diffuse origin. About half of the total sewage discharge and the corresponding P load of the Balaton region originate from the Zala watershed.

The treatment plant of the city of Zalaegerszeg is the largest point source on the Balaton watershed, amounting to one third of the total point source P inputs. Measurements described in this chapter show that about 60% of the total P load from Zalaegerszeg reaches Lake Balaton.

Based on our study the following main measures should be taken to control the eutrophication of the lake:

(1) The protective reservoir system of Kis-Balaton, aimed at removing pollutants carried by the Zala River, should be constructed. If the first phase of this system, to be completed by the end of 1984, results in the expected efficiency, then sufficient time will be gained to implement all the control measures planned for other regions of the lake.

(2) Retention reservoirs and filter zones of a similar type should be constructed on other appropriate sites.
(3) A final and complete solution of the sewage problem should be assured (P precipitation, diversion to other catchment areas, sewage and sludge disposal and utilization, etc.).
(4) Agricultural management strategies should be included, such as erosion control (especially along the northern and southern shorelines, then later, upstream of Zalaapáti), elimination of pollution loads from large-scale livestock breeding farms, appropriate determination of land uses, and the satisfactory solution of fertilizer application.

CHAPTER 15

The Influence of Watershed Development on the Long-Term Eutrophication of Lake Balaton

L. Dávid and L. Telegdi

15.1. Introduction

As discussed in Chapter 1 the acceleration of eutrophication in Lake Balaton is caused by the increasing socioeconomic and water resources development in the lake's contributing watersheds (watershed development) during the last few decades. Artificial eutrophication is the response of the lake ecosystem to man's interference. However, watershed development, which is *a priori* and spontaneous from the viewpoint of the lake, can be modified by human policies adopted to protect the lake.

Assuming there is a close connection between eutrophication of the lake and man-made watershed development (Dávid *et al.* 1979), then there are two basic approaches to model eutrophication (Figure 15.1). The first approach, A, relates watershed development to nutrient loading which, in turn, affects eutrophication, while the second approach, B, directly links watershed development with eutrophication.

In case A – as demonstrated in other chapters herein – the processes must be described with fairly detailed, structural models. Once developed, such models are extremely useful (in particular for understanding and short-term management). However, the construction of reliable models requires a good interdisciplinary knowledge about many physical and other processes and a well coordinated data collection program – time-consuming operations.

In approach B, a direct relationship is sought in terms of an integrated, multicriteria empirical model which can be used particularly for long-term

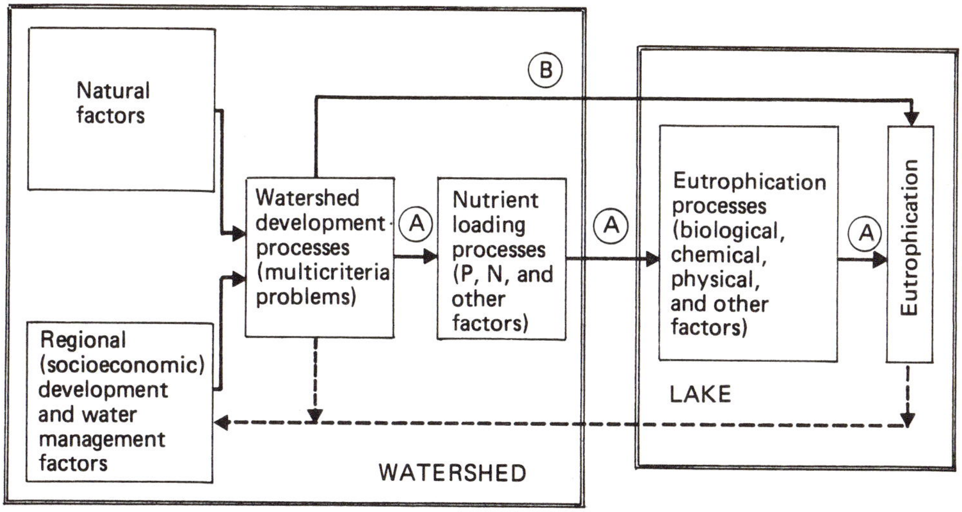

Figure 15.1. Approaches A and B to modeling eutrophication.

management of eutrophication. The long-term effects of human activities can be measured and simulated by this approach, since the required historical time-series of the basic factors (characterizing the stage of regional development of a watershed) are usually available prior to the study from regular statistical data. The development of such a model requires a relatively short time. However, considering the detailed processes involved, this approach can only be considered as a rough approximation. The integrated character does not supply information on the actual behavior of the individual processes that may play a role (e.g., even nutrient loads cannot be quantified, see Figure 15.1), but it is good for general and regional evaluation.

The purpose of this chapter is to present and adapt to Lake Balaton a multiregional and multicriteria watershed development model for the description of the long-range eutrophication process.

15.2. Watershed Development Model for Lake Balaton

In this section only a short description of the model is given for Lake Balaton; details on the general formulation of the approach and on its application are given in Dávid and Telegdi (1982).

As shown in previous chapters, for Lake Balaton four typical lake basins (or water bodies) can be distinguished, with seven subwatersheds (Figure 15.2). With this regionalization and taking into account that eutrophication of a basin depends on the development of connected subwatersheds and on the

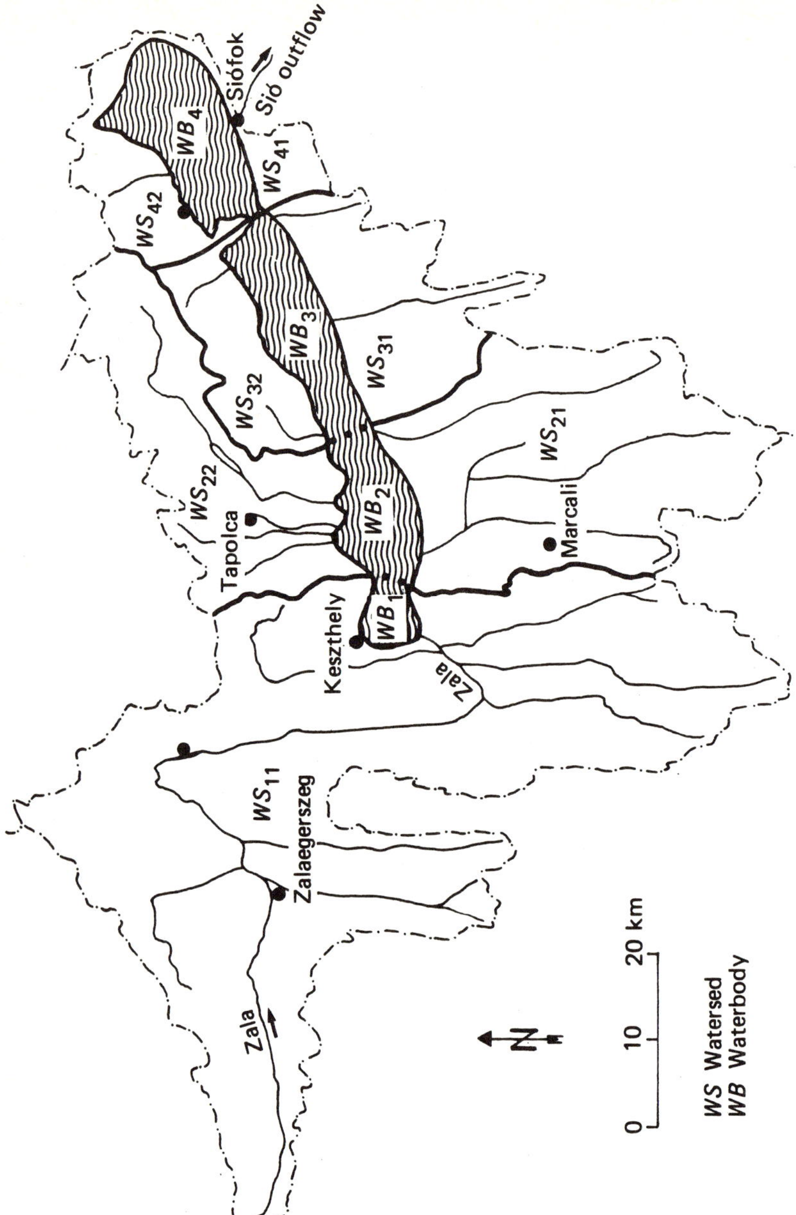

Figure 15.2. Regionalization of Lake Balaton and its watershed.

state of the western neighboring basin (see Chapter 4), the long-term eutrophication can be described by the following model:

$$E^*(l,j) = a_1 + b_1 D^*(l,j) \quad , \qquad \text{if } j = 1$$

and (15.1)

$$E^*(l,j) = a_2 + b_2 D^*(l,j) + c_2 E^*(l,j-1) \quad , \qquad \text{if } j = 2,3,4 \quad ,$$

where l is time in years; j is the water body, namely Keszthely basin (WB_1; j = 1), Szigliget Basin (WB_2; j = 2), Szemes Basin (WB_3; j = 3), Siófok Basin (WB_4; j = 4); E^* is the eutrophication index, a three-year moving average of the summer average of chlorophyll-a (Chl-a) concentration (E_j^*); a_1, b_1, a_2, b_2, c_2 are parameters subject to model calibration; and D^* is the integrated watershed development index corresponding to WB_j, calculated from

$$D^*(l,j) = \sum_{i=1}^{2} H(j,i) \times D(l,j,i) \quad . \tag{15.2}$$

In equation (15.2) i is the watershed (WS_i) associated with WB_j. For j = 1, i is only 1; for j = 2, 3, 4, i = 2, because there are two corresponding watersheds for these basins (Figure 15.2). $H(j,i)$ is the natural geographical indicator of WS_{ji}, indicating the role of WS_{ji} in the multiregional, hierarchical

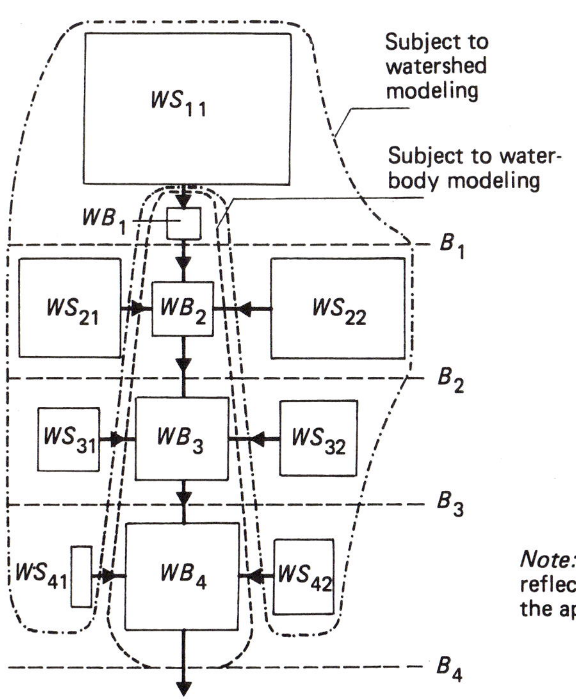

Note: The size of a block reflects the extension of the appropriate area

Figure 15.3. Multiregional, hierarchical system of Lake Balaton and its watershed.

system of the Balaton area (Figure 15.3). It is expressed by the ratio of the surface area of WS_{ji} and the total watershed area of the basin ($0 \leq H \leq 1$). $D(l,j,i)$ is the watershed development index for WS_{ji} in year l. It can be viewed as a multiattribute utility function (Keeney and Raiffa 1976) and written as

$$D(l,j,i) = \sum_{k=1}^{K} w_k I_k(l,j,i) \tag{15.3}$$

where I_k are indicator indices, each of which expresses a watershed development criterion influencing nutrient loads and has to be composed from the basic factors characterizing development of the watershed (natural, regional, and water management factors). The terms w_k are the weighting factors, constant in time and space ($w_k \geq 0$, $\sum_{k=1}^{K} w_k = 1$). I_k are dimensionless, most are time-dependent, and represent short-term or long-term management actions together with the associated influence on nutrient loads. They are simple functions of the basic factors.

Values of I_k range between 0 and 100. Zero means no influence on nutrient loads, while 100 shows a maximal one. From this, the definition of w_k, and equation (15.3) it follows that D also varies between 0 and 100; and a decrease in D means a reduction in nutrient loads.

The system of indicator indices involves $K = 25$ elements (composed from 50 basic factors, see Dávid and Telegdi 1982), listed in Table 15.1.

The value of weighting factors varies between 0.074 and 0.014. The visitor loading and the fertilizer use have the maximum and the ratio characterizing arable and forest land-use has the minimum weighting factors.

It is worth noting that equation (15.1) is based primarily on intuition and has a similar structure to that of the deterministic lake eutrophication model given by equations (4.1) and (4.2), which were derived from a detailed analysis following approach A [and then used in the frame of the eutrophication management optimization model (EMOM)]. Both models express the eutrophication index of a basin (summer average and annual peak Chl-a concentrations, respectively) as a linear function of the subwatershed characteristics of the basin considered and that of the neighbor, "upstream" basin. The subwatershed characteristics are, in the present model, the development indices, while in equation (4.1) they are the volumetric, biologically available P loads (the connection between these two groups of variables is represented by the indicator indices I_k).

The development and application of this model took three steps:

(1) The model was calibrated using time series of the watershed development index and summer average Chl-a concentrations (observations of VITUKI, see Figure 1.4) for the period 1973–80. Parameter values

$$\begin{array}{lll} a_1 = -215.3 & b_1 = 16.8 & \\ a_2 = 4.5 & b_2 = 1.3 & c_2 = 0.5 \end{array}$$

Table 15.1. System of indicator indices.

k	Indicator indices
1	Erosion potential
2	Quantity distribution of precipitation
3	Time distribution of precipitation
4	Density of natural water courses
5	Natural energy potential of the watershed
6	Population ratio involved in industry
7	Population ratio involved in agriculture
8	Visitor (tourist) loading
9	Density of possible point sources
10	Ratio characterizing arable and forest land-use
11	Urbanized part of the watershed
12	Ratio of vineyards and orchards
13	Fertilizer use
14	Density of animal population
15	Density of motoring roads
16	Use of available water resources
17	Amount of used water for infiltration into the soil
18	Ratio of untreated sewage discharge to the total
19	Ratio characterizing unregulated runoff
20	Ratio characterizing population supplied with drinking water works
21	Ratio characterizing population supplied with sewage works
22	Ratio of irrigation and drainage area to that of the arable land
23	Population density
24	Density of all water works in the watershed
25	Beach length indicator for direct recreation loading

were derived. The analysis of parameters shows that the eutrophication index of WB_j $(j > 1)$ depends on that of WB_{j-1} 1.4 times more than on the development index of the adjacent watersheds. This figure reflects the major role of the Zala watershed and Keszthely basin in the eutrophication of the entire lake.

(2) Model simulations were performed on the basis of development indices derived from statistical data for 1930–73. For this period no or scarce Chl-a data were available, so the model performance gives an idea of the past development of eutrophication.

(3) The model was applied to 1981–85 for various watershed development scenarios.

Results for the entire period 1930–85 are given in Figure 15.4 while the regional structure of eutrophication of the Balaton lake–watershed system is illustrated in Figure 15.5.

As shown in the figures the watershed development and water quality were rather uniform until 1960–65, both in time and space ($D \leq 25$ and $E \leq 8$). Later, eastern subwatersheds developed more intensively, but because of the specific topography of the system (see Table 1.2 and Figure 15.5) the water quality deterioration became much stronger in western segments of the lake.

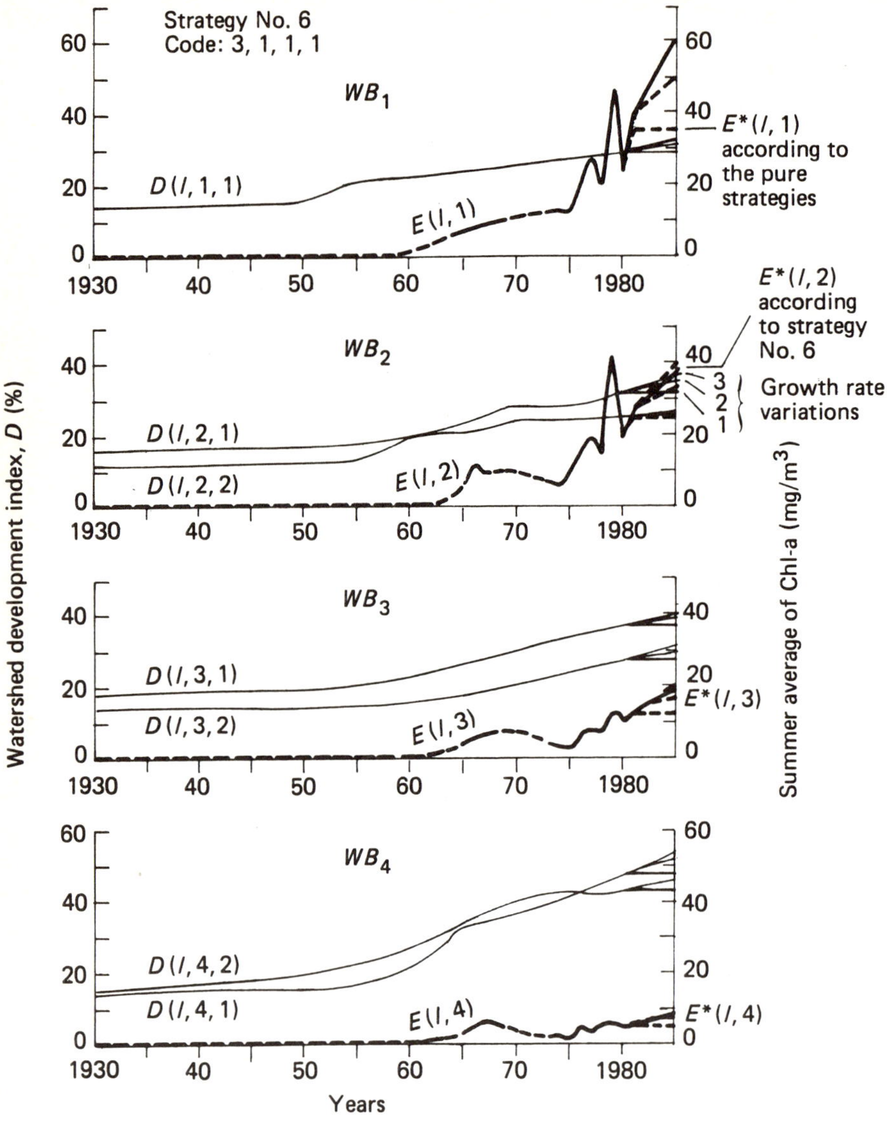

Figure 15.4. Past and future changes of watershed development and eutrophication indices.

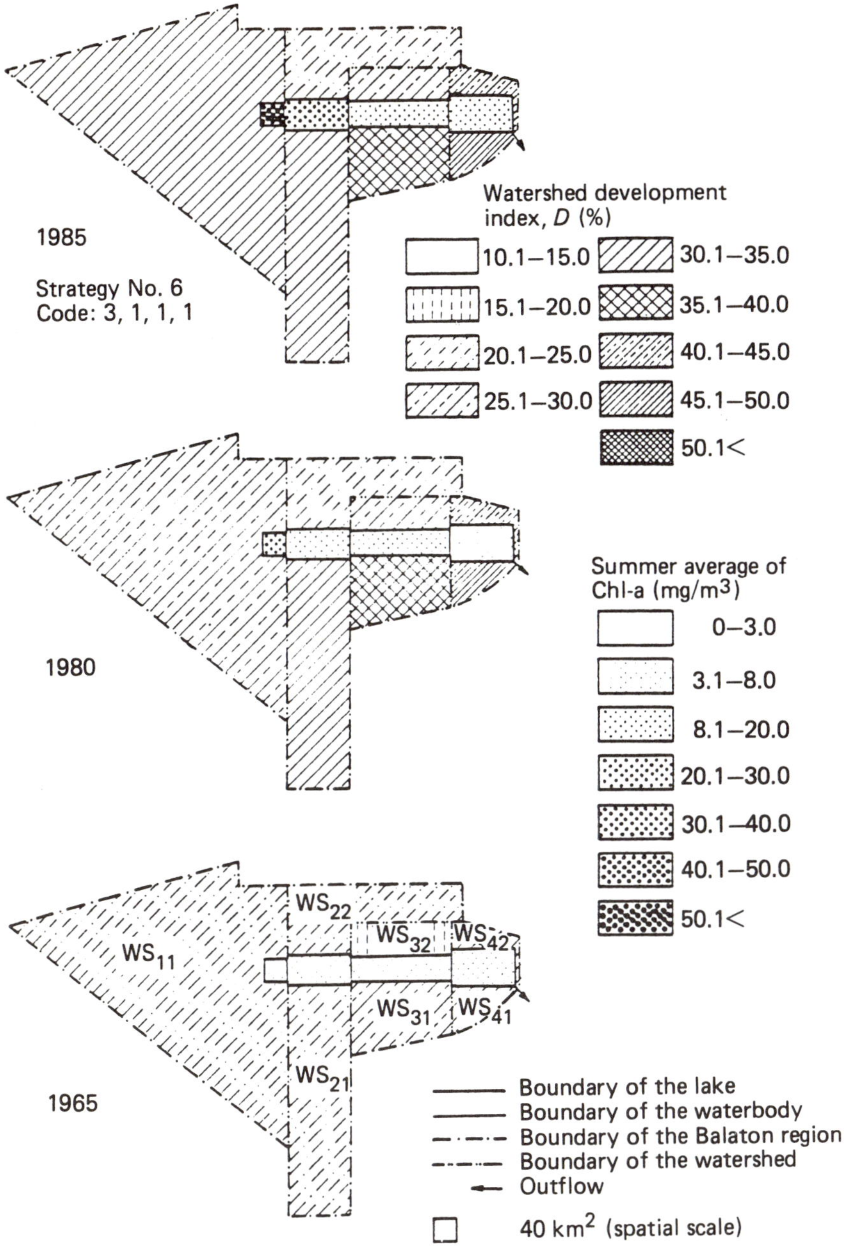

Figure 15.5. Regional structure of eutrophication for the Lake Balaton system.

15.3. Simulation of Eutrophication for 1981–85

The application of the calibrated model is presented in this section. First, realistic watershed development scenarios are established, which then make it possible to evaluate the future trend of eutrophication, assuming the validity of the calibrated model.

The simulation of eutrophication in the lake is based on three watershed development variations, all of which are the result of various human activities on the corresponding watersheds. In the case of strong growth-rate variation, indicated by the number 3 in Figure 15.4, the increase of the watershed development index D follows the previous trend between 1981–85. It assumes that there will be practically no effective lake protection policy to decrease the nutrient loading. On the other hand there is a zero growth-rate variation indicated by 1, which assumes a lake protection policy such that the watershed development index D remains constant at the 1980s level. That is, the consequences of negative changes in the watershed leading to an increase in nutrient discharges and positive measures for controlling nutrient loads will compensate each other. Finally, there is a medium growth-rate variation indicated by 2, which assumes that the lake protection policy will be gradually more and more effective and, therefore, it involves a decreasing growth rate of D. Considering the real conditions of decision making and implementation of lake protection measures, it is not yet feasible to assume a decrease in D between 1981–85. Therefore a variation like this has been considered only for the example of the Kis-Balaton project (see below).

The watershed development indices for the seven watersheds calculated according to the three variations outlined are presented in Table 15.2, as are the respective D^* values.

Based on these growth-rate variations of D, several lake protection strategies can be constructed according to the spatial distribution of these variations over the watersheds, WS_{ji}. For each strategy or scenario there exists a set of indicator indices $I_k(l, j, i)$ [see equation (15.3)], and a total of 33 lake protection strategies were developed. There are three pure strategies (No. 1, No. 14, and No. 25), in which the same growth-rate variation is applied for all of the seven watersheds (variations 3, 2, and 1, respectively); these are uniform development strategies. The other strategies are combined ones in which different growth-rate variations are used for the individual watersheds.

By the application of equation (15.1), the response of the lake basins to these lake protection strategies can be simulated.

The simulation results of the pure strategies are listed in Table 15.2 and also presented in Figure 15.4. In the case of strategy No. 1, which is the most pessimistic one, the lake eutrophication index (the average of the waterbody indices) will increase 78% by 1985 (as compared with 1980). In the case of strategies No. 14 and No. 25 the corresponding values are 46 and 5%, respectively (the last small increase of E^* is a consequence of the increase of D between 1979 and 1980).

The effectiveness of various strategies is measured by the percentage of decrease in the lake average eutrophication index compared to strategy No. 1.

Table 15.2. Results for three watershed development scenarios.

Indices and their growth rate variations (grv)			WB_1	WB_2		WB_3		WB_4	
			WS_{11}	WS_{21}	WS_{22}	WS_{31}	WS_{32}	WS_{41}	WS_{42}
Watershed development index (%)	1980		29.0	31.6	24.2	37.0	27.5	47.8	43.5
	1985	Zero (1)	29.0	31.6	24.2	37.0	27.5	47.8	43.5
		Medium (2)	30.7	35.0	24.4	39.5	29.5	52.5	44.0
		Strong (3)	32.0	37.0	25.3	40.5	31.5	54.5	46.5
Geographical indicator, $H(j,i)$			0.515	0.185	0.126	0.061	0.053	0.022	0.038
Integrated watershed development index, D^* (%)	1980		14.9	8.9		3.7		2.7	
	1985	Zero (1)	14.9	8.9		3.7		2.7	
		Medium (2)	15.8	9.5		4.0		2.8	
		Strong (3)	16.5	10.0		4.1		3.0	
Eutrophication index, E^* (mg/m^3)	1980		31.5	26.0		11.0		5.1	
	1985	Zero (1)	35.0	27.5		13.0		5.5	
		Medium (2)	49.5	32.8		17.0		7.3	
		Strong (3)	60.5	39.2		20.8		9.5	
Differences	D^*_{85} Strong–Zero		1.6	1.1		0.4		0.3	
	E^*_{85} Strong–Zero		25.5	11.7		7.8		4.0	

For illustration of the model application, the simulation results for one extreme strategy (No. 6) are indicated by continuous lines in Figure 15.4. In this case, all of the efforts are concentrated at $WS_{21}, WS_{22}, \ldots, WS_{42}$, but no action is taken at WS_{11} (for the spatial distribution of D and E see Figure 15.5). This strategy is contrasted below with the reverse situation: all the efforts are planned for a single watershed, WS_{11} (zero growth for WS_{11} and strong growth-rate variation for all the other WSs; strategy No. 26).

15.4. Evaluation of Results

Discussion

(1) If we consider the pure strategies, the three-year average values of eutrophication index (summer mean Chl-a) in 1985 will be between 60.5–35.0; 39.2–27.5; 20.8–13.0 and 9.5–5.5 mg/m^3 for the four basins (Table 15.2). This indicates that the same increase of watershed development in the various watersheds results in the greatest eutrophication in the Keszthely bay.

(2) The difference between the maximum and minimum values of simulated eutrophication indices is largest in the first basin, and its value decreases towards the eastern end of the lake (see Table 15.2). The ratio of these differences is 6.4:3:2:1 for $WB_1, \ldots, WB_4$, and indicates the range of effectiveness of possible lake protection measures in various watersheds of the four basins.

(3) The evaluation and comparison of simulation results for strategies No. 6 and No. 26 allow us to draw the following conclusions (see Figures 15.4 and 15.5 for strategy No. 6).

When applying strategy No. 6 the eutrophication will be only a little less advanced than that of the worst case (No. 1), and so the effectiveness is poor, 6%. This means that, although six watersheds are involved in the management actions planned, the efforts are concentrated at the wrong place, so due to the hierarchical character of the system the effectiveness will be low.

On the contrary, strategy No. 26 (zero growth for WS_{11}) results in favorable water quality conditions in WB_1 and at the same time only slight deterioration occurs in other basins. The effectiveness is 33%.

(4) Using a systematic analysis, the relative efficiency of protection measures taken at various watersheds can be evaluated. Accordingly, a protection measure realized at WS_{11} is 6.2 times more effective than at WS_{41}–WS_{42}. The corresponding values are 1.7 and 1.1 for WS_{21}–WS_{22} and WS_{31}–WS_{32}, respectively. These ratios are in agreement with conclusion (2).

(5) The model was used to evaluate the effectiveness of the Kis-Balaton reservoir ($\sim 75 \times 10^6$ m^3 volume) located at the mouth of the Zala river in WS_{11}. Assuming that only this protection measure will be established, nine indicator indices of the total of 25 are influenced. The nine indices reflect hydrologic, environmental, social, regional, economic, and technical impacts. Eight indices of the nine decrease (in particular, the reduction of I_{18} and I_{19} is remarkable) and only one increases. As a result of these changes, D^* of WS_{11} is reduced by about 10% leading to improvements in the eutrophication index E^* [see equation (15.1)] to 45, 40, 26, and 39% for $WB_1, \ldots, WB_4$, respectively, as compared with 1980 (see Table 15.2).

Keeping in mind that the construction and full operation of the reservoir requires 5–8 years and, meanwhile, a slight watershed development is expected, the improvement of water quality might be smaller. Therefore, the Kis-Balaton project should be completed as early as possible.

Conclusions

(1) The model introduced is able to describe the relationship between eutrophication and nutrient-influencing watershed development on a multiannual and multiregional basis by the consideration of a multicriteria utility function of watershed development. Following calibration, the model was used for characterizing the historical (long-term) evaluation of the eutrophication of Lake Balaton. Additionally, the model was applied to simulate the consequences of various lake protection policies and to compare their effectiveness.

(2) According to simulation results all of the limited, available control efforts and resources should concentrate on the Zala watershed in the next five years; protection measures taken on the Zala watershed are about six times more efficient from the viewpoint of the entire lake than those in eastern regions of the lake. The relative efficiency figures derived can serve as useful, general guidelines for elaborating water management – and regional development plans.

References

Dávid, L. and Telegdi, L. (1982) *The Influence of Watershed Development on the Eutrophication of Shallow Lakes; a Modeling Approach Applied to Lake Balaton*. Collaborative Paper CP-82-42 (Laxenburg, Austria: International Institute for Applied Systems Analysis).

Dávid, L., Telegdi, L., and van Straten, G. (1979) *A Watershed Development Approach to the Eutrophication Problem of Lake Balaton*. Collaborative Paper CP-79-16 (Laxenburg, Austria: International Institute for Applied Systems Analysis).

Keeney, R.L. and Raiffa, H. (1976) *Decisions with Multiple Objectives: References and Value Tradeoffs*. (New York: Wiley).

Index